Lecture Notes in Computer Science   11126

Commenced Publication in 1973
Founding and Former Series Editors:
Gerhard Goos, Juris Hartmanis, and Jan van Leeuwen

More information about this series at http://www.springer.com/series/7409

Josep Domingo-Ferrer · Francisco Montes (Eds.)

Privacy in Statistical Databases

UNESCO Chair in Data Privacy
International Conference, PSD 2018
Valencia, Spain, September 26–28, 2018
Proceedings

Springer

Editors
Josep Domingo-Ferrer (iD)
Universitat Rovira i Virgili
Tarragona
Spain

Francisco Montes (iD)
Universitat de València
Burjassot
Spain

ISSN 0302-9743   ISSN 1611-3349 (electronic)
Lecture Notes in Computer Science
ISBN 978-3-319-99770-4   ISBN 978-3-319-99771-1 (eBook)
https://doi.org/10.1007/978-3-319-99771-1

Library of Congress Control Number: 2018952341

LNCS Sublibrary: SL3 – Information Systems and Applications, incl. Internet/Web, and HCI

This Springer imprint is published by the registered company Springer Nature Switzerland AG
The registered company address is: Gewerbestrasse 11, 6330 Cham, Switzerland

Preface

Privacy in statistical databases is a discipline whose purpose is to provide solutions to the tension between the social, political, economic, and corporate demand of accurate information, and the legal and ethical obligation to protect the privacy of the various parties involved. In particular, the need to enforce of the EU General Data Protection Regulation (GDPR) in our world of big data has made this tension all the more pressing. Stakeholders include the subjects, sometimes also known as the respondents (the individuals and enterprises to which the data refer), the data controllers (those organizations collecting, curating, and to some extent sharing or releasing the data), and the users (the ones querying the database or the search engine, who would like their queries to stay confidential). Beyond law and ethics, there are also practical reasons for data controllers to invest in subject privacy: If individual subjects feel their privacy is guaranteed, they are likely to provide more accurate responses. Data controller privacy is primarily motivated by practical considerations: If an enterprise collects data at its own expense and responsibility, it may wish to minimize leakage of those data to other enterprises (even to those with whom joint data exploitation is planned). Finally, user privacy results in increased user satisfaction, even if it may curtail the ability of the data controller to profile users.

There are at least two traditions in statistical database privacy, both of which started in the 1970s: The first one stems from official statistics, where the discipline is also known as statistical disclosure control (SDC) or statistical disclosure limitation (SDL), and the second one originates from computer science and database technology. In official statistics, the basic concern is subject privacy. In computer science, the initial motivation was also subject privacy but, from 2000 onwards, growing attention has been devoted to controller privacy (privacy-preserving data mining) and user privacy (private information retrieval). In the past few years, the interest and the achievements of computer scientists in the topic have substantially increased, as reflected in the contents of this volume. At the same time, the generalization of big data is challenging privacy technologies in many ways: This volume also contains recent research aimed at tackling some of these challenges.

Privacy in Statistical Databases 2018 (PSD 2018) was held under the sponsorship of the UNESCO Chair in Data Privacy, which has been providing a stable umbrella for the PSD biennial conference series since 2008. Previous PSD conferences were held in various locations around the Mediterranean, and had their proceedings published by Springer in the LNCS series: PSD 2016, Dubrovnik, LNCS 9867; PSD 2014, Eivissa, LNCS 8744; PSD 2012, Palermo, LNCS 7556; PSD 2010, Corfu, LNCS 6344; PSD 2008, Istanbul, LNCS 5262; PSD 2006, the final conference of the Eurostat-funded CENEX-SDC project, held in Rome, LNCS 4302; and PSD 2004, the final conference of the European FP5 CASC project, held in Barcelona, LNCS 3050. The eight PSD conferences held so far are a follow-up of a series of high-quality technical conferences on SDC that started 18 years ago with Statistical Data Protection (SDP) 1998, held in

Lisbon in 1998 and with proceedings published by OPOCE, and continued with the AMRADS project SDC Workshop, held in Luxemburg in 2001 and with proceedings published by Springer in LNCS 2316.

The PSD 2018 Program Committee accepted for publication in this volume 23 papers out of 42 submissions. Furthermore, 11 of these submissions were reviewed for short oral presentation at the conference. Papers came from 15 different countries in four different continents. Each submitted paper received at least two reviews. The revised versions of the 23 accepted papers in this volume are a fine blend of contributions from official statistics and computer science. Topics covered include tabular data protection, microdata and big data masking, synthetic data, record linkage, and spatial and mobility data.

We are indebted to many people. First, to the Organizing Committee for making the conference possible and especially to Jesús Manjón, who helped prepare these proceedings. In evaluating the papers, we were assisted by the Program Committee and by Daniel Baena, Dimitrios Karapiperis, and José Antonio González Alastrué as external reviewers. We also wish to thank all the authors of submitted papers and we apologize for possible omissions.

Finally, we dedicate this volume to the memory of Prof. Stephen Fienberg, who was a Program Committee member of all past editions of the PSD conference.

July 2018   Josep Domingo-Ferrer
                                                                Francisco Montes

Organization

Privacy in Statistical Databases, PSD 2018

Program Committee

Jane Bambauer	University of Arizona, USA
Bettina Berendt	Katholieke Universiteit Leuven, Belgium
Elisa Bertino	CERIAS, Purdue University, USA
Aleksandra Bujnowska	EUROSTAT, European Union
Jordi Castro	Polytechnical University of Catalonia, Spain
Josep Domingo-Ferrer	Universitat Rovira i Virgili, Spain
Jörg Drechsler	IAB, Germany
Khaled El Emam	University of Ottawa, Canada
Mark Elliot	Manchester University, UK
Sébastien Gambs	Université du Québec à Montréal, Canada
Sarah Giessing	Destatis, Germany
Sara Hajian	Eurecat Technology Center, Spain
Alan Karr	CoDA, RTI, USA
Julia Lane	New York University, USA
Bradley Malin	Vanderbilt University, USA
Laura McKenna	Census Bureau, USA
Gerome Miklau	University of Massachusetts-Amherst, USA
Krishnamurty Muralidhar	University of Oklahoma, USA
Anna Oganyan	National Center for Health Statistics, USA
Christine O'Keefe	CSIRO, Australia
David Rebollo-Monedero	Polytechnical University of Catalonia, Spain
Jerome Reiter	Duke University, USA
Yosef Rinott	Hebrew University, Israel
Pierangela Samarati	University of Milan, Italy
David Sánchez	Universitat Rovira i Virgili, Spain
Eric Schulte-Nordholt	Statistics Netherlands, The Netherlands
Natalie Shlomo	Manchester University, UK
Aleksandra Slavković	Penn State University, UK
Jordi Soria-Comas	Universitat Rovira i Virgili, Spain
Tamir Tassa	The Open University, Israel
Vicenç Torra	University of Skövde, Sweden
Vassilios Verykios	Hellenic Open University, Greece
William E. Winkler	Census Bureau, USA
Peter-Paul de Wolf	Statistics Netherlands, The Netherlands

Program Chair

Josep Domingo-Ferrer   UNESCO Chair in Data Privacy,
                             Universitat Rovira i Virgili, Spain

General Chair

Francisco Montes   Universitat de València, Spain

Organizing Committee

Joaquín García-Alfaro   Télécom SudParis, France
Jesús Manjón   Universitat Rovira i Virgili, Spain
Romina Russo   Universitat Rovira i Virgili, Spain

Contents

Spatial and Mobility Data

Tabular Data Protection

Symmetric vs Asymmetric Protection Levels in SDC Methods for Tabular Data

Daniel Baena, Jordi Castro$^{(\boxtimes)}$, and José A. González

Department of Statistics and Operations Research,
Universitat Politècnica de Catalunya, Jordi Girona 1–3,
08034 Barcelona, Catalonia, Spain
danibaena@gmail.com, {jordi.castro,jose.a.gonzalez}@upc.edu

Abstract. Protection levels on sensitive cells—which are key parameters of any statistical disclosure control method for tabular data—are related to the difficulty of any attacker to recompute a good estimation of the true cell values. Those protection levels are two numbers (one for the *lower protection*, the other for the *upper protection*) imposing a *safety interval* around the cell value, that is, no attacker should be able to recompute an estimate within such safety interval. In the symmetric case the lower and upper protection levels are equal; otherwise they are referred as asymmetric protection levels. In this work we empirically study the effect of symmetry in protection levels for three protection methods: cell suppression problem (CSP), controlled tabular adjustment (CTA), and interval protection (IP). Since CSP and CTA are mixed integer linear optimization problems, it is seen that the symmetry (or not) of protection levels affect to the CPU time needed to compute a solution. For IP, a linear optimization problem, it is observed that the symmetry heavily affects to the quality of the solution provided rather than to the solution time.

Keywords: Statistical disclosure control · Tabular data
Cell suppression · Controlled tabular adjustment
Interval protection · Mixed integer linear optimization
Linear optimization

1 Introduction

The three statistical disclosure control methods for tabular data considered in this work (namely: cell suppression problem (CSP) [5,10], controlled tabular adjustment (CTA) [2,4,13], and interval protection (IP) [8,11]) belong to the family of *post-tabular* data protection methods, which modify or suppress table cells once the table have been built (in contrast to pre-tabular methods, which change microdata files, and therefore, although being faster, may not guarantee

Supported by grant MTM2015-65362-R of the Spanish Ministry of Economy and Competitiveness.

table additivity if the true values of marginal or total cells want to be preserved). More details can be found in the monograph [14] and the survey [6].

Each method protects sensitive cells in a different way. CSP removes sensitive cells; other additional cells have also to be removed to avoid recomputing the original value of sensitive cells. CSP results in a large and difficult mixed integer linear problem, which can be solved optimally (using Benders decomposition as done in [10]) or heuristically (e.g., using shortest paths for some hierarchical tables as in [5]). IP (or partial cell suppression, which was its original name coined in [11]) can be seen as a linear version of CSP, where cell values are replaced by intervals containing the true value. IP, unlike CSP, is a linear optimization problem, and therefore—at least theoretically—it can be solved in polynomial time by efficient interior-point methods [17]. CTA replaces sensitive values by safe values (i.e., outside the safety interval), thus forcing changes in other cells to preserve the table additivity. CTA is also formulated as a mixed integer linear optimization problem, which can be solved optimally by general purpose solvers [9], or heuristically [2,13]. This work provides a formulation of CSP, CTA and IP from the same set of parameters.

One of the key parameters for the optimization models for CSP, CTA and IP are the *lower and upper protection levels*: these two numbers define a protection interval around the cell value, such that no attacker should be able to obtain an estimation of the true value within such interval. When the lower and upper protection levels are equal, we have a *symmetric* interval around the true value; otherwise we refer to the *asymmetric* case. A priori, asymmetric intervals could benefit the solution of mixed integer linear optimization problems, such as CTA and CSP. Indeed, some results along these lines were obtained in [9] for CTA with quadratic objectives. Another objective of this work is to check if such behaviour is observed for CTA and CSP in the solution of a set of hierarchical tables.

For IP, being a linear optimization model, such symmetry is not expected to provide faster executions. However, as it will be shown in the computational results, the use of asymmetric protection levels is instrumental to avoid the disclosure of the true cell values.

This short paper is organized as follows. Section 2 shows a formulation of CSP, CTA and IP using a unified set of parameters. Section 3 reports and compares the results obtained on a set of generated hierarchical instances, using symmetric and asymmetric protection levels, for the three tabular data protection methods.

2   Formulation of CSP, CTA and IP for Tabular Data

The parameters that define any CSP, CTA or IP instance are:

- A general table, consisting of a set of n cells and a set of m linear relations $Aa = b$, where $A \in \mathbb{R}^{m \times n}$ is the matrix defining the table structure, $a = (a_1, \ldots, a_n)^\top \in \mathbb{R}^n$ is the vector of cell values, and the right-hand side $b \in \mathbb{R}^m$ is usually 0 if the table is additive.

- Upper and lower bounds $u \in \mathbb{R}^n$ and $l \in \mathbb{R}^n$ for the cell values, which are assumed to be known by any attacker: $l \leq a \leq u$ (e.g., $l = 0$, $u = +\infty$ for a positive table).
- Vector of nonnegative weights $w \in \mathbb{R}^n$, associated to either the cell suppressions for CSP, the cell perturbations for CTA, or the width of interval replacing cells for IP. That is, $w_i, i = 1, \ldots, n$ measures the cost (or data utility loss) associated to hiding the true value of cell i. If $w_i = 1$ for all $i = 1, \ldots, n$, the same cost is given to any cell; if $w_i = 1/a_i$ a relative cost is considered depending on the cell values; other options are possible, such as, for instance, $w_i = 1/\sqrt{a_i}$.
- Set $\mathcal{S} \subseteq \{1, \ldots, n\}$ of sensitive cells, decided in advance by applying some sensitivity rules.
- Lower and upper protection levels for each sensitive cell lpl_s and upl_s $s \in \mathcal{S}$ (usually either a fraction of a_s or directly obtained from the sensitivity rules). No sliding protection is considered, unlike in [10].

2.1 Formulation of Cell Suppression Problem (CSP)

CSP aims at finding a set $\mathcal{C}$ of complementary cells to be removed such that for all $s \in \mathcal{S}$

$$\underline{a_s} \leq a_s - lpl_s \quad \text{and} \quad \overline{a_s} \geq a_s + upl_s, \tag{1}$$

$\underline{a_s}$ and $\overline{a_s}$ being defined as

$$
\begin{aligned}
&\underline{a_s} = \min_x x_s \\
&\text{s. to } Ax = b \\
&\quad l_i \leq x_i \leq u_i \ \ i \in \mathcal{S} \cup \mathcal{C} \\
&\quad x_i = a_i \ \ i \notin \mathcal{S} \cup \mathcal{C}
\end{aligned}
\quad \text{and} \quad
\begin{aligned}
&\overline{a_s} = \max_x x_s \\
&\text{s. to } Ax = b \\
&\quad l_i \leq x_i \leq u_i \ \ i \in \mathcal{S} \cup \mathcal{C} \\
&\quad x_i = a_i \ \ i \notin \mathcal{S} \cup \mathcal{C}.
\end{aligned}
\tag{2}
$$

The classical model for CSP, originally formulated in [15], considers two sets of variables: (1) $y_i \in \{0, 1\}, i = 1, \ldots, n$, is 1 if cell i has to be suppressed, and 0 otherwise; (2) two auxiliary vectors $x^{l,s} \in \mathbb{R}^n$ and $x^{u,s} \in \mathbb{R}^n$, for all $s \in \mathcal{S}$, to impose as constraints that the solutions to problems (2) would satisfy (1). The resulting model is

$$
\min_{y, x^{l,s}, x^{u,s}} \sum_{i=1}^{n} w_i y_i
$$

s. to

$$
\left.
\begin{aligned}
Ax^{l,s} &= 0 \\
(l_i - a_i)y_i \leq x_i^{l,s} &\leq (u_i - a_i)y_i \quad i = 1, \ldots, n \\
x_s^{l,s} &\leq -lpl_s \\[1em]
Ax^{u,s} &= 0 \\
(l_i - a_i)y_i \leq x_i^{u,s} &\leq (u_i - a_i)y_i \quad i = 1, \ldots, n \\
x_s^{u,s} &\geq upl_s
\end{aligned}
\right\} \quad \forall \ s \in \mathcal{S} \tag{3}
$$

$$y_i \in \{0, 1\} \quad i = 1, \ldots, n.$$

When $y_i = 1$, the inequality constraints of (3) with both right- and left-hand sides impose bounds on the deviations $x_i^{l,p}$ and $x_i^{u,p}$ for cell i; these deviations are prevented when $y_i = 0$, that is, when the cell is published (non-suppressed). Formulation (3) gives rise to a mixed integer linear optimization problem of n binary variables, $2n|\mathcal{S}|$ continuous variables, and $2(m + 2n + 1)|\mathcal{S}|$ constraints.

2.2 Formulation of Controlled Tabular Adjustment (CTA)

Instead of suppressing cells, CTA computes an alternative safe table x: the closest to a using some particular distance $\ell_{(w)}$ based on cell weights w. In this context *safe* means that the values of sensitive cells are outside the protection interval $[a_s - lpl_s, a_s + upl_s]$ for all $s \in \mathcal{S}$. The optimization problem to be solved is:

$$
\begin{aligned}
\min_{x} \ & \|x - a\|_{\ell_{(w)}} \\
\text{s. to } & Ax = b \\
& l \leq x \leq u \\
& x_s \leq a_s - lpl_s \text{ or } x_s \geq a_s + upl_s \quad s \in \mathcal{S}.
\end{aligned} \tag{4}
$$

Defining cell deviations $z = x - a$, $l_z = l - a$ and $u_z = u - a$, (4) can be reformulated as:

$$
\begin{aligned}
\min_{z} \ & \|z\|_{\ell_{(w)}} \\
\text{s. to } & Az = 0 \\
& l_z \leq z \leq u_z \\
& z_s \leq -lpl_s \text{ or } z_s \geq upl_s \quad s \in \mathcal{S}.
\end{aligned} \tag{5}
$$

The "or" constraints of (5) can be modeled using binary variables $y_s \in \{0,1\}$, $s \in \mathcal{S}$, such that $y_s = 1$ if cell s is "upper protected" (i.e., $z_s \geq upl_s$), and $y_s = 0$ if it is "lower protected" ($z_s \leq -lpl_s$). For distance ℓ_1, the resulting mixed integer linear optimization formulation is

$$
\begin{aligned}
\min_{z^+, z^-} \ & \sum_{i=1}^{n} w_i(z_i^+ + z_i^-) \\
\text{s. to } & A(z^+ - z^-) = 0 \\
& 0 \leq z_i^+ \leq u_{z_i} \quad i \notin \mathcal{S} \\
& 0 \leq z_i^- \leq -l_{z_i} \quad i \notin \mathcal{S} \\
& upl_i y_i \leq z_i^+ \leq u_{z_i} y_i \quad i \in \mathcal{S} \\
& lpl_i(1 - y_i) \leq z_i^- \leq -l_{z_i}(1 - y_i) \quad i \in \mathcal{S} \\
& y_i \in \{0,1\} \quad i \in \mathcal{S}.
\end{aligned} \tag{6}
$$

where z_i, $i = 1, \ldots, n$, is split as $z_i = z_i^+ - z_i^-$, such that $|z_i| = z_i^+ + z_i^-$. Problem (6) has $|\mathcal{S}|$ binary variables, $2n$ continuous variables and $m + 4|\mathcal{S}|$ constraints.

2.3 Formulation of Interval Protection (IP)

The purpose of IP is to replace cell values a_i by feasible intervals $[lb_i, ub_i]$, $i = 1, \ldots, n$—where feasible means that $l_i \leq lb_i$ and $ub_i \leq u_i$, such that estimates of

a_s, $s \in \mathcal{S}$, computed by any attacker should be outside the protection interval $[a_s - lpl_s, a_s + upl_s]$. This means—similarly to what is was done for CSP—that

$$\underline{a_s} \leq a_s - lpl_s \quad \text{and} \quad \overline{a_s} \geq a_s + upl_s, \tag{7}$$

$\underline{a_s}$ and $\overline{a_s}$ being defined as

$$\underline{a_s} = \min_x x_s \qquad \qquad \overline{a_s} = \max_x x_s$$
$$\text{s.to } Ax = b \qquad \text{and} \qquad \text{s.to } Ax = b$$
$$lb_i \leq x_i \leq ub_i \ i = 1, \ldots, n \qquad \qquad lb_i \leq x_i \leq ub_i \ i = 1, \ldots, n. \tag{8}$$

Like in CSP, the previous problem can be formulated as a large-scale (linear in that case, instead of mixed integer linear) optimization problem. For each sensitive cell $s \in \mathcal{S}$, two auxiliary vectors $x^{l,s} \in \mathbb{R}^n$ and $x^{u,s} \in \mathbb{R}^n$ are introduced to impose, respectively, the lower and upper protection requirement of (7). The resulting optimization problem is:

$$\min_{lb,ub} \sum_{i=1}^{n} w_i(ub_i - lb_i)$$
$$\text{s.to}$$

$$\left.\begin{array}{l} Ax^{l,s} = b \\ lb_i \leq \ x_i^{l,s} \ \leq ub_i \qquad i = 1, \ldots, n \\ \qquad x_s^{l,s} \ \leq a_s - lpl_s \\ Ax^{u,s} = b \\ lb_i \leq \ x_i^{u,s} \ \leq ub_i \qquad i = 1, \ldots, n \\ \qquad x_s^{u,s} \ \geq a_s + upl_s \end{array}\right\} \ \forall \ s \in \mathcal{S} \tag{9}$$

$$l_i \leq lb_i \leq a_i \quad i = 1, \ldots, n$$
$$a_i \leq ub_i \leq u_i \quad i = 1, \ldots, n.$$

Problem (9) is very large, with $2n(|\mathcal{S}| + 1)$ continuous variables and $2(m + 2n + 1)|\mathcal{S}|$ constraints. On the other hand, unlike CSP, it is linear (no binary, no integer variables), and thus theoretically it can be efficiently solved in polynomial time by general or by specialized interior-point algorithms. As far as we know, no efficient implementation has been developed yet for IP, and there are only some preliminary prototypes [8]. Some related heuristics for variations of this problem were considered in [16].

3 Computational Experience

To study the effect of symmetric and asymmetric protection levels for CSP, CTA and IP we generated a set of six hierarchical tables using the generator introduced in [5]. Two versions of each table were considered: one with symmetric protection

Table 1. Instances dimensions.

Instance	n	$|\mathcal{S}|$	m	nnz
1	1444	135	171	2964
2	2544	237	303	5216
3	2108	196	243	4318
4	1444	216	152	2945
5	1488	220	173	3040
6	1411	209	168	2890

levels, the other with asymmetric ones. In the symmetric case a 20% of the cell value was considered, while a 5% and 20% were used for respectively the lower and upper protection levels for asymmetric instances. A priori bounds l and u were 0 and a large value, respectively (so they were always inactive). Table 1 reports the number of cells, sensitive cells, table linear relations, and number of nonzero entries of matrix A, for each instance. For CSP and CTA we used the efficient (C++) implementations described in [1,7], respectively. For IP the prototype code (a Benders decomposition implemented in AMPL [3,12]) of [8] was used.

Solution times for CSP and CTA appear in Table 2. Since the IP prototype is quite inefficient, their times are not reported. The optimality gap (i.e., the relative distance between the upper and lower bound of the optimization problem) required in both methods was 0.1%. Two symmetric tables exceeded the one hour time limit for CSP, so the final gap reached (in brackets) is notably higher than the required one. From Table 2, CTA is clearly more efficient for asymmetric than for symmetric instances; this is consistent with the results of [9], albeit they were for a quadratic version of CTA (i.e., using the ℓ_2 Euclidean instead of the ℓ_1 distance in the objective function). For CSP the pattern is not so definitive: asymmetric instances 4, 5 and, specially, 6 were slower than the corresponding symmetric variants. However, for the two largest instances (2 and 3) the symmetric cases were clearly outperformed by the asymmetric ones.

Table 2. Computation times, in seconds.

	CTA		CSP	
Instance	Symm.	Asymm.	Symm.	Asymm.
1	2.03	0.35	39.12	37.11
2	12.23	0.53	3600 (73%)	638.42
3	10.44	0.65	3600 (74%)	513.61
4	4.66	0.84	20.88	73.17
5	5.15	1.20	25.99	88.06
6	5.30	1.14	53.31	693.1

As for IP, not being a mixed integer linear problem, we do not expect differences in CPU times between symmetric and asymmetric instances (and, in addition, we would need an efficient IP code to check them, which is not the case). However we can perform a comparison between the quality of the intervals obtained for symmetric and asymmetric variants. In this respect, we first observed that most of the cells were not replaced by an interval, that is, lb_i and ub_i were the same value. Table 3 shows the number and percentage of cells which have been replaced in each instance by an interval, that is, one with different endpoints lb_i and ub_i. About 9.3% of cells are sensitive in instances 1 to 3, so one out of two interval-replaced cells is non-sensitive in the symmetric cases. Instances 4 to 6, with higher proportion of sensitive cells (about 15%), show lower rates of non-sensitive cells among all the interval-replaced cells. In all the instances, the number of cells replaced by an interval increases slightly for the asymmetric cases.

Table 3. Count and percentage of cells which have been replaced by an interval by IP.

	Symmetric		Asymmetric	
Instance	N. of cells	(%)	N. of cells	(%)
1	263	18.2	309	21.4
2	471	18.5	529	20.8
3	403	19.1	451	21.4
4	334	23.1	387	26.8
5	377	25.3	412	27.7
6	360	25.5	401	28.4

The quality of the protection is given by its difficulty to disclose the original cell values. In principle, an interval should be safe since any value inside it has the same chance to be the value sought by the attacker. However, we (somehow unexpectedly) found that an instance with symmetric protection levels is far more vulnerable. Table 4 describes the proportion of cells that have been replaced with an interval whose midpoint is exactly the original value (represented here by the zero value). The intervals have been standardized to have a width of 100. The five classes represented are given by the midpoint position: for instance, -50 means that the interval is $[-100, 0]$, that is, the rightmost value is equal to the original cell value; $(-50, 0)$ means that the original cell value is located somewhere strictly between the midpoint and the right endpoint. The proportion of cells lying in the midpoint (0) is very large among the symmetric cases, and represents a real risk of disclosure, since just taking the average of the interval has many chances to guess the original cell value. On the other hand, the proportion of such cases in instances with asymmetric protection levels is negligible. Figure 1 compares a typical instance (number 3), showing graphically the benefit of dealing with asymmetric protection levels. The other instances studied exhibited a similar behaviour.

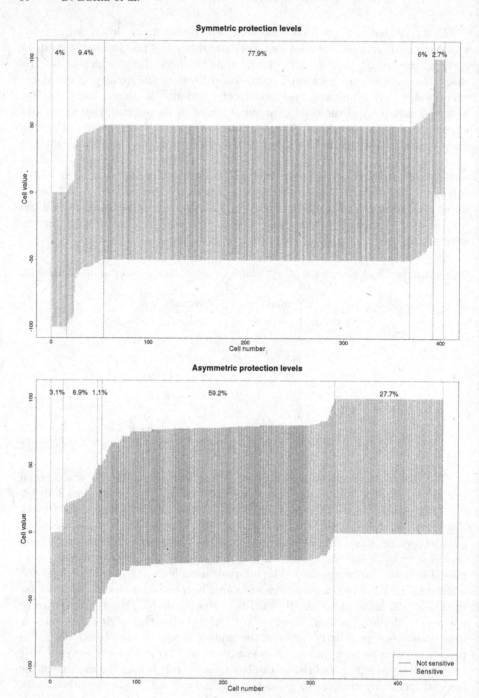

Fig. 1. Instance number 3, showing standardized intervals in both symmetric and asymmetric cases. The cells have been ranked in each case according to their interval (so the position along the x-axis is usually different in the two plots).

Table 4. Percentage distribution of intervals position.

	Instance	−50	(−50, 0)	0	(0, 50)	50
Symm.	1	2.7	5.7	87.1	4.6	0
	2	4	6.4	85.4	4	0.2
	3	4	9.4	77.9	6	2.7
	4	4.2	6.6	80.8	6.6	1.8
	5	6.4	7.4	77.5	6.9	1.9
	6	4.7	8.1	76.9	8.3	1.7
Asymm.	1	5.8	3.9	2.9	60.8	26.2
	2	2.3	4.5	2.3	59.2	31.4
	3	3.1	8.9	1.1	59.2	27.7
	4	1.6	4.4	2.1	68.7	23
	5	3.6	5.8	1	67.7	21.6
	6	2	6	3.5	67.3	20.9

4   Conclusions

From the computational results in the solution of a set of six hierarchical tables, using efficient implementations of CSP and CTA, and a prototype code for IP, we conclude:

- For the mixed integer linear problems CTA and CSP, symmetry of protection levels has an impact on the solution time. For CTA this assertion was always true: asymmetric instances were faster than symmetric ones. For CSP this fact was not so conclusive: only for the largest instances tested asymmetry provided faster executions.
- For IP asymmetric protection levels affected the quality of the solution, rather than solution times. In general, symmetric protection levels provided very poor intervals, and in most cases their midpoints disclosed the true cell value. Therefore, the use of asymmetric protection levels in IP should be highly recommended.
- Protection levels automatically provided by sensitivity rules are always symmetric: this practice should be reconsidered according to the results of this work.

A similar analysis to that done for IP could be performed for the intervals obtained by the auditing phase of the CSP; this is part of the future work to be done.

References

1. Baena, D., Castro, J., Frangioni, A.: Stabilized Benders methods for large-scale combinatorial optimization, with application to data privacy. Research report DR 2017/03, Department of Statistics and Operations Research, Universitat Politècnica de Catalunya, Barcelona, Catalonia (2017)
2. Baena, D., Castro, J., González, J.A.: Fix-and-relax approaches for controlled tabular adjustment. Comput. Oper. Res. **58**, 41–52 (2015)
3. Benders, J.F.: Partitioning procedures for solving mixed-variables programming problems. Comput. Manage. Sci. **2**, 3–19 (2005). English translation of the original paper appeared in Numerische Mathematik, 4 (1962) 238–252
4. Castro, J.: Minimum-distance controlled perturbation methods for large-scale tabular data protection. Eur. J. Oper. Res. **171**, 39–52 (2006)
5. Castro, J.: A shortest paths heuristic for statistical disclosure control in positive tables. INFORMS J. Comput. **19**, 520–533 (2007)
6. Castro, J.: Recent advances in optimization techniques for statistical tabular data protection. Eur. J. Oper. Res. **216**, 257–269 (2012)
7. Castro, J., González, J.A., Baena, D.: User's and programmer's manual of the RCTA package. Technical report DR 2009/01, Department of Statistics and Operations Research, Universitat Politècnica de Catalunya, Barcelona, Catalonia (2009)
8. Castro, J., Via, A.: Revisiting interval protection, a.k.a. partial cell suppression, for tabular data. In: Domingo-Ferrer, J., Pejić-Bach, M. (eds.) PSD 2016. LNCS, vol. 9867, pp. 3–14. Springer, Cham (2016). https://doi.org/10.1007/978-3-319-45381-1_1
9. Castro, J., Frangioni, A., Gentile, C.: Perspective reformulations of the CTA problem with L_2 distances. Oper. Res. **62**, 891–909 (2014)
10. Fischetti, M., Salazar, J.J.: Solving the cell suppression problem on tabular data with linear constraints. Manage. Sci. **47**, 1008–1026 (2001)
11. Fischetti, M., Salazar, J.J.: Partial cell suppression: a new methodology for statistical disclosure control. Stat. Comput. **13**, 13–21 (2003)
12. Fourer, R., Gay, D.M., Kernighan, D.W.: AMPL: A Modeling Language for Mathematical Programming. Duxbury Press, Duxbury (2002)
13. González, J.A., Castro, J.: A heuristic block coordinate descent approach for controlled tabular adjustment. Comput. Oper. Res. **38**, 1826–1835 (2011)
14. Hundepool, A., et al.: Statistical Disclosure Control. Wiley, Chichester (2012)
15. Kelly, J.P., Golden, B.L., Assad, A.A.: Cell suppression: disclosure protection for sensitive tabular data. Networks **22**, 28–55 (1992)
16. Robertson, D.: Automated disclosure control at Statistics Canada. In: Paper presented at the second international seminar on statistical confidentiality, Luxembourg (1994)
17. Wright, S.J.: Primal-Dual Interior-Point Methods. SIAM, Philadelphia (1997)

Bounded Small Cell Adjustments
for Flexible Frequency Table Generators

Min-Jeong Park[✉]

Statistical Research Institute, Statistics Korea, Daejeon, Korea
mjstat@korea.kr

Abstract. Statistics Korea has disseminated census data through the
Statistical Geographic Information Service (SGIS) system. Users can easily access the system on a web-site and obtain frequencies on the map
for diverse size-of-area units according to their selection of variables. In
order to control the disclosure risk for frequency tables, we thoroughly
examined the Small Cell Adjustments (SCA) method to find the reasons
for disclosures; we then suggested the Bounded Small Cell Adjustments
(BSCA) procedure in this paper. From the analysis on the census data of
a Korean city of approximately 1.5 million people, we demonstrated the
efficiency of BSCA, which reduces information loss under B in most cells
while maintaining B-anonymity in all cells as intended in the SCA idea.
The B denotes the criterion value defining a small cell. Furthermore, we
have discussed the relationship between disclosure risk and information
loss by BSCA.

Keywords: Frequency table · Table generating system
Small cell adjustments · Information loss · Disclosure risk
Risk-utility map

1 Introduction

One important mission for national statistical offices is to help people conveniently access data and easily find useful information without privacy violations.
Recently, statistical agencies have tried to develop a web-based platform for
offering statistics in a more user-friendly way. From 2009, Statistics Korea has
managed a user-friendly interface, called the Statistical Geographic Information
Service (SGIS), in order to provide frequency tables mapped out with information from several censuses.

For Statistical Disclosure Control (SDC) regarding tabular data, we have
many established methods or algorithms. The Small Cell Adjustment (SCA)
can be described as a semi-controlled rounding algorithm among post-tabular
methods [2]. The effects of SDC techniques for frequency tables are discussed
with suggestions regarding risk and information loss measures in [3]. However,
most methods have some drawbacks resulting in too much information loss, or
giving rise to disclosure accidents, or being hard to implement in a real system.

© Springer Nature Switzerland AG 2018
J. Domingo-Ferrer and F. Montes (Eds.): PSD 2018, LNCS 11126, pp. 13–27, 2018.
https://doi.org/10.1007/978-3-319-99771-1_2

In this paper, we suggest a practical SDC solution for a frequency table generator. We call it Bounded Small Cell Adjustments (BSCA). The suggested solution can achieve k-anonymity [4] for confidentiality. The value of k is equal to B that is previously select to define a small cell by the agency. Moreover, the BSCA controls information loss less than B in most situations. While developing the procedure, we propose to change the paradigm for thinking about risk and utility.

Section 2 explains the table structures in order to understand real platforms. In Sect. 3, we show our findings on information loss and disclosure accidents when we employ the existing SCA algorithms. Then we discuss the necessity to reform the risk-utility frame, illustrate the BSCA procedure, and show the analysis result for the census data of a Korean city of approximately 1.5 million people in Sect. 4. A conclusion is shown in Sect. 5.

2   Table Structure in a Real System

In this paper, we analyze the census microdata of a Korean city in order to show how the proposed algorithm works well. The data set consists of four variables, which are gender G (2), age A (18), marital status M (5) and education level E (8). The number in the parenthesis indicates the number of categories for each variable. The largest area unit, the city, is denoted as Local Area 1 (LA1). Smaller area units are denoted as LA2 and LA3, and the smallest area unit is called an Output Area (OA). In our data, the city LA1 consists of 5 LA2s, 78 LA3s, and 2,506 OAs. The sizes of area units, which we measure by number of individuals, are shown in Table 1. On average, each OA consists of about six hundreds individuals.

Table 1. The size (the number of individuals) of area units

Area level	OA	LA3	LA2	LA1
Average	586	18,841	293,920	1,469,599
Range	58–4,125	2,578–47,734	200,371–491,320	1,469,599

Now, we can define the hierarchy between tables according to the number of variables and the area level. Assume that the microdata set has N individuals, P variables, and four area levels. We denote the true frequency table as T and the masked table as $\tilde{T}$. For example, $\tilde{T}^{vp}_{i.LA2}$ indicates that the masked frequency table has p variables at the i-th LA2 unit. Table 2 shows all types of frequency tables for $P = 4$. Note that Table 2 also treats all margins. For example, the margins of T^{v4}_{OA} can be found in T^{v3}_{OA} or T^{v4}_{LA3}. The details of the frequency table are written in Appendix A with examples in Table 11.

T^{vP}_{OA} should be obtained by aggregating the given microdata file. Others can be produced from T^{vP}_{OA}. Moreover, in order to run the system faster, we have to use T^{vP}_{OA} instead of microdata within the system. Therefore, if we apply a random technique only to T_{OA}^{vP} and obtain other tables from T_{OA}^{vP} already saved in the system, then we can provide consistent frequencies through the system. We call T_{OA}^{vP} a full table.

Table 2. The types of true frequency tables for $P = 4$ variables

T	OA	LA3	LA2	LA1
$p = 4$	T_{OA}^{v4}	T_{LA3}^{v4}	T_{LA2}^{v4}	T_{LA1}^{v4}
$p = 3$	T_{OA}^{v3}	T_{LA3}^{v3}	T_{LA2}^{v3}	T_{LA1}^{v3}
$p = 2$	T_{OA}^{v2}	T_{LA3}^{v2}	T_{LA2}^{v2}	T_{LA1}^{v2}
$p = 1$	T_{OA}^{v1}	T_{LA3}^{v1}	T_{LA2}^{v1}	T_{LA1}^{v1}

On the other hand, we need to consider that users can aggregate the disseminated tables for comparison. So we add 'ag' to the subscript(superscript) for denoting aggregation according to area(variable) levels. For example, $\tilde{T}_{LA3.ag}^{v4}$ at a LA3 area unit is obtained from the tables $\tilde{T}_{OA}^{v4}$ of the corresponding OA units.

Finally, we can discuss the properties for the table generating system. If there exist some differences between $\tilde{T}$ and $\tilde{T}_{ag}$ (or $\tilde{T}^{ag}$), we say that the system does not have additivity. The difference between $\tilde{T}$ and hidden true T can be called Information Loss (IL). When small cells can be disclosed, we say that Disclosure Risk (DR) is high. If the frequency of a specific cell could have different values, the system does not have consistency. In order to disseminate $\tilde{T}$ efficiently, we have to consider additivity, IL, DR and consistency within the system.

3   Small Cell Adjustments

In order to maintain confidentiality by achieving k-anonymity in tables, we have employed the SCA method. Small cells, which have lower frequency values than pre-defined criterion B, are regarded as high risk for disclosure. Zero cells (no information), boundary and large cells (enough anonymity) are considered safe. SCA is a semi-controlled rounding method since it changes only small cells into zeros or boundary cells having frequency B in a probabilistic manner so that the totality of the table is expected to be preserved. The SCA method can have two versions of its algorithm in a real system. The first one is as follows:

(SCA Algorithm 1). Small cell adjustments controlled to OA totals

1. For given P variables, construct true full tables T_{OA}^{vP} at each OA.
2. Adjust small cells in each table of T_{OA}^{vP}, which apply a random rounding technique to the frequency f_k and obtain the masked frequency $\tilde{f}_k$ in tables $\tilde{T}_{OA}^{vP}$. The random rounding with base B can be described as follows:
For $b = 1, \ldots, (B - 1)$,
$$f_k = b \longrightarrow \tilde{f}_k = 0 \text{ with probability } (B - b)/B$$
$$f_k = b \longrightarrow \tilde{f}_k = B \text{ with probability } b/B$$

3. Obtain $\tilde{T}_{LA}^{vp}$ by aggregating $\tilde{T}_{OA}^{vP}$ (represented as $\rightarrow$, $\downarrow$ or $\searrow$ in Table 3) where $p = 1, \ldots, P$ and $LA = OA, LA3, LA2, LA1$. The bold in Table 3 indicates that frequencies are randomly rounded as described in Step 2. Since we get $T \neq \tilde{T} = \tilde{T}^{ag} = \tilde{T}_{ag}$, the additivity is achieved by SCA1.

Table 3. Two versions of SCA algorithm

$\tilde{T}$	SCA1				SCA2			
	OA	LA3	LA2	LA1	OA	LA3	LA2	LA1
$p = 4$	$\tilde{\boldsymbol{T}}_{OA}^{v4}$	$\rightarrow \tilde{T}_{LA3}^{v4}$	$\rightarrow \tilde{T}_{LA2}^{v4}$	$\rightarrow \tilde{T}_{LA1}^{v4}$	$\tilde{\boldsymbol{T}}_{OA}^{v4}$	$\tilde{\boldsymbol{T}}_{LA3}^{v4}$	$\tilde{\boldsymbol{T}}_{LA2}^{v4}$	$\tilde{\boldsymbol{T}}_{LA1}^{v4}$
$p = 3$	$\downarrow \tilde{T}_{OA}^{v3}$	$\searrow \tilde{T}_{LA3}^{v3}$	$\searrow \tilde{T}_{LA2}^{v3}$	$\searrow \tilde{T}_{LA1}^{v3}$	$\downarrow \tilde{T}_{OA}^{v3}$	$\downarrow \tilde{T}_{LA3}^{v3}$	$\downarrow \tilde{T}_{LA2}^{v3}$	$\downarrow \tilde{T}_{LA1}^{v3}$
$p = 2$	$\downarrow \tilde{T}_{OA}^{v2}$	$\searrow \tilde{T}_{LA3}^{v2}$	$\searrow \tilde{T}_{LA2}^{v2}$	$\searrow \tilde{T}_{LA1}^{v2}$	$\downarrow \tilde{T}_{OA}^{v2}$	$\downarrow \tilde{T}_{LA3}^{v2}$	$\downarrow \tilde{T}_{LA2}^{v2}$	$\downarrow \tilde{T}_{LA1}^{v2}$
$p = 1$	$\downarrow \tilde{T}_{OA}^{v1}$	$\searrow \tilde{T}_{LA3}^{v1}$	$\searrow \tilde{T}_{LA2}^{v1}$	$\searrow \tilde{T}_{LA1}^{v1}$	$\downarrow \tilde{T}_{OA}^{v1}$	$\downarrow \tilde{T}_{LA3}^{v1}$	$\downarrow \tilde{T}_{LA2}^{v1}$	$\downarrow \tilde{T}_{LA1}^{v1}$

In the SCA procedure, the IL is intended to be in $(-B, B)$ ($= [-2, 2]$ for $B = 3$). However, IL can not be controlled when the tables are aggregated. Table 4 shows that unintended IL larger than B is obtained in many cells since small ILs at OA are compounded.

Table 4. Distribution of absolute IL ($|\tilde{f} - f|$) by SCA1

	OA				LA3			
|IL|	0	1	2	**3+**	0	1	2	**3+**
$p = 4$	93.6%	4.3%	2.1%	**0.0%**	73.0%	16.7%	8.1%	**2.2%**
$p = 3$	81.8%	10.4%	5.9%	**1.8%**	54.4%	18.9%	12.8%	**13.9%**
$p = 2$	55.7%	19.1%	12.8%	**12.4%**	28.5%	17.5%	14.2%	**39.7%**
$p = 1$	19.0%	19.2%	15.9%	**45.9%**	5.7%	10.2%	9.0%	**75.1%**
	LA2				LA1			
|IL|	0	1	2	**3+**	0	1	2	**3+**
$p = 4$	50.1%	11.9%	10.3%	**27.8%**	41.1%	8.2%	7.7%	**43.0%**
$p = 3$	36.2%	9.4%	8.3%	**46.2%**	30.8%	5.0%	4.8%	**59.4%**
$p = 2$	17.4%	5.7%	6.1%	**70.8%**	15.8%	1.2%	2.1%	**81.0%**
$p = 1$	1.8%	1.2%	3.6%	**93.3%**	3.0%	0.0%	0.0%	**97.0%**

In order to reduce IL, we can try to modify the SCA procedure as follows:

(SCA Algorithm 2). Small cell adjustments controlled to LA totals

1. For the given P variables, make full tables T_{LA}^{vP} at all area units.
2. Adjust small cells in each table of T_{LA}^{vP} by a random rounding technique with a base value B and obtain $\tilde{T}_{LA}^{vP}$.

3. Obtain $\tilde{T}_{LA}^{vp}$ by aggregating $\tilde{T}_{LA}^{vP}$ (denoted as $\downarrow$ in Table 3). Note that we get $T \neq \tilde{T} = \tilde{T}^{ag} \neq \tilde{T}_{ag}$.

It is reasonable to expect that SCA2 will reduce IL more than the SCA1 procedure. However, additivity is lost in the system. We might be able to explain the broken additivity in the system as the action for privacy protection while recommending users not to aggregate frequencies since we already provided all frequencies. The more serious drawback of SCA2 is the possibility of disclosure accidents.

Assume that a LA3 has six OAs with a frequency value of 1 for each specific cell. Let us employ the procedure SCA2 with $B = 3$ and assume that the small frequencies at the OA level are all changed into 3 accidentally. At LA3 level, the true frequency value 6 in this cell is not smaller than 3, so it is not rounded and the masked frequency is 6. Finally, users can conclude that each true value at OA is 1 since they know that the candidates are $\{1, 2, 3\}$ and that the sum of the candidates is 6. Table 5 describes this disclosure example. The bold means that the random rounding technique is applied to the frequencies.

Table 5. A disclosure example by SCA2

	OA$_1$	OA$_2$	...	OA$_6$	LA3
$\tilde{f}$	**3**	**3**	... **3**	6
Candidates of f	1,2,3	1,2,3	...	1,2,3	6
Inferred f	1	1	...	1	6

Consequently, the SCA procedures can not be properly employed because of the considerable IL or disclosure possibility of small cells. Therefore, we need to find an alternative and efficient solution for the frequency table generating system.

4   Bounded Small Cell Adjustments

Usually, we have put more stress on privacy than data utility. In a paper published in PSD 2014, this priority was discussed as follows:
"A widely accepted paradigm is that protection has priority over utility. This means that a minimum level of protection is a-priori decided and set in the optimization problem through constraints. Then an output maximizing the utility is searched among all solutions with an acceptable level of protection [1]".

In this viewpoint, the paradigm that we have had so far for disseminating tables can be described as follows:

1. **First consider an acceptable level of disclosure risk** and find a method to reduce DR measures.

2. Calculate IL using distance measures and select a masking method minimizing IL among the ones satisfying DR criteria.

In this paper, we suggest a shift in thinking on this paradigm. Eventually, we may derive the same conclusion by adopting a solution that minimizes IL with an acceptable level of protection. However, a more efficient solution can be found from putting the priority on IL reduction over DR control. Here is the new paradigm, under which we have constructed an efficient algorithm in this paper.

1. **First consider the permissible level of IL.** In a frequency table, IL values are certain in $\{0, 1, 2, \ldots, N\}$. When we employ the SCA technique, which changes frequencies $1, 2, \ldots, (B - 1)$ to 0 or B, we specify the permissible level of IL as $(B - 1)$. We therefore define $max\ IL = (B - 1)$.
2. Next consider DR measures. In the SCA, the disclosure probability of a masked small cell is $1/B$, which means that the small cell achieves B-anonymity or uncertainty. We have decided to employ k-anonymity as the DR measure and have studied how to achieve anonymity in the algorithm.

Under this paradigm, we can connect IL and DR with B. The ideal relationship between IL and DR can be defined by a rational function as follows:

$$y = \frac{1}{x + 1}, \quad x = max\ IL, \quad y = max\ DR.$$

Figure 1 illustrates the optimal risk-utility relationship. The right panel shows the range of IL and DR if we can apply an ideal solution to the system for $B = 3$.

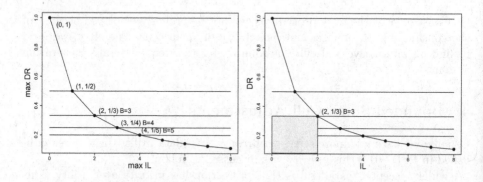

Fig. 1. The optimal relationship between information loss and disclosure risk

Now we will propose a new algorithm named the Bounded Small Cell Adjustments (BSCA). We will describe the notations first, and then discuss the main ideas and the algorithm of the BSCA along with IL results in our data.

4.1 Notations

Let LA_L represent a lower-dimensional or larger area level, such as T_{OA}^{vp} $(p < P)$ or T_{LA}^{vp} $(p \leq P,\ LA = LA3, LA2, LA1)$, while let LA_S be the level of full table T_{OA}^{vP}. Denote the number of full tables that belong to the ith LA_L table as K_i. Then, for a specific cell, the set of frequency values at LA_S in the ith LA_L table can be represented by $\{f_{ij} : j = 1, \ldots, K_i\}$. Table 6 illustrates how the frequency f_i at the LA_L level can be obtained from the frequencies $\{f_{ij}\}$ in the full tables.

Table 6. Notations of frequencies according to hierarchy

			LA_S			LA_L
Id	1	$\ldots$	j	$\ldots$	K_i	i
Frequency	f_{i1}	$\ldots$	f_{ij}	$\ldots$	f_{iK_i}	f_i

We can adjust risky small cells at LA_S using the random rounding technique with a base value B. Without loss of generality, we assume that there is no large cell at the LA_S level $(f_{ij} \leq B)$ in order to make the discussion simpler. The large cells can just be added at the final step in any frequency calculation case.

4.2 Main Ideas

In order to construct a new, more efficient solution, we have considered two main issues, first, how to reduce IL, and second, how to prevent unexpected disclosures.

First, we suggest the use of a Fixed Grid Rule (FGR) when providing frequency f_i at LA_L by adding up the corresponding small cells $\{f_{ij}\}$ at LA_S, so that we can efficiently control IL. Note that FGR is applied when f_i at LA_L is large $(f_i > B)$.

The idea of the FGR is simple. We divide integers with intervals and make fixed grids with length of B. Then we can use a representative value for each interval. When the frequency value f_i is in the grid of $[nB + 1,\ nB + B]$, we can assign the medium of the gird, which is $nB + 1 + [B/2]$ for an odd integer B, to $\tilde{f}_i$. If so, then IL can be less than $B/2$. Meanwhile, all masked frequency values can achieve B-anonymity since $\tilde{f}_i$ has B candidates. Figure 2 shows how FGR works. For better illustration, we just use a plane instead of a line.

Conceptually, the FGR might be similar to microaggregation. The important difference is that the grid is fixed and therefore the disseminated values are stable for changes in data distribution. That is, microaggregation needs a process of grouping neighbors so that a particular value can belong to different groups according to changes in other values, whereas fixed grids do not allow such change to that particular value. FGR is also very similar to conventional rounding except that the representative values are different.

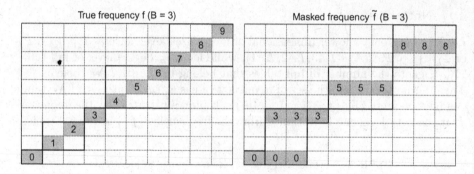

Fig. 2. Fixed grid rule for $B = 3$

Secondly, we have to examine all information given to users by the system since we want to avoid unexpected disclosure accidents. Users know the process of SCA and also can observe how many boundary cells ($\tilde{f}_{ij} = B$) are in the disseminated tables at LA_S. Let us denote the number of those boundary cells ($\tilde{f}_{ij} = B$) as k_i. Then the information given to users for $\{f_{ij}\}$ and f_i can be summarized as in Table 7.

Table 7. Users' information for true frequencies

$\tilde{f}_{ij}$	# (cells)	$\sum \tilde{f}_{ij}$	f_{ij}	$f_i = \sum f_{ij}$
0	$K_i - k_i$	$k_i B$	$0, \dots, (B-1)$	$[k_i, u_i]$
B	k_i		$1, \dots, B$	
Observation	Observation	Observation	Conjecture	Conjecture
$u_i = (K_i - k_i)(B - 1) + k_i B = K_i(B - 1) + k_i$				

Users can infer that the smallest value of true frequency f_i ($= \sum f_{ij}$) can be k_i if all cells disseminated as B at LA_S actually have true values of 1, while all zero cells in the disseminated tables are actually zero. Users can guess that the largest value of f_i is $(K_i - k_i)(B - 1) + k_i B$ if all cells disseminated as B have a true value of B and the zero cells are all actually $(B - 1)$. Finally, the set D of candidates of true frequencies can be inferred as follows:

$$f_i = \sum f_{ij} \in [k_i, \ (K_i - k_i)(B - 1) + k_i B] = [k_i, u_i] = D.$$

In Appendix B, there is a discussion on DR if we only use FGR or conventional rounding for the dissemination of hierarchical frequency tables.

In conclusion, the disseminator has to consider the three constraints in Table 8 in order to properly provide $\tilde{f}_i$ at the ith LA_L. First, the masked frequency can not be in $(0, B)$ since we employed the SCA. Next, we want to control IL under B as intended in the SCA. Lastly, for avoiding unexpected disclosures, we wish to prevent additional information leakage from the system. Therefore, the candidates set C directly provided by the system, which is the gird given to $\tilde{f}_i$, should be included in $D = [k_i, u_i]$, which is the interval derived from the system. That is, the frequency f_i has to achieve anonymity B despite the information of D.

Table 8. Three constraints for masking frequencies at LA_L level

Source	Constraints on $\tilde{f}_i$
SCA	$\tilde{f}_i = 0$ or $\tilde{f}_i \geq B$
IL control	$\|\tilde{f}_i - \sum f_{ij}\| < B$
DR control	$\tilde{f}_i \in C \subset D = [k_i, u_i]$

4.3 The Algorithm

In order to decide $\tilde{f}_i$, first we compare $\sum_j \tilde{f}_{ij}$ with $f_i = \sum_j f_{ij}$. If IL is less than B ($\|f_i - \sum_j \tilde{f}_{ij}\| < B$), $\sum_j \tilde{f}_{ij}$ satisfies the three constraints in Table 8 because $\sum_j \tilde{f}_{ij} = k_i B$. Therefore, we can directly use $\sum_j \tilde{f}_{ij}$ as $\tilde{f}_i$ for dissemination. Note that $\|f_i - \sum_j \tilde{f}_{ij}\| < B$ always holds when $f_i = 0$ or $K_i = 1$.

For all cases of $\|f_i - \sum_j \tilde{f}_{ij}\| \geq B$, we suggest two more steps for both reducing IL and preventing disclosures. Step 1 is applying FGR in order to control IL. For this, $f_i(= \sum_j f_{ij})$ should be represented with *mod* of B as follows:

$$f_i - 1 = a_i B + b_i, \qquad a_i \in Z^+, \text{and } b_i \in \{0, 1, 2, \ldots, (B-1)\}. \qquad (1)$$

Note that we employ the *mod* function for $f_i - 1$ instead of f_i in order to properly apply the FGR to the frequency values as illustrated in Fig. 2. For example, if $B = 3$, we would like to make a grid not for $\{3, 4, 5\}$ but for $\{4, 5, 6\}$. Now we choose $\tilde{f}_i - 1 = a_i B + [B/2]$, which is the median integer on the corresponding grid. Then IL can be controlled under $B/2$.

On the other hand, the grid C given to $\tilde{f}_i$, which is the set of candidates officially provided by the system, is $[\tilde{f}_i - [B/2], \tilde{f}_i + [B/2]]$. Disclosures (or decreasing anonymity) can happen if C exists out of the interval $D = [k_i, u_i]$, which represents the information that users can learn from the system. That is, disclosures can happen when $\tilde{f}_i - [B/2] < k_i$ or $u_i < \tilde{f}_i + [B/2]$.

The left panel in Fig. 3 shows the case of $\tilde{f}_i - [B/2] < k_i$, in which the anonymity decreases. From the value of $\tilde{f}_i$, users can learn that f_i must be in $[a_i B + 1, a_i B + B]$. Users also have information that $f_i \geq k_i$. Therefore, the set of candidates for f_i shrinks to $[k_i, a_i B + B]$ from $[a_i B + 1, a_i B + B]$ and anonymity decreases. The right one displays the case of $u_i < \tilde{f}_i + [B/2]$. In these two cases, users can obtain additional information by analyzing the system.

In Step 2 controlling DR, for the case of $\tilde{f}_i - [B/2] < k_i$, we suggest adding B to $\tilde{f}_i$. For the case of $u_i < \tilde{f}_i + [B/2]$, we propose to subtract B from $\tilde{f}_i$ in

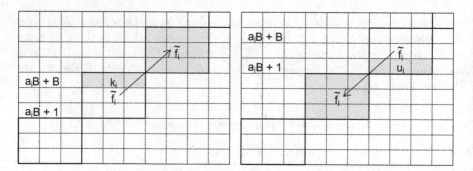

Fig. 3. Moving grid for maintaining anonymity

order to make anonymity more than B. Since the number of integers in D is large enough, that is $\#(D) = (K_i - 1)(B - 1) + B \geq B + B - 1$ for $K_i \geq 2$, the moved grid can stay in D.

However, IL can increase up to $[B/2] + (B - 1)$ from $[B/2]$ since f_i is considered to be in $[\tilde{f}_i - [B/2] - (B - 1), \tilde{f}_i + [B/2] + (B - 1)]$ instead of $[\tilde{f}_i - [B/2], \tilde{f}_i + [B/2]]$. Note that Step 2 is unnecessary when $\tilde{f}_i - [B/2] = k_i$ or $u_i = \tilde{f}_i + [B/2]$ so that IL does not increase up to $[B/2] + B$.

Through Step 2, these two disclosive cases causing $C \not\subset D$ are selectively processed with. The probability of these cases happening are very small since most small frequencies have to be changed into zero (or B) at the same time. Only in these rare disclosive situations does the IL increase while absolute IL in most cells is still under B as intended in the SCA.

For example, assume that $B = 3$, $\{f_{ij} = 1, j = 1, 2, \ldots, 6\}$ and $\{\tilde{f}_{ij} = 3, j = 1, 2, \ldots, 6\}$ by SCA, which is the same example in Table 5. After applying FGR in Step 1, $\tilde{f}_i$ is decided as 5 and $C = [4, 6]$. Users know that f_i should be in $[k_i, u_i] = [6, 18]$, so the set of candidates reduces to $\{6\}$ and the anonymity decreases to 1 from 3. However, by Step 2, we add $B = 3$ to the result of Step 1 and obtain $\tilde{f}_i = 8$. The cell at LA3 achieves anonymity greater than $B (= 3)$ since the set of candidates by the BSCA is $[\tilde{f}_i - [B/2] - (B - 1), \tilde{f}_i + [B/2] + (B - 1)] = [5, 11]$. Users can be informed that the true frequency is mostly in $\{7, 8, 9\}$ and rarely in $\{5, 6, 7, 8, 9, 10, 11\}$. In fact, IL is 2, which is actually $B - 1$, in this case. Table 9 summarizes this example.

Finally, the algorithm can be summarized as follows:

(BSCA Procedure). Bounded small cell adjustments controlled to OA totals

1. For the given P variables, make full tables T_{OA}^{vP} at each OA.
2. Adjust small cells in each table of T_{OA}^{vP} by a random rounding technique with a base value B and obtain $\tilde{T}_{OA}^{vP}$.
3. Construct $\tilde{T}_{LA}^{vp}$ by aggregating $\tilde{T}_{OA}^{vP}$, that is $\tilde{f}_i = \sum_j \tilde{f}_{ij}$.
(Steps 1–3 are the same to SCA1.)
4. For the cells that $|\tilde{f}_i - f_i| \geq B$ (in which $K_i \geq 2$),
   A. When $a_i > 0$, which means that f_i is the frequency value of a large cell,

Table 9. Example of applying the BSCA algorithm (B = 3)

	OA_1	OA_2	...	OA_6	LA3
f	1	1	...	1	6
SCA	3	3	...	3	18 or 6
(Step 1) FGR	3	3	...	3	5
candidates	1,2,3	1,2,3	...	1,2,3	$[4,6] \cap [6,18]$
(Step 2) Moving Grid	3	3	...	3	8
candidates	1,2,3	1,2,3	...	1,2,3	$[5,11] \cap [6,18]$

① update $\tilde{f}_i = a_i B + [B/2] + 1$ by applying FGR to $f_i - 1 = a_i B + b_i$.
② if $C \not\subset D$, update $\tilde{f}_i$ in order to prevent information leakage as follows:

$$\tilde{f}_i - [B/2] < k_i \longrightarrow \tilde{f}_i = \tilde{f}_i + B$$
$$u_i < \tilde{f}_i + [B/2] \longrightarrow \tilde{f}_i = \tilde{f}_i - B$$

B. When $a_i = 0$, which implies a small or a boundary cell,
① Assign $\tilde{f}_i = B$.
② if $C \not\subset D$, update $\tilde{f}_i$ in order to prevent information leakage as follows:

$$1 < k_i \longrightarrow \tilde{f}_i = \tilde{f}_i + [B/2]$$
$$u_i < B \longrightarrow \tilde{f}_i = \tilde{f}_i - [B/2] \quad \text{(can not happen)}$$

The only difference between SCA1 and BSCA is Step 4 in the above procedure. Note that Step 4 does not deal with large cells among $\{f_{ij}\}$ at LA_S level. The large cells at LA_S are added to $\tilde{f}_i$ after the BSCA procedure. Therefore, the final frequency can have any type of value because of large cells, even though the small cell aggregation results by the BSCA procedure are usually given in a form of $aB + 1 + [B/2]$ that is the representative value of a grid.

We have applied the BSCA algorithm to our census microdata and constructed masked frequency tables. Table 10 shows the distribution of IL by BSCA. In our data, the full table T_{OA}^{v4} has 6.4% small cells. Since the random rounding technique is only applied to the full table without any aggregation process, 93.6% cells can have zero IL in T_{OA}^{v4}. Absolute IL in most cells is smaller than or equal to $(B-1)(= 2 \text{ for } B = 3)$ as intended in SCA. Only in under 1% of cells do the absolute IL values increase up to $[B/2] + (B-1)(= 3, \text{for } B = 3)$. On the other hand, the anonymity $B(= 3)$ is guaranteed in all cells.

4.4 Comparison of SCA Algorithms on a Risk-Utility Map

Figure 4 shows the relationship between IL and DR according to SCA algorithms on a risk-utility map. If we employ SCA2, DR can increase up to 1. IL can be increased up to a considerable amount when constructing $\tilde{T}_{LA}^{vp}$ from $\tilde{T}_{LA}^{vP}$ for $p = 1, \ldots, (P - 1)$. The shaded area in the left panel illustrates those increases.

Table 10. Distribution of absolute IL ($|\tilde{f} - f|$) by BSCA ($B = 3$)

|IL|	OA				LA3			
	0	1	2	3	0	1	2	3
$p = 4$	93.6%	4.3%	2.1%	**0.0%**	73.2%	18.1%	8.4%	**0.4%**
$p = 3$	82.0%	11.5%	6.3%	**0.2%**	57.5%	28.8%	13.3%	**0.4%**
$p = 2$	58.2%	27.3%	14.0%	**0.5%**	39.7%	45.4%	14.7%	**0.3%**
$p = 1$	31.5%	51.8%	16.5%	**0.2%**	30.0%	61.0%	9.0%	**0.0%**
|IL|	LA2				LA1			
	0	1	2	3	0	1	2	3
$p = 4$	57.6%	31.7%	10.3%	**0.4%**	55.4%	36.8%	7.6%	**0.1%**
$p = 3$	50.7%	40.7%	8.4%	**0.2%**	49.6%	44.4%	5.8%	**0.2%**
$p = 2$	39.4%	55.1%	5.4%	**0.1%**	44.4%	53.3%	2.4%	**0.0%**
$p = 1$	34.6%	63.6%	1.8%	**0.0%**	39.4%	60.6%	0.0%	**0.0%**

However, BSCA can maintain anonymity, which means efficiently control DR, as is planned in the SCA. BSCA additionally increases IL just by $\lceil B/2 \rceil$ in a very few cells. For $B = 3$, the right panel in Fig. 4 shows that IL can increase only up to $(B - 1) + \lceil B/2 \rceil = 3$, which shows the conclusion in Table 10.

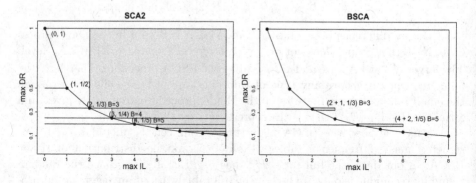

Fig. 4. Comparison of SCA algorithms on RU map

5　Concluding Remarks

In this paper, we have proposed an alternative SDC solution for frequency table generating systems. In Sect. 2, we introduced a data structure in order to deal with diverse shapes of tables according to the number of variables and the levels of area units. Then Sect. 3 has discussed the existing SCA procedures, which were originally designed to have a *max* IL of $(B-1)$ and a *max* DR of $1/B$ in any single table, along with their large IL and DR in a real system.

In Sect. 4, we tried putting the priority on reduction of IL instead of DR. Then we have examined how to reduce IL and maintain anonymity. In order to reduce IL efficiently, we have suggested employing the fixed grid rule. For preventing information leakage, we have studied the information given to users and then proposed to selectively move one gird for the risky frequencies.

Through the proposed BSCA algorithm, we have found that IL in most cells can be controlled less than or equal to $(B-1)$ while maintaining the anonymity of B in all cells. When disseminating frequencies, we have only to announce the ratio of cells whose IL increases up to $(B-1+B/2)$. In our city example, the ratio of IL over $(B-1)$ is under 1% for $B=3$. In addition, zero cells can have some uncertainty caused by the small cells, which occupy 6.4% of our data as shown in Table 10.

In the future, we will apply the BSCA algorithm to another dataset in which P is larger than 4 and the area units are a grid scale. Utility measures such as correlation should also be checked. However, the BSCA algorithm would be useful at least for publishing frequencies as descriptive statistics.

Appendix A

Multi-dimensional tables can be represented in a flattened form. Table 11 shows a three-dimensional flattened frequency tables in our data. The categories of each variable are listed from lexicographic ordering. We can represent any p-dimensional frequency table of an area unit in this kind of flattened form. In general, we have $_PC_P = 1$ table of T^{vP}, $_PC_{P-1}$ tables of $T^{v(P-1)}$, ..., $_PC_1 = P$ tables of T^{v1} and totally $2^P - 1$ tables at each area unit for P variables.

Table 11. Three dimensional flattened frequency tables, T^{v3}

Case	G	A	M	Frequency	Case	G	A	E	Frequency
1	1	1	1	$f_{1,1,1} = f_1$	1	1	1	1	$f_{1,1,1} = f_1$
		...					...		
$K = 180$	2	18	5	$f_{2,18,5} = f_K$	$K = 288$	2	18	8	$f_{2,18,8} = f_K$

Case	G	M	E	Frequency	Case	A	M	E	Frequency
1	1	1	1	$f_{1,1,1} = f_1$	1	1	1	1	$f_{1,1,1} = f_1$
		...					...		
$K = 80$	2	5	8	$f_{2,5,8} = f_K$	$K = 720$	18	5	8	$f_{18,5,8} = f_K$

Appendix B

Here, we would like to discuss how disclosures can occur by applying FGR only to hierarchical tables. According to Eq. (1), frequencies having hierarchy is denoted as

$$f_j = a_j B + b_j + 1, \qquad j = 1, \ldots, K$$
$$f = aB + b + 1 = \sum_{j=1}^{K} a_j B + \sum_{j=1}^{K} b_j + K.$$

Note that $a = [(f-1)/B]$ and $b \in \{0, 1, \ldots, B-1\}$. We omit subscript i for convenience. If we apply FGR to the frequencies at each hierarchical level, the masked values are as follows with $Q(\{b_1, \ldots, b_K\}) = [(\sum_{j=1}^{K} b_j + K - 1)/B]$:

$$\tilde{f}_j = a_j B + [B/2] + 1, \qquad j = 1, \ldots, K$$
$$\tilde{f} = aB + [B/2] + 1 = \sum_{j=1}^{K} a_j B + Q(\{b_1, \ldots, b_K\}) \cdot B + [B/2] + 1,$$

From the masked frequencies $\{\tilde{f}_j\}$ and $\tilde{f}$ provided by the system, users can directly obtain the information of

$$f_j \in C_j = [a_j B + 1, a_j B + B]$$
$$f \in C = [aB + 1, aB + B]$$

On the other hand, from $\{\tilde{f}_j\}$, users can also infer that the true f should be in $D = [\sum_{j=1}^{K} a_j B + K, \sum_{j=1}^{K} a_j B + KB]$.

Therefore, disclosures can happen (or anonymity can decrease) if $C \not\subset D$:

$$(1) \quad \sum_{j=1}^{K} a_j B + K > aB + 1 = \sum_{j=1}^{K} a_j B + Q(\{b_1, \ldots, b_K\}) \cdot B + 1 \quad \text{or}$$
$$(2) \quad \sum_{j=1}^{K} a_j B + KB < aB + B = \sum_{j=1}^{K} a_j B + Q(\{b_1, \ldots, b_K\}) \cdot B + B.$$

An example can be found when $\{f_1, f_2\} = \{4, 4\}$, $f = 8$ and $B = 3$. Note that (1) and (2) do not hold at the same time.

If we use FGR at each hierarchical level, we may not be able to avoid disclosures. However, we use FGR once when we aggregate the small cells masked by SCA in the full table and then have a moving grid step to avoid disclosures in our algorithm.

References

1. Hernández-García, M.-S., Salazar-González, J.-J.: Further developments with perturbation techniques to protect tabular data. In: Domingo-Ferrer, J. (ed.) PSD 2014. LNCS, vol. 8744, pp. 24–35. Springer, Cham (2014). https://doi.org/10.1007/978-3-319-11257-2_3
2. Hundepool, A., et al.: Statistical Disclosure Control. Wiley, Hoboken (2012)
3. Shlomo, N., Antal, L., Elliot, M.: Measuring disclosure risk and data utility for flexible table generators. J. Official Stat. **31**(2), 306–324 (2015)
4. Sweeney, L.: Achieving k-anonymity privacy protection using generalization and suppression. Int. J. Uncertainty, Fuzziness Knowl.-Based Syst. **10**(5), 571–588 (2002)

Designing Confidentiality on the Fly
Methodology – Three Aspects

Tobias Enderle[(✉)], Sarah Giessing, and Reinhard Tent

Federal Statistical Office of Germany, 65180 Wiesbaden, Germany
{Tobias.Enderle,Sarah.Giessing,
Reinhard.Tent}@destatis.de

Abstract. In the development of so called "Confidentiality on the fly" methodology building on random noise implemented with a cell key method, a number of issues have to be addressed. First, there is the choice of the probability distributions for the noise. Of course parameter sets yielding a low loss of information are desirable, but the disclosure risk avoidance potential of a parametrization should also be taken into account. This requires benchmarking of the risk avoidance potential of candidate settings.

Another issue is the communication of the potential effects of the noise on published results. The paper looks at the effect noise may have on estimates resulting from a division of two noisy counts.

Thirdly, a cell key method produces in the first place consistent, but non-additive results which might be difficult to communicate. One is tempted to restore additivity which – amongst other challenges – raises the issue of a technical solution.

1 Introduction

The online tool Table Builder of the Australian Bureau of Statistics (ABS) which allows users to define and download tables implements a disclosure control technique on the basis of additive noise that is applied 'on the fly' while the output is generated [9, 15]. A cell key technique ensures consistency of the output, i.e. the random noise added to a specific cell is assigned as a "function" of the cell key. For logically identical cells the noise will always be exactly the same, because cell keys are assigned consistently. This method, sometimes referred to as "cell key method", is one of the SDC methods addressed in EU project "Open Source tools for perturbative confidentiality methods".

The present paper studies some aspects in the context of the cell key method. A common issue with any perturbative SDC method for counts data is the reliability of results computed as a ratio of noisy counts involving smaller counts. Some considerations in this context will be the subject of Sect. 3.

The other two issues we look at in this paper are connected to the non-additivity of methods like the cell key method: perturbed interior cells of a table or hypercube generally do not add up exactly to the perturbed margins. Connected to any non-additive SDC method there is always a certain risk of disclosure by differencing. This risk should be taken into account when deciding on the noise parameters. When

J. Domingo-Ferrer and F. Montes (Eds.): PSD 2018, LNCS 11126, pp. 28–42, 2018.
https://doi.org/10.1007/978-3-319-99771-1_3

choosing noise parameters, we should be able to compute a measure of the differencing risks to benchmark risk avoidance potentials of different candidate settings. In this paper, we look for an efficient way of organizing a risk avoidance benchmarking, c.f. Sect. 4.

While the disclosure risk problem of non-additivity can be limited by suitable choice of noise parametrization, non-additivity may still be difficult to communicate to users of the data. [9] suggested to restore additivity. Following up on earlier work (c.f. [5]), Sect. 5 presents functionalities of a wrapper tool developed in R for control and execution of the CTA package [3], used successfully to restore additivity without manual intervention for a large set of small tables.

2 Recalling Issues of Noise Design for Protection of Counts Data

In this paper we assume an additive noise method to be defined by a transition matrix P (sometimes also referred to as "*p-table*"). Following a suggestion of [11], [14] describes an algorithm utilizing the NLopt-package for non-linear optimization [13] to compute transition matrices based on a maximum entropy approach[1].

Restoring to the denotation of [10], P is the $L \times L$ transition matrix[2] containing conditional probabilities: $p_{ij} = P$ (perturbed cell value is j | original cell value is i), where p_i refers to the i [th] row-vector of matrix P. Let v_i the column vector of the noise which is added, if an original value of i is turned into a value of j. I.e. the jth entry of v_i is $(j - i)$. As suggested in [9], the noise distributions resulting from the algorithm described in [11] ensure that the perturbations take integer values and that the following criteria hold:

1. the mean of the perturbation values is zero;
2. the perturbations have a fixed variance σ^2;
3. the perturbations will not produce negative cell values or positive cell values below a specified threshold j_s; and
4. the absolute value of any perturbation is less than a specified integer value D.

When using the algorithm from [11] to compute a transition matrix (or p-table), we must supply the parameters σ^2 and D, thus fixing the noise variance (typical choices are between 0.8 and 3) and its maximum absolute value. In addition, we can require no small, non-zero counts up to a certain threshold j_s to appear in the perturbed data ($j_s :$ = 2 prevents counts of 1 or 2 in the perturbed data, for example).

As explained in [11], for rows i from $i_s := D + j_s + 1$ onwards, it makes sense to compute the row vector p_i using the row vector p_{i_s} of conditional probabilities for the

---

[1] A first release of the R package "ptable" (Perturbation Table Generator) implementing the algorithm is planned for end of 2018 and will be available at https://github.com/sdcTools/ptable.

[2] As index j may take a value of zero (when a cell value is changed to zero), in the following we start counting matrix and vector indices at 0, enumerating rows and columns of the $L \times L$ matrix by 0, 1, 2, ..., $L - 1$. The number of rows and columns L, which we assume - without loss of generality (w.l.g.) - to be the same, differs for different set ups.

original count i_s and simply shift the set of non-zero entries $i - i_s$ places to the right. The algorithm therefore only computes the first i_s rows of P.

3 Ratios of Noisy Counts

So, the noise for the counts data has a fixed variance (identical for all counts) and the noise distribution is identical for all counts from count i_s onwards. Unfortunately, these nice statistical properties do not hold for estimates computed by dividing two perturbed counts. The ratio $\hat{R}$ of two noisy counts $\hat{X}$ and $\hat{Y}$ is an estimate of the true ratio $R := X/Y$ which is only approximately unbiased (see Appendix A.1 for formal proof). The conditional distribution (conditional on a particular realization $X = x$ for the enumerator and $Y = y$ for the denominator of the ratio R) of the estimate $\hat{R}$ and its conditional variance and coefficient of variation (CV) strongly depend on the size of counts X and Y. For illustration of what can happen, assume an extremely unfortunate case with small original counts $x = 4$ and $y = 4$, i.e. $R = 100\%$. Assume further $D = 3$ and a perturbation of $U = +D$ for the enumerator and $V = -D$ for the denominator[3]. Then $\hat{R} = 700\%$ – instead of the original 100%!

A positive point is: we can show that for large counts in the denominator (Y) the conditional variance of $\hat{R}$ tends to zero (see Appendix A.2). We can also show that the conditional CV of the estimate tends to zero, when both counts are large, the denominator Y and the enumerator X (c.f. Appendix A.3).

On the other hand, for small counts in either the denominator or the enumerator (or both) the ratio estimate is relatively imprecise. See Table 4 in Appendix A.3 for illustration. Assuming an illustrative but common noise distribution, the table presents the CV of the estimate $\hat{R}$ for a selection of counts x and y.

In particular, for large x, but small y, and also the other way round for large y, but small x, the standard deviation of the estimate is quite high, at an order of magnitude of about $0.1\ R^4$ and should thus not be ignored by users.

Therefore, up to certain thresholds for the counts in enumerator and denominator, it might be useful to compute and provide confidence intervals (or approximations thereof) for the ratio estimates. This might be interesting especially for internal users within an agency and avoid misinterpretations or release of data at too low level of detail.

3.1 Interval for $\hat{R}$

As the noise distributions for enumerator and denominator are discrete random distributions defined by the data provider, it is a straightforward exercise to compute the (conditional) probability density function for the deviations $|\hat{R} - R|$ given an observation (x, y) by computing the potential outcomes along with the respective

---

[3] In this example we assume w.l.g. the case $j_s := 0$, i.e. all counts allowed in the perturbed data.

[4] See the last line in Table 3 (Appendix A.1) for illustration of the case of large x. For formal evidence see [6], A.1.4.

probabilities. Cumulating the probabilities (after sorting them by size of $|\hat{R} - R|$) leads to the cumulative distribution function which could be used to look up confidence intervals. A simpler, rule-of-thumb like approach is to focus on an event with high probability and to compute the maximum deviation on this event. To this end we use $b_1(x, y) := \max_{(u,v) \in M} \left| \frac{x+u}{y+v} - \frac{x}{y} \right|$, where $M := \left\{ (k, l) \in \{-1, 0, 1\}^2 : |k| + |l| \leq 1 \right\}$. Looking up the entries p_{ij} in the transition matrix in row $i = i_s$ with the highest probabilities, i.e. for $j = i - 1, i, i + 1$ we compute $\sum_{(k,l) \in M} p_{i,i+k} p_{i,i+l} = p_{i,i}^2 + 4p_{i,i+1} p_{i,i}$, (because of the symmetry of non-zero entries in p_{i_s}) to obtain a lower bound for the probability of $\left\{ |\hat{R} - R| \leq b_1(x, y) \right\}$. With the probabilities of the illustrative example from Table 3 in the appendix the bound is above 60%. A slightly wider bound is $b_2(x, y) := \max_{(u,v) \in \{-1,0,1\}^2} \left| \frac{x+u}{y+v} - \frac{x}{y} \right|$. The corresponding lower bound for the probability of $\left\{ |\hat{R} - R| \leq b_2(x, y) \right\}$ is then $\sum_{k=-1}^{1} \sum_{j=-1}^{1} p_{i,i+j} p_{i,i+k} = p_{i,i}^2 + 4p_{i,i+1} p_{i,i} + 4p_{i,i+1} p_{i,i-1}$, and about 75% for the data from the numerical example.

4 Benchmarking Disclosure Risks of Different Parameter Settings

As explained in Sect. 2, random noise is defined by its parameters, like f.i. the maximum deviation D. As mentioned in the introduction, there is always a certain risk of disclosure by differencing connected to any non-additive SDC method like cell key based noise. When deciding on the noise parameters, one should therefore take into account the risk avoidance potential of a candidate set of parameters. In order to compare, i.e. benchmark different candidate settings regarding risk avoidance, first of all we need a method or tool to compute such risks. [11] suggests computation of feasibility intervals (c.f. [12], 4.3.1) for the original counts. These are typically computed by means of linear programming (LP) methods (see for example [8]). Solutions are obtained for each cell of interest by minimizing and maximizing a variable representing its cell value, subject to a set of constraints expressing the "logical" table relations, and some a priori upper and lower bounds on unknown cell values. In our scenario we assume all original non-zero cells to be "unknown". Rather conservatively, our scenario assumes the intruder to know the maximum noise deviation D leading to an a priori lower and upper bound for each cell. As explained in [11], for practical reasons, instead of an LP tool we rely on an implementation of the shuttle algorithm [2] in SAS, even though in some situations we may underestimate the true risks (c.f. [11], fn. 6). If the upper and the lower bound of a feasibility interval computed for a cell coincide, we consider the cell as disclosable.

Secondly, we need a dataset suitable for observing the risks. If the strategy is to use a fixed dataset and apply the risk assessment method to different noisy versions of this dataset, depending on risk averseness of an agency, especially when we need to compare low risk parameter settings, it may indeed take a major data set to observe a significant number of disclosable cells, at least for the variants with the higher risk avoidance potential, to get statistically valid results from the experiments. Setting up a suitable dataset for the experiments may thus require some effort. Once constructed, using f.i. Census data and topic breakdowns[5], one might be tempted to use it over and over again, no matter to which data the noise is supposed to be applied to after all.

Here, we propose an alternative strategy. The goal is to observe comparable risk indicators for different noise parameterizations also using data from collections much smaller than a population census. The general idea is to "replace" extensiveness of the test data set by concentrating on sparsely populated tables on one hand, and on some randomization on the other hand. Regarding the latter, the idea is to randomize in two ways, i.e. randomize (1) the data, and (2) the perturbation applied to the data. As further research option, randomizing the variable breakdowns might also be interesting to look at in future work.

A randomization of the data can be achieved by drawing subsamples from the original (micro-)data. To obtain m different versions of our tables or hypercubes, we draw m independent subsamples, construct the tests tables for each subsample and compute the risk measure for each "realization" of the test tables. Using a low sampling fraction in this step offers an additional advantage: it leads to less densely populated tables and thus (in our experience, see Sect. 4.1) typically increases the share of cells that will be disclosed by the method used to compute the risk measure[6]. We expect this in turn to improve the stability of the comparison of parameter settings. Randomizing the perturbation is straightforward for a stochastic method like random noise. To obtain n perturbed versions of each table, we draw record keys for each microdata subsample n times.

4.1   Test Application and Results

For demonstrating the suggested methods we basically use an extract from the synthetic dataset used for the testing in [5] which corresponds to a particular NUTS3[7] area. To this dataset we have added one more level of detail in the geography by "inventing" 7 small municipalities (the smallest with only 30 inhabitants) making up this synthetic small region on NUTS3 level. The data set partly implements the definitions of EU Census 2021 hypercube 9.2[8], i.e. a cross combination of the variable breakdowns SEX, AGE.M, and YAE.H (Year of arrival in the country since 1980)[9]. For good reasons – from SDC point of view – the geography of the original hypercube is defined only

---

[5] In previous experiments we successfully used a data set involving about 30 million cells.

[6] The drawn subsamples in our experiment have the following relative frequency distribution in average: 45% zeroes, 17% small cell counts (i.e. 1s and 2s), 12% cell counts between 3 and 10 and 26% larger cell counts (i.e. >10).

[7] The classification NUTS is a hierarchical system that is used to divide the territory of the EU into smaller regions. The level NUTS3 represents the smallest territorial unit.

[8] Details on the content of the hypercubes can be found in the Census 2021 draft implementing regulation (c.f. [7]).

[9] Some variables have more than one breakdown, each with different levels of detail. In the terminology of the draft implementing regulation, 'H' identifies breakdowns with the highest level of detail, 'M' identifies breakdowns with a medium level of detail, and 'L' identifies breakdowns with the lowest level of detail and 'N' identifies the breakdown that refers to the national level.

Table 1. Benchmarking risk vs. utility - indicators for disclosure risk and information loss

	(a) All original non-zero cells		(b) Original small cells (1's, 2's)	
	Disclosed cells (in %)	Mean absolute deviation	Disclosed cells (in %)	Mean absolute deviation
	Random noise			
D = 1	62.51	0.60	82.10	0.60
D = 2	1.64	0.27	4.82	0.29
D = 5	0.00	1.15	0.00	1.33
	Deterministic rounding			
base 3	95.70	0.75	99.00	1.00
base 5	11.91	1.25	21.37	1.31
base 7	0.50	1.68	0.66	1.31
base 9	0.02	2.08	0.03	1.31

down to NUTS3, not including the municipality level. The breakdown of our extended test hypercube is GEO.H x SEX x AGE.M x YAE.H. Exchanging GEO.M by GEO.H increases disclosure risks tremendously and thus facilitates risk observation and comparison. The synthetic area has a population of about 33,000 people, the hypercube consists of about 20,000 cells (thus an extremely low mean cell size of 1.5 people) and about 18,000 relations between interior and marginal cells.

. Table 1 presents the results of an evaluation of the disclosure risk vs. information loss (mean absolute deviation) for the synthetic hypercube for three variants of noise with maximum deviation of $D = 1$ vs. $D = 2$ vs. $D = 5$ and, for sake of comparison, for four variants of deterministic rounding to, e.g. rounding bases 3, 5, 7 and 9.

The table presents the averaged simulation results[10] of the rate of disclosed cells and the absolute deviation between original and perturbed cells. Both measures are computed (a) for all (original) non-zero counts, and (b) for only the cells with small counts (i.e. 1's, 2's). Obviously, rounding to base 3 offers little protection only. For this method, the algorithm discloses the table almost completely, especially the small counts. Rounding to base 5 improves the rates to about 11% in general and 21% of the small counts; rounding to bases 7 or 9 leads to further improvement at the expense of increasing information loss. The rates of disclosed cells for the noise variant with $D = 1$ are also quite high and range between those of the rounding variants with base 3 and 5. Rates for $D = 2$ tend to be higher than for rounding to base 7, but much lower than for rounding to base 5. For the noise variant with $D = 5$ no cases of disclosure were observed[11].

Notably, figures for the risk indicator obtained for just one "realization" of the noise and one subsample would not be very reliable, as we observed after the randomization

---

[10] Both indicators presented are derived stepwise. First, a measure within each subsample is computed by averaging over all n perturbed realizations. Finally, these m estimates of an indicator are averaged to an overall result.

[11] Besides the maximum deviation other specifics of a noise distribution affect both indicators. However, the maximum deviation usually has the strongest impact on disclosure risk

experiment where we replicated the perturbations $n = 100$ times for each of $m = 10$ subsamples drawn from the synthetic hypercube. Especially the different perturbations within each subsample produce large deviations of the disclosure risk indicator. Within the first subsample, for instance, we observe a mean disclosure rate of 63% for the noise variant $D = 1$. The maximum rate is 78%, the minimum 44%, with a standard deviation of 0.07.

Finally, since we may underestimate disclosure risks using the shuttle algorithm, we compared for two of the noise variants the result of the shuttle algorithm to results obtained using the LP based audit tool of τ-Argus (c.f. [4]). For noise with $D = 2$ the shuttle algorithm disclosed the same rate of 15.4% of the original non-zero counts within a minute while the LP tool takes 90 min. For noise with $D = 1$, the shuttle algorithm slightly underestimates the risk. The LP based tool discloses 69.12% percent taking 8 h for the computation, the shuttle algorithm 68.53% within a minute.

5   A Tool to Restore Additivity

As seen in the previous section, the disclosure risk due to non-additivity of random noise can be controlled by suitable parametrization. Non-additivity of this SDC method may nevertheless be difficult to communicate. [5] has presented a variety of heuristics how to use the CTA algorithm for restoring additivity to EU Census hypercubes after an initial perturbation using the cell key method. Naturally, an approach like CTA which "balances" the adjustments within the hypercube relations leads to much less perturbation in hypercube margins than simple summation of the noisy lowest level counts. Also not surprising was the result that information loss due to restoring additivity is generally the smaller, the smaller the "instance" considered for adjustment. [5] therefore recommended either not to restore additivity at all, or to restore additivity to separate low level geographic area hypercubes.

In this paper we follow up on this recommendation. For testing we use a number of small, municipality level one- and two-way tables without hierarchical structure[12] from the German Census 2011 standard publication for about 15% of the German municipalities with a very high share of small municipalities in the selection.

5.1   CTA Algorithm as Additivity Module

Leaning to the denotation of [3], a CTA instance is represented by (i) a set of cells y_i; $i = 1,...,n$, that satisfy m linear relations $Ay = b$ (y being the vector of y_i's; matrix A and vector b imposing the tabular constraints, expressing for example that the cell values of some set of cells must be identical to the cell value of another (marginal) cell); (ii) a lower and upper *a priori* bound for each cell $i = 1,...,n$, respectively l_i and u_i. In addition to that we can also define (iii) a set $P = \{i_1, i_2, ..., i_p\} \subseteq \{1,...,n\}$ of indices of "sensitive cells" and require (iv) for each sensitive cell $i \in P$ a lower and upper

---

[12] We only consider links due to a common population total. In such cases we state tables as set of linked tables in the problem setup.

protection level, respectively lpl_i and upl_i, such that the adjusted values satisfy either $x_i \geq y_i + upl_i$ or $x_i \leq y_i - lpl_i$.

Given these settings, the purpose of CTA is to find the set of closest feasible adjusted values x_i; $i = 1,...,n$ satisfying these conditions. This is expressed as the following optimization problem (in terms of the deviations $z_i =: x_i - y_i$ and w_i being a vector of cell weights)[13]:

$$
\begin{aligned}
&\min_{z} \quad \sum_{i=1}^{n} w_i |z_i| \\
&s.t. \quad Az = 0 \\
&\qquad l_i \leq z_i \leq u_i, \quad i = 1,\ldots,n \\
&\qquad z_i \leq -lpl_i \text{ or } z_i \geq upl_i, \quad i \in P
\end{aligned}
\tag{1}
$$

5.2 Problem Setup for the Use Case with Sensitive Cells

As explained in [5], in a context where non-zero counts up to j_s (viz., 1's and 2's) are not allowed we need a strategy involving the following two phases.

Initial Phase: As in this context there will be no noisy cell values below j_s, no cells need to be declared sensitive in CTA problems defined in the initial CTA phase (t = 1). Some cells $y_i^{(0)}$ may nevertheless be adjusted to $y_i^{(1)}$, where $0 < y_i^{(1)} \leq j_s$.

Second Phase: For CTA execution with index $t = 2$, define as set of sensitive cells $P^{(2)} := \left\{ i; 0 < y_i^{(1)} \leq j_s \right\}$. For $i \in P^{(2)}$ define protection levels $lpl_i := y_i^{(1)}$ and $upl_i := j_s + 1 - y_i^{(1)}$. Then, in a feasible solution $y_i^{(2)} > j_s$ or $y_i^{(2)} = 0$ for all $i \in P^{(2)}$.

In order to avoid that some other cells ($i \notin P^{(2)}$) are adjusted in this phase to $y_i^{(2)}$, where $0 < y_i^{(2)} \leq j_s$, we define special lower a priori bounds $l_i < y_i^{(1)} - j_s$. For all zero-cells, i.e. where $y_i^{(1)} = 0$, define upper a priori bounds $u_i = 0$. To avoid infeasibility problems we define those special a priori bounds only for cells not appearing as margin cell in any of the relations defined by A.

5.3 A Stable Implementation

The aim of the work described here is a stable implementation executing the CTA algorithm without manual intervention on large sets of small tables, guaranteeing a fully feasible solution, in particular in the use case with sensitive cells. We also want solutions where the maximum of the deviations z_i is "as small as possible", but if we define very tight a priori bounds, CTA problems may turn out to be infeasible. The CTA version of [3] offers many configuration options to enforce to a successful execution (to some degree) even of a basically infeasible problem. Unfortunately, none of them is universally ideal for every instance. With infeasible instances, in practice we face the problem that the CTA algorithm may

---

[13] For the exact mathematical statement and the linearity issue of the optimization problem see [3].

a. not return any result since the given LP is considered infeasible,
b. return an additive solution still containing sensitive cell values, or
c. return a solution that is still not exactly additive.

In order to automate the procedure, we developed a wrapper algorithm in R that not only applies CTA to a given instance, but automatically checks the output and takes suitable measures, preparing a follow up execution of the instance if needed. To this end, at first CTA is executed with the lower and upper bounds (l_i and u_i) as initially defined by the user, and the attainment of additivity as the procedures main objective. The result is checked then for additivity as well as for the occurrence of sensitive cells (non-zero counts up to j_s). If the result is additive and does not contain small counts, the procedure ends. If, however, the result is additive but still does contain small counts, the corresponding cells are set to sensitive and CTA is executed again on the new instance, but this time with elimination of sensitive cell values as new main objective. If the CTA algorithm did not return any result at all, the lower and upper bounds get expanded and CTA will be executed again.

Since in the first step the main objective is set to the restoration of additivity, theoretically case c. should not occur. However, due to certain specifications of the LP in reality it did. Repeating the application with restoration of additivity as main objective will of course not change the result. So in this case our algorithm falls back to changing the main objective to the elimination of small counts. Even though this may not lead to an accepted solution directly, it provides a new starting point for the next iteration.

Now, if in a second execution of the CTA algorithm the main objective is set to elimination of small counts, the new result might be non-additive again. In this case CTA is run again with the attainment of additivity as main objective. While in some cases this leads to a feasible solution, we can imagine that one might as well end up in an infinite loop, where the same values are changed back and forth, alternating between an additive solution or a solution without small counts. To solve this problem it turned out to be useful to store a history of sensitive cells and the corresponding cell values, in order to check if such a loop occurs. In this case an automatic manipulation of the hypercube happens: If the cell value that jumps in and out of the sensitive domain came from above the given threshold, it is set to zero, and if otherwise it was zero before it jumped into the sensitive domain, its value is increased such that it is not sensitive anymore. Subsequently CTA will be executed on the new data.

In case, however, this strategy still leads to an infinite loop, if a cell is set to an unsafe value three times, we set it back to its initial value and set the parameter *fixdir* of the CTA algorithm which fixes the direction of the cells perturbations to "no" in the next call of CTA. While actually "no" is the default setting of the CTA algorithm, due to better results in tests we regularly set *fixdir* to "random". Since a change of this parameter does alter the results returned by CTA, this way we can break from loops that cannot be handled otherwise.

While testing this algorithm a further problem occurred. Due to specifications of an instance it may happen that the CTA output is identical to its input. In this case our algorithm extends the lower and upper bounds and the main objective is changed to restoration of additivity, if it was set to elimination of small counts in the previous step and *vice versa*. If nevertheless the data still remain unchanged after three iterations, the parameter *fixdir* of the CTA algorithm is set to "no" again in the next call of CTA.

5.4 Test Application and Results

The wrapper algorithm was tested using a collection of small, one- and two-way tables without hierarchical structure[14] from the German Census 2011 standard municipality level publication. The test data originate from 1,585 municipalities with varying size and a high share of small municipalities.

Table 2. Restoring additivity - indicators for quality and performance

	Random noise (non-additive)	Random noise after CTA (considering sensitive cells)	Random noise after CTA (not considering sensitive cells)
No. of cells without deviation	47.06%	47.99%	49.10%
Mean absolute distance[a]	0.784	0.763	0.726
Cumulated relative absolute distance[a]	108,679	113,055	102,007
Maximum absolute distance	6	18	15
Mean number of CTA runs per table	–	2.44	1.00

[a]See [1], Table 1 for definition of the indicator.

First, random noise was added to the data in a parametrization with $j_s = 2$, i.e. the perturbed data did not contain any 1's and 2's. When set to preserve this absence of small counts, the recovery of additivity took the algorithm about two hours and ran without complications. Alternatively we applied the algorithm without the condition that no small counts should occur. This time it ran for only about 1.2 h.

Since naturally the original tables before perturbation were additive, it is not a surprising result that the restoration of additivity decreased the number of modified cells, as can be seen in Table 2. While also the mean absolute distance to the original data decreases after restoring additivity one should not ignore that the maximum absolute distance to the original data increased by a factor of (almost) two and that such extreme deviations cannot be predicted or even controlled in advance.

---

[14] We however consider links due to a common population total in the problem setup.

6 Summary and Conclusions

This paper has discussed three issues that should be considered when setting up a cell key based random noise method for disclosure control of counts data "on the fly". In Sect. 3 (and the appendix) it has examined relevant characteristics of ratio estimates computed by division of noisy counts: the conditional bias, variance, and CV. We have seen that with sufficiently large denominator the estimate for the ratio of two counts approximates unbiasedness, but the CV tends to zero only, if both, enumerator and denominator are large. To communicate possible effects of the noise on a ratio of counts, we have suggested an easy to calculate, simple kind of confidence interval.

In Sect. 4, the paper has stressed the importance of benchmarking disclosure risk avoidance potential of different noise parameterizations before selecting a particular parametrization for the noise. Aiming at the possibility to implement a reliable benchmarking procedure able to observe and compare also lower risk potentials even on the basis of relatively small datasets, the paper has looked into several options of randomization, in particular to randomize the noise, and to subsample the data, and has presented test results obtained with a SAS implementation of the shuttle algorithm [2] as tool for the risk assessment.

Finally, in the context of the non-additivity issue of the cell key method, the paper has explained some tricks implemented in a wrapper tool (in R) for the CTA package, exploiting the configuration options of that package. The tool has been used successfully to run CTA – if necessary with iterations – on a large number of small, noisy tables, making them additive while keeping the maximum of the deviation introduced in this step as low as possible. The paper concentrates here purely on the aspect of technical implementation. The question, if it is worthwhile to give up the consistency property of the cell key method in exchange for additivity of some tabular outputs – and which ones - is not discussed but needs further attention and research in the future.

Acknowledgements. The research leading to these results has partially received funding from the EU project "Open Source tools for perturbative confidentiality methods" (Specific grant agreement N° 2018.0108) under the Framework partnership agreement n° 11112.2014.005-2014.533

Appendix

This appendix provides some evidence for the claims made in Sect. 3. In addition to the denotation from Sect. 3 we define u and v the respective column vectors of possible outcomes of the noise U added to X and V added to Y, i.e. $u = v = (-D, -D+1, \ldots, -1, 0, 1, \ldots D-1, D)^T$ and the noise distribution defined by row vector p_{i_s} of transition matrix P for both, U and V[15]. So for any count $y, y \geq i_s$ denote p_y

---

[15] To keep it simple, we do not consider ratios that involve very small counts below i_s here, ignoring the distributions defined by the first rows of the matrix.

Table 3. Illustrative example for $c_1(y)$, $c_2(y)$ and $c_3(y) := \sqrt{c_2(y) - (c_1(y))^2}$, for selected values of y with $p_y = (0.009, 0.059, 0.182, 0.5, 0.182, 0.059, 0.009)$ (For large x, the standard deviation of $\hat{R}$ tends to $c_3(y)R$.)

y	4	5	9	10	20	50	100
c_2	1.35	1.17	1.04	1.03	1.01	1.00	1.00
c_1	1.09	1.08	1.01	1.01	1.00	1.00	1.00
c_3	0.41	0.24	0.12	0.11	0.05	0.02	0.01

the vector of non-zero elements of p_{i_s}. The $2D+1$ entries of p_y define the probabilities $p_{y,y+v_j}$ for original count y to change into $y+v_j$.

Given noise distribution V, we also define two functions c_1 and c_2:

$$c_1(y) := yE\left(\frac{1}{y+V}\right) \text{ and} \tag{2}$$

$$c_2(y) := y^2E\left(\frac{1}{y+V}\right)^2. \tag{3}$$

With $e \in \mathbb{R}^{2D+1}$ a column vector of 1's, we write $(ye+v)^{-1}$ for the vector with entries $\left(\frac{1}{y+v_j}\right)_{j=1,...,2D+1}$, and get $c_1(y) = yE\left(\frac{1}{y+V}\right) = yp_y(ye+v)^{-1} = y\sum_{j=1}^{2D+1} \frac{p_{y,y+v_j}}{(y+v_j)} = y\sum_{j=1}^{2D+1} \frac{p_{y,y+v_j}}{y\left(1+\frac{v_j}{y}\right)} = p_y\left(e+\frac{1}{y}v\right)^{-1}$. For large y, obviously, $e+\frac{1}{y}v$ approximates e. So, in that case $c_1(y)$ approximates $p_ye = 1$, because p_y defines a discrete probability distribution and hence its entries sum up to 1. With analog argument it is easy to see that the same holds for $c_2(y)$. For illustration, Table 3 provides figures for $c_1(y)$ and $c_2(y)$, computed for some selected values of y.

A.1 Conditional Bias of the Estimate $\hat{R}$

The Conditional Bias of $\hat{R}$ is $E(\hat{R} - R|X = x, Y = y)$. We write this as $E\left(\frac{x+U}{y+V}\right) - \frac{x}{y}$. Because U and V are independent and identically distributed with $EU = EV = 0$ we have

$$E\left(\hat{R}|X = x, Y = y\right) = E\left(\frac{x+U}{y+V}\right) = xE\left(\frac{1}{y+V}\right) = \frac{x}{y}c_1(y), \tag{4}$$

because of (2). As shown above, $c_1(y)$ tends to 1 for large y. Therefore the bias $E\left(\frac{x+U}{y+V}\right) - \frac{x}{y} = \frac{x}{y}c_1(y) - \frac{x}{y} = \frac{x}{y}(c_1(y) - 1)$ approximates zero, which means $\hat{R}$ is an approximately (for large y) unbiased estimate of R.

A.2 Conditional Variance of the Estimate $\hat{R}$

The conditional variance of the estimate $\hat{R}$, i.e. is defined as $Var(\hat{R}|(X.Y)) = E\left(\left(\hat{R} - E(\hat{R}|(X.Y))\right)^2|(X.Y)\right) = E(\hat{R}^2|(X.Y)) - (E(\hat{R}|(X.Y)))^2$. Conditioning on a particular realization $X = x$ for the enumerator and $Y = y$ for the denominator of the ratio R, we write

$$\mathrm{Var}(\hat{R}|(X.Y)) = E\left(\frac{x+U}{y+V}\right)^2 - \left(E\left(\frac{x+U}{y+V}\right)\right)^2. \tag{5}$$

Because U and V are independent and identically distributed with $EU = EV = 0$ and $Var U = Var V = \sigma^2$, and with $c_1(y)$ and $c_2(y)$ according to (2) and (3), we have

$$E\left(\frac{x+U}{y+V}\right)^2 = E(x+U)^2 E\left(\frac{1}{y+V}\right)^2 = \frac{x^2+\sigma^2}{y^2}c_2(y), \text{ and}$$

$$\left(E\left(\frac{x+U}{y+V}\right)\right)^2 = \left(E(x+U)E\left(\frac{1}{y+V}\right)\right)^2 = \left(x \cdot E\left(\frac{1}{y+V}\right)\right)^2 = \left(\frac{x}{y} \cdot c_1(y)\right)^2.$$

So (5) is equal to $\mathrm{Var}(\hat{R}|(X.Y)) = \frac{x^2+\sigma^2}{y^2}c_2(y) - \left(\frac{x}{y} \cdot c_1(y)\right)^2 = \frac{x^2}{y^2}(c_2(y) - (c_1(y))^2) + \frac{\sigma^2}{y^2}c_2(y)$, i.e. we have

$$\mathrm{Var}(\hat{R}|(X.Y)) = \frac{x^2}{y^2}\left(c_2(y) - (c_1(y))^2\right) + \frac{\sigma^2}{y^2}c_2(y). \tag{6}$$

As shown above, both, $c_1(y)$ and $c_2(y)$ tend to 1 for large y. Hence, for large y, $\mathrm{Var}(\hat{R}|(X.Y))$ tends to zero. Obviously, we also have $\mathrm{Var}(\hat{R}|(X.Y)) = \frac{x^2}{y^2}\left(\left(c_2(y) - (c_1(y))^2\right) + \frac{\sigma^2}{x^2}c_2(y)\right)$ which means that for large x, the standard deviation of $\hat{R}$ tends to $R\sqrt{c_2(y) - (c_1(y))^2}$. See the last line of Table 3 for illustration of the effects.

A.3 Coefficient of Variance (CV) of the Estimate $\hat{R}$

The conditional CV of $\hat{R}$ is defined as $\mathrm{CV}(\hat{R}|X,Y) := \frac{\sqrt{\mathrm{Var}(\hat{R}|(X.Y))}}{E(\hat{R}|X,Y)}$. With denotation from above, and because of (4) and (6) we can write this as $\mathrm{CV}(\hat{R}|X,Y) =$
$$\frac{\sqrt{\frac{x^2}{y^2}(c_2(y)-(c_1(y))^2)+\frac{\sigma^2}{y^2}c_2(y)}}{\frac{x}{y}c_1(y)} = \sqrt{\frac{(c_2(y)-(c_1(y))^2)}{(c_1(y))^2} + \frac{\sigma^2}{x^2}\frac{c_2(y)}{(c_1(y))^2}}, \text{ i.e.}$$

$$CV(\hat{R}|X, Y) = \sqrt{\frac{c_2(y)}{(c_1(y))^2}\left(1 + \frac{\sigma^2}{x^2}\right)} - 1. \tag{7}$$

As for large y, both, $c_1(y)$ as well as $c_2(y)$ tend to 1, from (7) we see that for large y, $CV(\hat{R}|X, Y)$ approximates $\frac{\sigma}{x}$, thus tending to zero, if (and only if) both, y and x become large.

Table 4. Numerical example for $CV(\hat{R}|X, Y)$ (in %) for selected combinations of x and y, calculated with $p_y = (0.009, 0.059, 0.182, 0.5, 0.182, 0.059, 0.009)$

$y\downarrow\ x\rightarrow$	4	5	9	10	15	20	50	100
4	46	43	40	39	38	38	38	38
5	36	32	27	27	26	25	25	25
9	28	23	16	15	14	13	12	12
10	27	23	15	14	12	12	11	10
15	26	21	13	12	10	8	7	7
20	26	21	12	11	8	7	5	5
50	25	20	11	10	7	5	3	2
100	25	20	11	10	7	5	2	1

References

1. Antal, L., Enderle, T., Giessing, S.: Statistical disclosure control methods for harmonised protection of census data (2017). https://ec.europa.eu/eurostat/cros/system/files/methods_for_protecting_census_data.pdf
2. Buzzigoli, L., Giusti, A.: An algorithm to calculate the lower and upperbounds of the elements of an array given its marginals. In: Statistical Data Protection (SDP 1998) Proceedings, pp. 131–147. Eurostat, Luxembourg ((1998)
3. Castro, J., Gonzalez, J.A., Baena, D., Jimenez, X.: User's and programmer's manual of the RCTA package (v.2). Technical report DR 2013-06 (2013). http://www-eio.upc.es/ ~jcastro
4. De Wolf, P.-P., Hundepool, A., Giessing, S., Castro, J., Salazar, J.J.: t-ARGUS User's Manual (2014). http://neon.vb.cbs.nl/casc/Software/TauManualV4.1.pdf
5. Enderle, T., Giessing, S.: Testing CTA as additivity module for perturbed census 2021 EU Hypercube Data. In: Joint UNECE/Eurostat Work Session on Statistical Data Confidentiality, Skopje, 20–22 September 2017 (2017). http://www.unece.org/fileadmin/DAM/stats/documents/ece/ces/ge.46/2017/1_testing_cta.pdf
6. Enderle, T., Giessing, S., Tent, R.: Designing confidentiality on the fly methodology – some aspects. Unpublished manuscript (2018)
7. Eurostat Unit F2: Commission implementing Regulation laying down rules for the application of Regulation (EC) No 763/2008 of the European Parliament and of the Council on population and housing censuses as regards the technical specifications of the topics and of their breakdowns, Item 2 of the agenda. In: 30th Meeting of the European Statistical System Committee, 28th September 2016, ESSC 2016/30/3/EN (2016)
8. Fischetti, M., Salazar-González, J.J.: Models and algorithms for optimizing cell suppression problem in tabular data with linear constraints. J. Am. Stat. Assoc. **95**, 916–928 (2000)

9. Fraser, B., Wooton, J.: A proposed method for confidentialising tabular output to protect against differencing. In: Monographs of Official Statistics. Work Session on Statistical Data Confidentiality, Eurostat-Office for Official Publications of the European Communities, Luxembourg, pp. 299–302 (2006)
10. Giessing, S., Höhne, J.: Eliminating small cells from census counts tables: some considerations on transition probabilities. In: Domingo-Ferrer, J., Magkos, E. (eds.) PSD 2010. LNCS, vol. 6344, pp. 52–65. Springer, Heidelberg (2010). https://doi.org/10.1007/978-3-642-15838-4_5
11. Giessing, S.: Computational issues in the design of transition probabilities and disclosure risk estimation for additive noise. In: Domingo-Ferrer, J., Pejić-Bach, M. (eds.) PSD 2016. LNCS, vol. 9867, pp. 237–251. Springer, Cham (2016). https://doi.org/10.1007/978-3-319-45381-1_18
12. Hundepool, A., et al.: Statistical Disclosure Control. Wiley, Chichester (2012)
13. Johnson, S.G.: The NLopt nonlinear-optimization package (2015). http://ab-initio.mit.edu/nlopt
14. Marley, J.K., Leaver, V.L.: A method for confidentialising user-defined tables: statistical properties and a risk-utility analysis. In: Proceedings of 58th World Statistical Congress, pp. 1072–1081 (2011)
15. Thompson, G., Broadfoot, S., Elazar, D.: Methodology for the automatic confidentialisation of statistical outputs from remote servers at the Australian Bureau of Statistics. Paper presented at the Joint UNECE/Eurostat Work Session on Statistical Data Confidentiality, Ottawa, 28–30 Oktober 2013 (2013). http://www.unece.org/fileadmin/DAM/stats/documents/ece/ces/ge.46/2013/Topic_1_ABS.pdf

Protecting Census 2021 Origin-Destination Data Using a Combination of Cell-Key Perturbation and Suppression

Iain Dove[✉], Christos Ntoumos, and Keith Spicer

Office for National Statistics, Titchfield PO15 5RR, UK
{iain.dove, Christos.ntoumos, Keith.spicer}@ons.gov.uk

Abstract. The UK Office for National Statistics (ONS) is intending to produce outputs involving travel to and from different locations (origins and destinations) in 2021, as they have done for previous Censuses. This data poses a particular challenge for protecting against disclosure risk, as categorising respondents on multiple geographical variables yields very sparse tables. This paper explores the disclosure risk and data utility of one option for protecting this data: applying cell-key perturbation (noise), and suppressing the remaining disclosive values. It finds that these methods provide good protection for the data with considerable loss of utility for outputs at low geographies. Whether this is an acceptable approach will be determined by user feedback.

Keywords: Origin destination · Flow data · Cell-key perturbation
Suppression

1 Introduction

National Statistical Institutions (NSIs) have a duty to protect respondents' information from being learned by others. Responsible use of information, including protecting against disclosures, is vital in protecting the reputation of any organisation and helping maintain survey response rates amongst members of the public, as well as being an administrative [8] and legal requirement (Data Protection Act) (1998 UK) [6], the Statistics and Registration Services Act (2007 UK) [7] and General Data Protection Regulation (2016 EU) [9]). This protection requirement leads to the disclosure control paradigm, in which NSIs try to reduce the risk of disclosure within statistical outputs, without compromising the utility of the data to researchers [3]. Origin-destination (flow) data, which describe movements between two different locations poses a particular challenge, as the data are inherently very sparsely distributed.

One of the most frequently used flow data in the UK is known as 'travel to work' data, which details where respondents commute to and from, alongside other characteristics. Origin-destination data is an important output used for informing transport infrastructure policy, (traffic projections, potential public transport routes) population-demographic projections (migration patterns), housing policy (second homes, migration by age groups) and others (Fig. 1).

© Crown 2018

J. Domingo-Ferrer and F. Montes (Eds.): PSD 2018, LNCS 11126, pp. 43–55, 2018.
https://doi.org/10.1007/978-3-319-99771-1_4

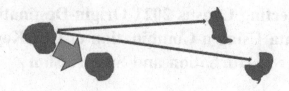

Fig. 1. Nearby areas often have large 'flows', which can be broken down by other variables without disclosing information about individuals, though many distant areas have very small, often disclosive flows

Nearby areas are much more likely to have larger flows than very distant areas. These large flows are of interest to researchers as they describe the characteristics of large groups of people. However, given a reasonable level of detail in the geography variables, many of the flows across longer or even moderate distances will be small, and potentially disclosive.

In the previous UK Census in 2011, restricting access was the main form of protection used, with less detailed tables available publicly at high geographic levels, and most outputs available under a licensed agreement whereby the user can access the data providing they agree to certain conditions (they will not attempt to learn about individuals, pass on the data to others etc.). Other more detailed outputs were accessible only within in a secure environment, specifying that although users have full access to the microdata within a secure location, all aggregated outputs or analysis need to be checked before release. This approach provided sufficient protection to the data, but access was difficult to obtain for some users. In particular, access was more readily available for academics and researchers than for business users or members of the public. ONS' main aim for origin-destination outputs in 2021 is to improve in this aspect by providing more publicly available data on origin destination, available to all users.

2   Disclosure Control Methods

2.1   Targeted Record Swapping

In the UK, the main form of protection applied to census data is targeted record swapping [4], where respondents (households) from different areas are swapped with each other. Some households appear in a different area in the data, other than their true location. All households have the potential to be swapped with another, introducing uncertainty that potentially identified records or households may have been swapped, and apparent identified data is false. Random swapping was applied in 2001 with all households equally likely to be swapped. In 2011, swapping was targeted towards more risky records, with households deemed at greater risk of identification more likely to be selected for swapping (Fig. 2).

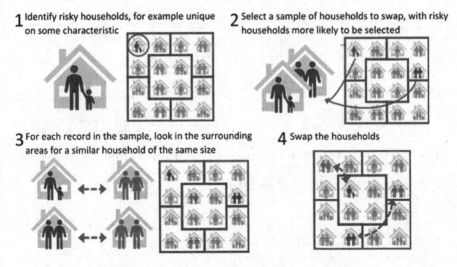

1 Identify risky households, for example unique on some characteristic

2 Select a sample of households to swap, with risky households more likely to be selected

3 For each record in the sample, look in the surrounding areas for a similar household of the same size

4 Swap the households

Fig. 2. An illustration of targeted record swapping

Record swapping provides good overall protection, though is designed to protect regular outputs rather than origin destination data. The targeting of risky records is based on households at risk within regular outputs, which may be a different set of households to those at risk in origin destination outputs. Because of the level of sparsity found in origin-destination data, it is also necessary to consider the perception of disclosure (the appearance of disclosure risk, even if protection has been applied), which will be much higher than usual. The perception of individuals information being disclosed can be damaging to potentially affected respondents, and affect the reputation of ONS, irrespective of whether the information gained is true, so will be considered here.

2.2 Cell-Key Perturbation

In Census 2021, ONS aims to provide more flexible outputs, give users more choice about the detail they would like, make the data much more accessible through a simple system, and provide the data more quickly than in previous censuses. This flexible approach should lead to more tables being provided to users than in 2011, and potentially a greater variety of detail or breakdowns. This increases the risk of differencing (taking the difference between two similar tables which differ slightly in one aspect to gain more information than the NSI had intended). For example, providing data by two different breakdowns, 12–15 year olds and 11–15 year olds, independently these data can be considered low risk, but when considered together (specifically the difference between them) they could provide a disclosive level of detail on the characteristics of 11 year olds.

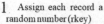

1 Assign each record a random number (rkey)

Record	rkey
r_1 ⟶	54
r_2 ⟶	4
r_3 ⟶	93
⋮	
r_N ⟶	26

2 For each cell, sum rkey and apply a function to get a cell key

Age by sex	Male	Female
0-15	.	.
16-24	.	4
25-34	.	.
⋮		

Record	rkey
r_2 ⟶	4
r_4 ⟶	61
r_{56} ⟶	7
r_{72} ⟶	90
Sum =	162

For example, take the last two digits → **ckey = 62**

3 Use a look up table to get a perturbation value

Cell Key ⟶

		1	2	3	...	61	62	63	...	99
Cell Value	1		+1							
	2			+1				-1		
	3									+1
	4	-1					+1			
	5			-1		-1				
	⋮									

4 Apply the perturbation value to the cell

Age by sex	Male	Female
0-15	.	.
16-24	.	5
25-34	.	.
⋮		

Fig. 3. Illustration of the cell-key perturbation method, first developed by the Australian Bureau of Statistics (ABS). Every record is assigned a random record key, after which the method is deterministic, allowing a repeatable procedure using the same record keys. The noise or perturbation applied to a cell is a function of the sum of the record keys of all records contained within the cell, and the cell value (number of records in the cell). The perturbation lookup table determines the perturbation value to apply to each cell, though most combinations are set to apply noise of +0

To protect against differencing, low levels of noise can be applied to outputs. If intruders attempt to difference two similar tables, the result will be the true values plus the noise added to both tables. This noise should ideally be small enough to leave large counts and analyses unaffected, whilst introducing uncertainty in any small counts obtained through differencing. ONS is investigating the use of cell-key perturbation on outputs, which involves the addition of noise in a pseudorandom mechanism (Fig. 3). This method was first developed by the Australian Bureau of Statistics (ABS) [1, 2, 5]. Every record is given a random 'record key'. The sum of record keys of all records contained in a cell determines the noise given to that cell. The record keys are permanently assigned, so that repeated queries will receive the same noise. It also has the effect that a group of records will receive the same noise, even if appearing within different tables.

2.3 Suppression

Very sparse tables lead to attribute disclosures (learning an attribute of an individual, using a reasonable level of prior knowledge). Given the nature of these outputs, the size of flows between geographies is considered to be a priority for users over the accompanying characteristic information. To prevent disclosures, we can consider suppressing the characteristic variable whilst maintaining basic information about the flow.

Table 1. Illustration of the outputs before applying the suppression method

Area of residence	Area of workplace	Mode of transport	Before suppression
Area A	Area C	Car	1
Area A	Area C	Bike	0
Area A	Area C	Walk	0
Area B	Area D	Car	5
Area B	Area D	Bike	2
Area B	Area D	Walk	0

Table 2. Illustration of the outputs after applying the suppression method

Area of residence	Area of workplace	Mode of transport	After suppression
Area A	Area C	All	1
Area B	Area D	Car	5
Area B	Area D	Bike	2
Area B	Area D	Walk	0

In Table 1, given knowledge of a respondent who lived in 'Area A' and worked in 'Area C', an intruder would learn that the individual must have travelled by car. The same process could be easily repeated for any other variable as there will remain only one respondent travelling from 'Area A' to 'Area C'. After applying suppression, Table 2 reveals that one person travelled from Area A to Area C, but no other information, the intruder does not learn any new information about the individual. The information revealed for flows from 'Area B' to 'Area D' is unaffected.

This disclosure risk can apply to large groups in the same way, e.g. if 10 respondents had travelled from 'Area A' to 'Area C', all of which used the same form of transport the same information could still be learned by an intruder. Though, when large groups of people have disclosures in this way, the information is often less sensitive, and could be learned by an intruder by other means (they learn that an individual travels by car not because they are identified in the output, but because a large number of similar respondents travel in this way). Section 3 investigates the impact on data utility of allowing disclosures on varying sizes of groups of respondents, specifically suppressing disclosures on any size of group of respondents (known as no threshold), suppressing group disclosures on fewer than 5 records (a group disclosure threshold of 5), and suppressing disclosures on fewer than 3 records (a group disclosure threshold of 3).

3 Application of Methods

3.1 Application of Perturbation

Using a combination of methods attempts to minimise the level of suppression necessary, potentially allowing more detail in larger flows to be released. As the main form of protection for census data, record swapping will have been applied to the microdata records before the production of origin-destination outputs.

The post-tabular methods described (perturbation and suppression) were applied on a number of univariate origin-destination tables using 2011 Census data. These could, in 2021, potentially be released publicly given this level of protection. The majority of these tables were not publicly available for the previous census. The data covered 3 different types of "flow", commuting flows to work, migrations flows from address one year ago to current address, and flows from main address to a listed second address. The data were aggregated at 4 different geographical levels, Local Authority (avg. 160,000 residents), Middle Super Output Area (MSOA, avg. 8000 residents), Wards (avg. 6500 residents) and Output Area (OA, avg. 300 residents). A number of popular characteristic variables were used: age, sex, marital status, ethnic group, religion, country of birth, occupation, industry, economic activity, approximated social grade, number of hours worked, method of travel, and distance travelled to work.

Note that these tables and results do not include zero cells. Origin destination tables inclusive of zero cells would be unmanageably large, with the vast majority of cells containing uninformative 'empty flows' from distant parts of the country.

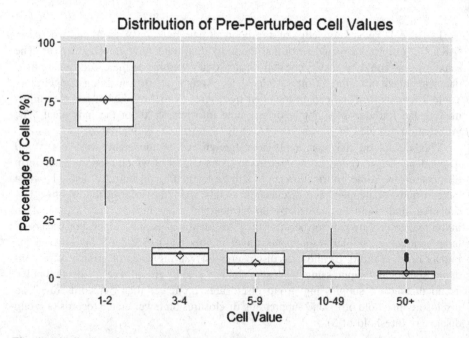

Fig. 4. Distribution of pre-perturbation cell values (excluding zeros). The majority of the table cell values are 1 or 2. One observation is recorded for each table.

As expected the vast majority of these tables consisted of very low counts across all geographic levels (Fig. 4). For tables at Output Area level, 95% or more of non-empty cells were value 1 or 2, an indicator of severe sparsity. Amongst higher geography tables, such as Local Authority, between 30 and 60% of cells were size 1 or 2.

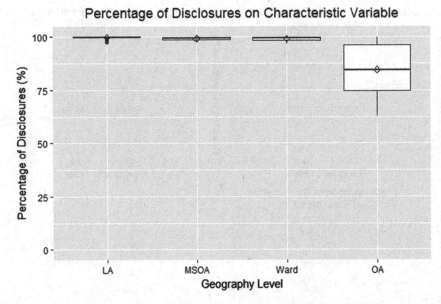

Fig. 5. Proportion of disclosures that reveal characteristic information, rather than geographic

Of all possible disclosures revealing information about the origin of a respondent, their destination or the value of their characteristic variable, a large majority of the disclosures were found to reveal the value of the characteristic variable (Fig. 5). Intuitively this is a result of the origin and destination variables having many more categories than the characteristic variable, even at higher geographies, respondents were likely to be unique on origin and destination. For MSOA, Ward, and Local Authority levels, these consistently accounted for more than 95% of disclosures. At Output Area level, disclosures on origin or destination formed a larger proportion of disclosures, but the characteristic variable still accounted for between 65 and 100%.

The focus of this paper from this point on will be disclosures on the characteristic variable, for the following reasons:

1. For any level of geography, disclosures on the characteristic variable form the majority of disclosures
2. The information in the characteristic variable is considered to be more sensitive than geographic information, and less likely to be previously known by an intruder
3. Given the structure of origin-destination outputs, a disclosure on the characteristic variable can lead more readily to other disclosures; one isolated flow from one geography to another can be broken down to disclose any other characteristic variable, an isolated flow from one geography by a characteristic variable can only be broken down by other geographical information, which may already have been disclosed.

Given these selected outputs, applying perturbation can be shown to not significantly address this disclosure risk (Fig. 6):

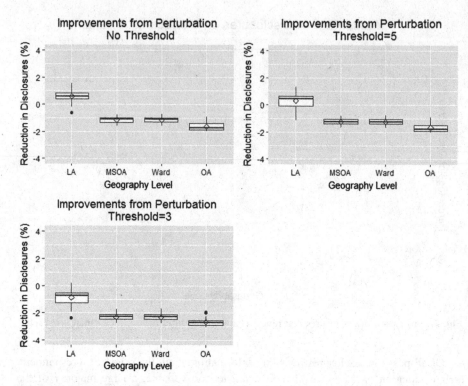

Fig. 6. Reduction in number of attribute disclosures by geography, after applying perturbation

Comparing the number of attribute disclosures before and after perturbation, very little difference is found. Perturbation was intended to protect against disclosure by differencing, rather than this type of attribute disclosure. It is clear that perturbation on its own is not sufficient to protect the data, it tends to reduce the number of disclosures by between 0 and 4%, though in Local Authority tables it often increases the number of disclosures (produces more apparent disclosures than it protects).

The impact of perturbation was also observed across different types of origin-destination table (commuting flows to work addresses, migrations from addresses one year ago to current address, and flows from main address to a listed second address). No significant relationship was found between type of flow table and the protection provided by perturbation, all tables received very low levels of protection, between 1–3% reduction.

It should be noted that a relatively low level of perturbation was used for this test case, applying more noise could be used to provide more protection, along with the associated further reduction in utility of the data. The level of perturbation that ONS may apply to Census 2021 data has not been fixed.

3.2   Application of Suppression

Suppression directly protects against disclosures on the characteristic variable by removing information deemed to have been disclosed. Given a threshold for a size of group disclosures to allow, it guarantees that information is not learned on characteristic variable (for groups smaller than the set threshold).

As with all forms of suppression, totals could be used to attempt to unpick the protection, and estimate the true values. In this case this form of attack is prevented by the perturbation. Consider flows within a large area, consisting of two small areas. It's possible that only one small area flow would be suppressed. Usually totals can be used to 'unpick' the suppression, in this case perturbation adds uncertainty to this unpicking (Table 3):

Table 3. Uncertainty in unpicking values provided by perturbation

Origin	Destination	True count	Perturbed count
Large area	Large area	20	19
Small area 1	Small area 1	7	7
Small area 1	Small area 2	4	4
Small area 2	Small area 1	1	*
Small area 2	Small area 2	8	9

Using the perturbed counts provided, with the flows within the large area as the totals, an intruder may estimate that the flow from small area 1 to small area 2 would be −1, which is clearly incorrect. This estimate is the true flow from small area 1 to small area 2 (1) plus noise added to the table. This prevention depends heavily on the outcome of perturbation, but comfortably provides uncertainty around values found in this way. The same mechanism applies to subsets of data broken down by other characteristic variables. This is a considerable benefit of applying perturbation, as secondary suppression can be difficult to implement and result in high utility loss.

Suppression provides strong protection against disclosure on the characteristic variable, though the resulting impact on the utility of the data must be considered [3]. A good measure of lost utility would be the level of suppression applied. Though the direct number of suppressions applied may be misleading, as larger flows could be considered to contain more data than smaller flows, the percentage of records contained in suppressed cells is also considered (Fig. 7).

The percentage of cells suppressed, against the percentage of records suppressed is plotted, using different thresholds for minimum group size (no threshold, threshold of 5, or threshold of 3). Ideal observations for utility would be in the bottom left of the graph, representing very few cells and very few records having been suppressed. Observations towards the top right of the graph represent high levels of suppression/ utility loss.

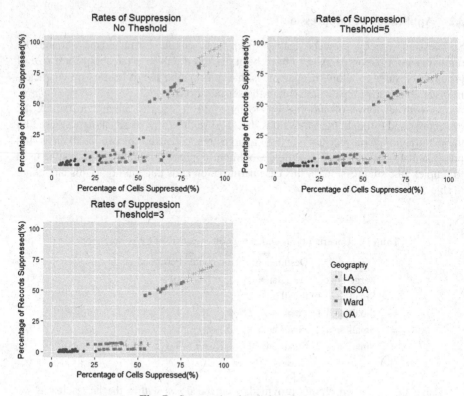

Fig. 7. Suppression levels by geography

A clear relationship between geography and suppression can be seen here, with lower geographies having much higher rates of suppression required as expected. High geography tables have much higher counts, and breakdowns are usually available with few attribute disclosures requiring suppression. Low geographies can be very sparse with up to 90% of cells resulting in disclosures. As is always the case, larger flows over shorter distances tend to be safe, long distance flows tend to be disclosive but small, and make up a small proportion of the table. Observations of MSOA and Ward (green triangles ad blue squares) are very close to each other, a likely result of their very similar size.

The association between proportion of cells and proportion of records suppressed can also be seen to be positive, and relatively flat when using the lower thresholds for disclosures, and higher geographies. Using a lower threshold reduces the necessary suppression, which causes a considerable improvement in utility of the lower geography tables (Output Area). Tables at Local Authority consistently have very few records suppressed (average 6%), with MSOA and Ward quite varied, depending heavily on the variables used.

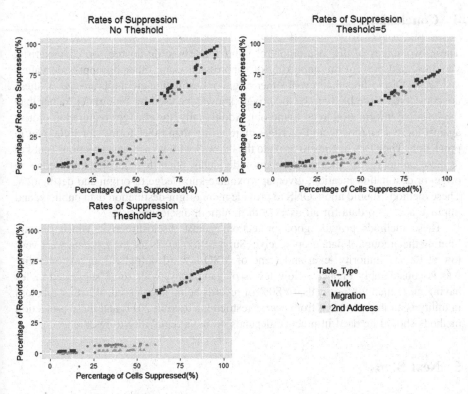

Fig. 8. Suppression levels by table type (Color figure online)

Suppression rates by table type is also plotted (Fig. 8). Migration tables (denoted by green triangles) consistently have lower level of records suppressed. This may be a result of the many respondents who have not migrated, which are very unlikely to be suppressed. Second address tables generally require more suppression than work tables, but clearly size of geography areas used has a greater impact than the type of table, observations are very clustered by geography and much more mixed by table type.

Perturbation has no overall effect on the univariate distributions as the method is designed to be unbiased for each cell count. (Even if certain categories are all very low counts whilst others are all very large, both categories will receive on average zero noise). Suppression however may affect the distribution of the characteristic variable, it's possible that some categories will be suppressed more often others (small flows are more likely to contain common categories on the characteristic variable, so overall the more common categories are more likely to have values suppressed/removed). To counteract this effect the overall univariate distributions will be released separately. Researchers can use this information to adjust for the missingness when necessary.

4 Conclusions

These two stages of disclosure control remove attribute disclosures on characteristic variables, and prevent the unpicking of suppression using other unsuppressed values and totals. Disclosures that reveal either origin or destination are not protected, as these are considered more likely to be the private knowledge used by an intruder, rather than the sensitive information to be found out. Additionally, the changes required to protect against these scenarios would sacrifice more utility than other forms, for little gain in protection. The ONS intends to provide the "flows" with no access restrictions (public) as these cannot reveal information about individuals, other than detailed knowledge on origin or destination location, given approximate knowledge of origin and destination. These methods should allow ONS to provide more origin-destination data publicly, and improve access to data for all users with minimum disclosure risk.

These methods provide good protection against disclosure risk, whilst directly limiting the amount of data users receive. Suppression levels for the test data were very low at Local Authority level and some of MSOA/Ward outputs. Suppression/utility loss is considerably greater at low levels of geography, with some suggested tables having more than 90% of cells, or 80% of records being suppressed. Whether this loss in utility is an improvement from access restrictions used in 2011, and these protection methods should be used in practice depends on its acceptability to researchers.

5 Next Steps

User research will ultimately decide whether this approach is deemed acceptable and is implemented in 2021. The next focus of this research should be to gauge users views on their needs for Origin-Destination data and whether this form of protection will be sufficient for their research purposes.

It is also worth investigating the impact of using a mix of geographic levels, which may significantly reduce the level of suppression needed, e.g. tabulating flows originating from relatively large areas (MSOA) but maintaining high level of detail in destinations (Output Area), or vice versa. This could be greatly beneficial if one geography is of much more interest to users than the other, this may allow more detail to be kept in priority variables by sacrificing detail in low priority variables without an increase in disclosure risk.

References

1. Fraser, B., Wooton, J.: A proposed method for confidentialising tabular output to protect against differencing. In: Joint UNECE Eurostat Work Session on Statistical Data Confidentiality, Geneva, Switzerland, 9–11 November 2005
2. Leaver, V.: Implementing a method for automatically protecting user-defined Census tables. In: Joint UNECE/Eurostat Work Session on Statistical Data Confidentiality, Bilbao, Spain (2009)

3. Hundepool, A., et al.: Statistical Disclosure Control. Wiley Series in Survey Methodology. Wiley, Hoboken (2012)
4. Shlomo, N., Tudor, C., Groom, P.: Data swapping for protecting census tables. In: Domingo-Ferrer, J., Magkos, E. (eds.) PSD 2010. LNCS, vol. 6344, pp. 41–51. Springer, Heidelberg (2010). https://doi.org/10.1007/978-3-642-15838-4_4
5. Shlomo, N., Young, C.: Invariant post-tabular protection of census frequency counts. In: Domingo-Ferrer, J., Saygın, Y. (eds.) PSD 2008. LNCS, vol. 5262, pp. 77–89. Springer, Heidelberg (2008). https://doi.org/10.1007/978-3-540-87471-3_7
6. Data Protection Act (1998). http://www.legislation.gov.uk/ukpga/1998/29
7. Statistics and Registration Service Act (2007). http://www.legislation.gov.uk/ukpga/2007/18/section/39
8. UK Statistics Authority Code of Practice for Official Statistics (2009). https://www.statisticsauthority.gov.uk/wp-content/uploads/2015/12/images-codeofpracticeforofficialstatisticsjanuary2009_tcm97-25306.pdf
9. GDPR legislation. https://ec.europa.eu/commission/priorities/justice-and-fundamental-rights/data-protection/2018-reform-eu-data-protection-rules_en

Synthetic Data

On the Privacy Guarantees of Synthetic Data: A Reassessment from the Maximum-Knowledge Attacker Perspective

Nicolas Ruiz[1(✉)], Krishnamurty Muralidhar[2],
and Josep Domingo-Ferrer[1]

[1] UNESCO Chair in Data Privacy,
Department of Computer Science and Mathematics,
CYBERCAT-Center for Cybersecurity Research of Catalonia,
Universitat Rovira i Virgili, Av. Països Catalans 26,
43007 Tarragona, Catalonia, Spain
nicolas.ruiz@oecd.org, josep.domingo@urv.cat
[2] Department of Marketing and Supply Chain Management,
Price College of Business, University of Oklahoma,
308 Brooks Street, Norman, OK 73019, USA
krishm@ou.edu

Abstract. Generating synthetic data for the dissemination of individual information in a privacy-preserving way is an approach that is often presented as superior to other statistical disclosure control techniques. The reason for such claim is straightforward at first glance: since all records disseminated are synthetic and not actual observed values, no individual can reasonably claim to face a privacy threat. Thus, and if the synthesizer used is good enough, synthetic data will potentially always offer a high level of information with low disclosure risk attached. Building on recent advances in the literature regarding the conceptualization of an intruder, this paper aims at challenging this claim by reassessing the privacy guarantees of synthetic data. Using the concept of a maximum-knowledge intruder, we demonstrate that synthetic data can in fact be always expressed as a re-arrangement of the original data and that, as a result, they may lead to configurations where disclosure risk may be higher than for non-synthetic disclosure control approaches. We illustrate the application of these results by an empirical example.

Keywords: Statistical disclosure control · Synthetic data
Maximum-knowledge attacker

1 Introduction

Data on individual subjects are increasingly collected and exchanged. By their nature, they provide a rich amount of information that can inform statistical and policy analysis in a meaningful way. However, due to the legal obligations surrounding these data, this wealth of information is often not fully exploited in order to protect the confidentiality of respondents. In fact, such requirements shape the dissemination policy of microdata

© Springer Nature Switzerland AG 2018
J. Domingo-Ferrer and F. Montes (Eds.): PSD 2018, LNCS 11126, pp. 59–74, 2018.
https://doi.org/10.1007/978-3-319-99771-1_5

at national and international levels. The issue is how to ensure a sufficient level of data protection to meet releasers' concerns in terms of legal and ethical requirements, while offering users a reasonable richness of information. Moreover, over the last decade the role of microdata has changed from being the preserve of National Statistical Offices and government departments to being a vital tool for a wide range of analysts trying to understand both social and economic phenomena. As a result, more parties, often very heterogeneous in their privacy and information requirements, are now involved in microdata transactions. This has opened a new range of questions and pressing needs about the privacy/information trade-off and the quest for best practices that can be both useful to users but also respectful of respondents' privacy.

Statistical disclosure control (SDC) research has a rich history in addressing those issues, by providing the analytical apparatus through which the privacy/information trade-off can be assessed and implemented. SDC consists in the set of tools that can enhance the level of confidentiality of any data while preserving to a lesser or greater extent their level of information (see [7] for an authoritative survey). Over the years, it has burgeoned in many directions. In particular, techniques applicable to microdata, which are the focus of this paper, offer a wide variety of tools to protect the confidentiality of respondents while maximizing the information content of the data released, for the benefits of society at large.

While generally considered as part of the SDC literature, the publication of synthetic data is an appealing alternative to, but also a significant departure from, pure SDC methods. The idea is simple: instead of disseminating an anonymized version of a dataset, i.e. the original data altered by the application of an SDC method, some data are instead created by drawing from a model fitted to the original data (hereafter called a synthesizer). At first glance it is clear that, since all values are synthetic and none of the individuals in the original data are included, disclosure risk must be practically nonexistent [14]. The original data are used to build the synthesizer, and thus the contribution of an individual to a data set is not pointless but is in fact used only as an informational basis. As a result, synthetic data seem to offer a clear and almost definitive advantage compared to other SDC methods: it would seem that synthetic data can be made as close as possible to the original data without any strong concern for privacy, while for non-synthetic SDC methods similarity to original data must be traded off against disclosure risk in a more stringent way (and hence utility is necessarily limited).

However, further scrutiny appears to weaken the advantage offered by synthetic data. For the sake of illustration, assume a dystopian society in possession of a perfect synthesizer, i.e. one that is able to perfectly replicate the statistical information observed over its population. In this case, an intruder using the synthetic data to conduct his attack may be able to re-identify some individuals or learn some sensitive information about them. From the point of view of the individuals, the fact that the information gained by the intruder is synthetic does not change much the situation: the right to privacy has been violated. While from a legal perspective this situation may not be unlawful [19], from an ethical perspective this can be clearly qualified as a negative outcome. Of course, in real life the perfect synthesizer does not exist. But the better the job done by the data releaser to create the synthetic data, the closer can be an attacker to gaining valuable information about some respondents in the original data. Thus, it can be reasonably argued that, ultimately, synthetic data are somehow subject to the same kind of risk/information trade-off faced by non-synthetic SDC methods.

It is based on these considerations that the privacy guarantees of synthetic data need to be explicitly considered. In [18], a privacy model to produce synthetic data with *ex ante* privacy guarantees was proposed. Here, we take an *ex post* approach, as previously performed in e.g. [6, 12], but based on a new, encompassing definition of an attacker, in order to reassess the privacy guarantees of synthetic data, regardless of how they have been obtained. The definition of an attacker on individual data has always been a thorny issue in the literature, not least because one must postulate how much background knowledge the attacker has. As a result, a variety of scenarios can be constructed, all based on *ad hoc* assumptions that may not comply with the views and the constraints faced by the data releasers, and that will also remain very context-specific. A recent proposal in the SDC literature has tried to circumvent these difficulties by proposing the concept of a maximum-knowledge attacker [1]. This attacker is based on a rather radical setting because he is entitled with the knowledge of *both* the original and the anonymized data set. While this may appear as unrealistic at first glance, such a scenario is conceptually powerful: any anonymized data set judged as sufficiently safe in term of disclosure risk under this scenario will in fact be safe under any kind of other possible scenario. As a result, this concept is a way to unify the comparison of the various performances of SDC techniques by using a common benchmark. It is also a way to ease the dissemination of individual data in the sense that, if a data releaser agrees with the level of disclosure risk contained in his anonymized data set under the maximum-knowledge attacker configuration, then he can be reassured that the release will be safe whatever the malicious attempts that could take place on his data.

Now, if the notion of a maximum-knowledge attacker seems valuable to gauge non-synthetic SDC techniques, it seems fair to submit synthetic data to the same kind of test. This is the purpose of this paper, structured as follows. Section 2 gives some background concepts on synthetic data and the maximum-knowledge attacker model needed later on. Section 3 characterizes the consequences of having some synthetic data submitted to a maximum-knowledge attack, and subsequently derives some new tools to assess their privacy guarantees. Section 4 presents some empirical results based on these tools. Conclusions and future research directions are gathered in Sect. 5.

2 Background Concepts

2.1 Synthetic Data

Synthetic data rely on a principle that is by nature similar to the imputation of missing values in a data set. The idea is to fit a model, called a synthesizer, to the original data; then values are drawn from the synthesizer to replace original data rather than merely imputing missing data. Three types of synthetic data can be distinguished [7]:

- *Fully synthetic data:* no original data are released and the values of all attributes across all records are synthetic.

- *Partially synthetic data:* across some if not all records, only sensitive attributes are synthesized while for example quasi-identifiers are original values.
- *Hybrid data:* original and fully synthetic data are combined, and the resulting data can be more or less similar to the original or fully synthetic data.

The above distinction will not have any consequences in what follows in this paper, so we will use the term synthetic data indistinctively to point to any of the three types. However, what is common to them is obviously the pivotal role of the synthesizer. Generating synthetic data worth disseminating is work-intensive, not least because creating a synthesizer that can replicate the intricate features of a micro data set necessitates some time and an involved level of expertise. It is beyond the scope of this paper to discuss the relative merits of the several approaches available to create a synthesizer, as well as the criteria that can be used to gauge it (see [4] for an extensive discussion), but a general principle is that the level of information offered by a synthetic data set can be only as good as the quality of the underlying synthesizer used to generate it. In this paper, we will simply assume that the data releaser did a good enough job so that the resulting synthetic data are worth disseminating and being analyzed by the users.

Regarding the practical characteristics of synthetic data, let us emphasize that they do not always come under the same format than the original data. First of all, they do not have to be of the same size, although having the same number of synthetic records than the number of original records seems a natural choice. To the best of the authors' knowledge, no firm guideline exists in the literature on this criterion (see however [13] for an empirical discussion). Depending on the context, an argument can be made for releasing synthetic data smaller than, same size as, or larger than the original data. Given this, we will assume that the number of synthetic records is the same as the original data. However, we will not restrict to the case of equal number of synthetic and original records, as one of the appeals of synthetic data is that they can come under any size. Specifically, we will outline below a pre-sampling procedure that can be applied before undertaking the evaluation of the privacy guarantees of synthetic data; this will in fact allow gauging synthetic data sets of any size.

A second difference with non-synthetic SDC methods is that synthetic data generally lead to the dissemination of several data sets, while for the former methods only one set is released. This practice is motivated by the goal of capturing the different designs of the original data [15]. Clearly, such a feature can quickly become cumbersome for the users (as well as for the releasers who need to generate the sets under various design configurations) and thus has to balance cost and accuracy [13]. Moreover, in the case where the original data are numerical and approximately multivariate normal, the sufficiency-based perturbation approach will perform at least as well as synthetic data for the preservation of information, while at the same time necessitating the release of only a single data set, which eases the tasks of the users [10].

Here again, no firm guideline exists on the right number of data sets to be released. The original proposal of releasing multiple data sets postulates as a rule-of-thumb a typical number between 3 and 10 [15], but later contributions outlined that this number is in fact context-dependent and may vary according to the analytical needs of the user and the properties of the employed synthesizer [13]. In this paper, we will assume that an arbitrary number M of synthetic data sets is released. As we will demonstrate, this number will turn out to be critical for the privacy guarantees of synthetic data.

Finally, in the introduction of this paper we briefly touched upon the fact that disclosure risk in fully synthetic data must always be by nature almost non-existent. Such claim has been made at various occasions in the literature, e.g. [4, 5, 13, 14], albeit it must be mentioned that: (i) this conclusion is less clear-cut for partially synthetic or hybrid data [5, 7] (which by construction will contain some of the original data), (ii) as far as pure synthetic data are concerned, some attempts to evaluate disclosure risk have also been previously proposed [6, 12]. In these last two cases however, it is again generally assumed that the risk is very low. The recent advances in the SDC literature on the notion of intruder cast a new light on this crucial feature of synthetic data.

2.2 The Maximum-Knowledge Attacker Model

The issue of an attacker's background knowledge has been recently pushed further in the literature through the proposal of the maximum-knowledge attacker model [1, 9]. This model defines an attacker who knows both the original data set and its entire corresponding anonymized version. This is a rather extreme configuration, unlikely to be mirrored by concrete situations, but it remains however conceptually very insightful, as an anonymized data set that can pass the test of such a situation will in fact be able to pass any test. It also has the advantage of completely resolving the issue of which background knowledge is to be assumed for operationalizing an attack on individual data [9]. Moreover, this model has as a consequence that, while it legitimates an exclusive focus on re-identification disclosure, it can be easily adapted to attribute disclosure assessment by excluding from the maximum-knowledge background a specific attribute [2]. In that case, the attacker's objective is to learn about the specific attribute's values as precisely as possible. This possibility, of particular interest for synthetic data, will be exploited in the empirical section of this paper.

The concept of a maximum-knowledge attacker is rooted in the known-plaintext attack defined in cryptology. While other types of attack can be conceived, they carry less meaning in the context of individual data [1]. A ciphertext-only attack, where only the anonymized data set is available, is less realistic than a known-plaintext attack: the attacker is likely to know at least a few original records or attributes, as part of his background knowledge. Regarding chosen-plaintext or chosen ciphertext-attacks, they are relevant only in the case in which the attacker can interact with the anonymization procedure; this may occur when protecting the answers to interactive queries to an on-line database, but it does not happen when releasing an anonymized data set. Note that a maximum-knowledge attacker, observing both the original data set and its anonymized version, has nothing to gain in terms of information. One can view his attempt as being purely slanderous, trying to discredit the data releaser by revealing his anonymization procedure.

Given the assumption that such a powerful person might exist, this leads to one question: what is exactly the perspective of that intruder? In fact, the reply relies on the record tracking numbers. Generally, and after having applied a non-synthetic SDC technique, data releasers can track which anonymized record derives from which original record through a number that does not carry any information of any sort and is unaffected by the anonymization procedure. Moreover, when the data are released, all numbers can be modified or deleted. But these numbers, known for practical purposes by the data releaser but not by the maximum-knowledge attacker, act in fact as a mask. Contrary to the statement made in [16], record tracking numbers in fact set the limit of the maximum-knowledge attacker. To make this clear, Table 1 illustrates the attacker's perspective on a toy example. In this example, the intruder has to retrieve the mapping between records in X and records in Y. His task is equivalent to retrieving some permutation structures. Note that permutation is in fact the overarching principle governing non-synthetic SDC methods [1, 11, 16, 17]. In what follows, we will also demonstrate that actually such principle applies to synthetic methods too.

Table 1. Point of view of a maximum-knowledge attacker

		Original dataset X				Anonymized dataset Y		
ID	X_1	X_2	X_3		ID	Y_1	Y_2	Y_3
1	13	135	3707		8	160	3248
2	20	52	826		20	57	822
3	2	123	-1317		-1	122	248
4	15	165	2419		18	135	597
5	29	160	-1008		29	164	-1927

Using synthetic data does have some implications for the maximum-knowledge attacker model. For non-synthetic SDC methods, the releaser has the advantage over the maximum-knowledge attacker of knowing the mapping between the tracking numbers in X and Y. The releaser can use this knowledge for example to assess how an individual has been protected; even the individual herself can verify her protection, if she can identify her own record in the non-synthetic data set. But *for synthetic methods the mapping between original and synthetic records does not make much sense*: a synthetic record does not derive from any specific single original record. Thus, *the advantage of the releaser over the maximum-knowledge attacker vanishes: both are at the same level of knowledge*. The privacy risk in synthetic data is not tied to a mapping: it is rather connected with knowing that synthetic records exist that are very close to some original records. In fact, real and synthetic individuals are linked by information. This can be assessed by a multivariate version of a rank-based record linkage procedure that will be developed below.

3   Synthetic Data from the Maximum-Knowledge Attacker Perspective

3.1   Multiple Reverse Mapping of Synthetic Data

We first start by observing that a synthetic data releaser can always transform the data such that each attribute in each synthetic data set can be expressed as a permutation of the original data. This procedure, called reverse mapping, has been recently proposed in the literature for non-synthetic SDC methods [1, 11]. To the best of our knowledge, this is the first time that it is developed for synthetic data.

Assume that a releaser generates $m = 1, \ldots, M$ synthetic data sets $Y^m = \left(Y_1^m, \ldots, Y_p^m\right)$ based on an original data set $X = (X_1, \ldots, X_p)$; denote by $X_j = (x_{1,j}, \ldots, x_{n,j})$ and $Y_j^m = \left(y_{1,j}^m, \ldots, y_{n_m,j}^m\right)$ the values of attribute $j = 1, \ldots, p$ over n records in the original data and n_m records in the mth synthetic data set, respectively. No further assumptions are made, except that the values of an attribute can always be ranked, which is obvious in the case of numerical or categorical attributes, but also feasible in the case of nominal ones [3].

In particular, the synthetic data sets need not be of the same size as the original data set. However, in order to perform reverse mapping, we need to compare sets of the same size. This issue can be fixed as follows: when the synthetic data have more (resp. less) records than the original data, synthetic data can be randomly sub-sampled (resp. super-sampled):

- When $n_m > n$, a subset Q^m of size n is randomly selected;
- When $n_m < n$, a superset Q^m of size n is created by randomly generating $n - n'$ additional records from the original n' ones;
- When $n_m = n$, the synthetic data are not modified and $Q^m = Y^m$.

Such a preliminary sampling procedure is viable provided that the original data set is large enough for it to be analytically interesting and representative. In the remainder of this paper, we will assume that $n_m = n$, $\forall m = 1, \ldots, M$, keeping in mind that the pre-sampling procedure can be eventually used to align the sizes of every synthetic data sets with the size of the original data. The multiple reverse mapping of synthetic data is then performed as follows:

Algorithm: multiple reverse mapping of synthetic data
Require: original data set X, with attributes $X_j = \left(x_{1,j}, \ldots, x_{n,j}\right)$, for j=1,...,p
Require: synthetic data sets Y^m, for m=1,..., M, where Y^m has attributes $Y_j^m = \left(y_{1,j}^m, \ldots, y_{n,j}^m\right)$, for j=1,...,p
For m=1, ..., M do
   For j=1,...,p do
      For i=1,...,n do
         Compute $k=Rank(y_{i,j}^m)$
         Set $z_{i,j}^m = x_{(k,j)}$ (where $x_{(k,j)}$ is the value of X_j of rank k)
      Next i
      Let $Z_j^m = (z_{1,j}^m, \ldots, z_{n,j}^m)$
   Next j
   Let data set $Z^m = (Z_1^m, \ldots, Z_p^m)$
Next m
Return data sets, $Z^1, \ldots, Z^M$

The resulting reverse-mapped attribute j in the mth synthetic data set Z_j^m expresses Y_j^m as a permutation of X_j. Since the point values of a synthetic attribute are unlikely to be the same as the point values of the original data, particularly in the case of numerical attributes, one must also add E_j^m, the difference between Y_j^m and Z_j^m, to get an exact recomposition of Y_j^m as a function of X_j. Then, and since Z_j^m is a permutation of X_j, it always holds that (with P_j^m denoting a permutation matrix):

$$Y_j^m = P_j^m X_j + E_j^m, \ \forall j = 1,\ldots,p \ \text{ and } \ \forall m = 1,\ldots,M \tag{1}$$

Equation (1) shows that, conceptually, a synthetic data set is functionally equivalent to (i) permuting the original data; (ii) adding some noise to the permuted data. But, since the noise added has to be necessarily small, as it cannot by construction alter ranks, it does not offer protection of any sort against disclosure risk. In fact, it represents an information loss (as it modifies the marginal distributions of a data set) that is not matched by a decrease in disclosure risk: if, for example, an attacker learns from a data set that the income of an individual is 102 while in reality it is 100, privacy has been violated in the same way as if the intruder was able to retrieve the exact value. Thus, the imprecision due to the small noise is not relevant for privacy. But any anonymization method, synthetic or not, must intuitively comply with the basic principle that any information loss triggered by anonymization must have a counterpart in terms of improved protection. Clearly, the small noise addition does not comply with this principle and can thus be discarded. As a result, the anonymized version of a data set always has an underlying structure that exactly preserves the marginal distributions of the original data (as they are simply a permutation of the original ones), but alters the relative ranks across attributes [15]. Stated otherwise, what ultimately brings protection (and also information loss) are the changes in relationships between attributes.

At first glance, viewing synthetic data as a rank permutation may seem counter-intuitive. After all, and as mentioned above, there is no mapping between the synthetic records and the original records. However, the synthetic data set tries to mimic the information in the original data set. In turn, this mimicked information can be expressed as a function of the original data, but with a different rank structure. Thus, at a fundamental level of functioning, a synthesizer can be viewed as a generator of different permutation structures of the original data, or equivalently as a way to generate some permutation matrices for anonymization. The generation of M synthetic data sets is thus equivalent to the generation of M permutation matrices. As it has been previously characterized in the literature that any non-synthetic SDC method is also equivalent to the generation of specific permutation matrices [16, 17], *the distinction between synthetic and non-synthetic approaches to anonymization does not seem a fundamental one. As a consequence, synthetic methods must undergo a disclosure risk scrutiny just like their non-synthetic counterparts.*

The ramifications of the above conclusion can further be grasped by recalling the example of a perfect synthesizer. In that case, with a perfect mimic of the information, all multivariate relationships must be exactly preserved. As a result, the permutation matrix has to be the identity matrix (which is a particular case of a permutation matrix where no permutation takes place) and the synthetic data set is the same as the original data set. More realistically, *the better is a synthesizer, the closer to the identity matrix will be each of the underlying permutation patterns contained in the multiple synthetic data sets begin generated.*

Finally, and while the scope of this paper is to investigate the privacy guarantee of synthetic data, it must be noted that the results developed above have broader implications. A releaser could for example decide to release only reverse-mapped synthetic data sets. This solution would not entail additional privacy risks as we saw, but will always offer superior information quality due to the exact preservation of the marginal distributions. Each synthetic data set will thus convey a different rank structure according to the targeted design feature of the original data. Such a possibility is a path for future research.

3.2   Multiple Rank-Based Record Linkage Attack

The multiple reverse mapping procedure can be easily engineered by the data releaser because he has at his disposal both the original and the synthetic data sets, as in the case of non-synthetic SDC techniques [11]. But as we have argued, in the case of synthetic data, the releaser and the maximum-knowledge attacker are at the same level of knowledge. Thus the attacker, who tries to perform the equivalent of a known-plaintext attack in cryptography, can also reverse map each synthetic data set, eliminate the small noise addition and ultimately be confronted with a collection of data sets that contain only the original data but with different permutation structures. Here, a fundamental departure from non-synthetic anonymization is that the attacker is entitled to several attempts to perform his attack. For instance, if trying to learn say the level of income of an individual, the attacker will try on the M data sets to retrieve the value. Intuitively, one can then see that the question of privacy in synthetic data may be trickier than previously thought: the attacker, by retrieving M values of income during his attack, could be confused (if the values are very different), comforted (if the values are close), or most likely be helped by narrowing the range of potential values. That is, it is in fact possible that synthetic data may entail a higher degree of privacy risk than non-synthetic anonymized data (in the latter type of data, only one anonymized data set is typically released).

To mount the attack against synthetic data, the recently developed procedure of rank-based record linkage [8] can be repeated M times. We consider this specific linkage type better than other types (such as distance-based linkage or probabilistic linkage), because, as outlined above, data anonymization can be basically described as rank perturbation. Thus, rank-based record linkage appears to be the overarching procedure for evaluating disclosure risk (see [17] for a detailed explanation).

Denote by $O = (o_{ij})$ and $S^m = \left(s_{lj}^m \right)$ the rank matrices of the original data set and of the mth synthetic data set, respectively[1]. The procedure of multiple rank-based record linkage on synthetic data is as follows:

---

[1] Using these notations, o_{ij} is the rank of attribute j in original record i and s_{lj}^m is the rank of attribute j in synthetic record l of the mth synthetic data set.

Algorithm: *multiple rank-based record linkage*
Require: *rank matrix O of the original data set*
Require: *rank matrices* $S^1, ..., S^M$ *of the M synthetic data sets* $Y^1, ..., Y^M$
For m=1,..., M do
   For i=1,...,n do
      For l=1,...,n do
         Compute $d_{il}^m = Criterion[abs(o_{i1} - s_{l1}^m), ..., abs(o_{ip} - s_{lp}^m)]$
         Next l
         Linked index of i in $Y^m = arg\,min_l(d_{il}^m)$
   Next i
Next m
Return *linked indices of i in the M synthetic data sets*

This procedure is the multi-data set version of the procedure outlined in [8]. It reports the M possible matches of an original record with the M synthetic data sets. Several criteria can be selected, such as the sum or the minimum of rank differences. To evaluate the privacy guarantees of non-synthetic methods, the criterion will generally depend on the method, e.g. the sum for noise addition or the maximum for data swapping [8]. In the context of synthetic data, this choice is less clear and several criteria should ideally be considered.

4 Empirical Illustrations

The objective of this section is to illustrate the concepts of multiple reverse mapping of synthetic data and multiple rank-based record linkage. The experiment is based, without loss of generality, on a small data set of 20 observations and three attributes, and proceeds as follows:

- The assumed original data set is generated by sampling $N(50, 10^2)$, $N(500, 50^2)$ and $N(2500, 250^2)$ distributions, respectively. The correlation coefficient between the first and the second attribute is 0.56, 0.25 between the first and the third, and 0.16 between the second and the third.
- $M = 3$ synthetic data sets are generated using a similar sampling procedure. The synthetic data are directly generated with the same size as the original data, albeit one can use the pre-sampling procedure developed above to eventually align the sizes of the former with the size of the latter.
- For the sake of illustration, we consider three different levels of closeness to the original data. As stated previously, the goal of this paper is not to discuss the issue of how to generate a satisfying synthesizer. Rather, by using three different sets, we try to account for the difficulty of generating a satisfying synthesizer:
  - The first synthetic data set is very close to the original data (but does not replicate them perfectly). It was sampled from the same normal distributions from which the original data set was sampled. As a result, the joint relationships

between the three attributes are slightly altered (the correlation coefficient between the first and the second synthetic attribute is 0.52, 0.18 between the first and the third and 0.21 between the second and the third).
- The second synthetic data set has also the joint relationships between the three attributes slightly altered (the correlation coefficient between the first and the second synthetic attribute is 0.44, 0.25 between the first and the third and 0.21 between the second and the third) but with also the properties of the marginal distributions not exactly preserved, i.e. the attributes are sampled from N $(45, 8^2)$, $N(450, 40^2)$ and $N(2200, 200^2)$ distributions, respectively.
- The third synthetic data set has its marginal distributions sampled from the same as the second one. However, no particular effort is made to preserve the joint relationships (the correlation coefficient between the first and the second synthetic attribute is 0.17, 0.12 between the first and the third and 0.09 between the second and the third).

Table 2 shows the multiple reverse-mapping procedure for the first attribute in the three synthetic data sets[2]. It can be seen that each synthetic data set is expressed as a permutation of the original data. As outlined in the last section, these versions do not entail more disclosure risk than the first generated synthetic data sets, but offer an improved level of information by exactly preserving marginal distributions.

Now, a maximum-knowledge attacker can exactly perform reverse mapping for all attributes and can attempt to recreate the correct linkage. A releaser can also do the same to gauge the privacy of his synthetic data sets before release. Of course, identity disclosure may appear as a peculiar notion for synthetic data but it is still conceivable: an attacker may try to identify which synthetic individuals are the most similar to real individuals, i.e. trying to retrieve some clones. However, we believe that more interesting in the context of synthetic data is attribute disclosure, i.e. when a confidential information contained in the synthetic data sets can be revealed and will closely or exactly correspond to the information of a real individual.

A maximum-knowledge attacker can conduct an attack on a specific attribute by ignoring his knowledge of this attribute in the original data; this is part of the flexibility offered by the maximum-knowledge attacker model (see above and also [2]). The maximum-knowledge attacker can then use the multiple rank-based record linkage procedure to see how well he can recreate the ranks of the ignored attribute; that would simulate a partial-knowledge attacker who did not know the third original attribute and wanted to guess it. Table 3 shows the result of such an attack when knowledge of the third attribute of the original data set is ignored and the sum of rank differences criterion is used to perform multiple rank-based record linkage on the first and second attributes.

In this example, one can see that the outcome of an attack on synthetic data can either create confusion to a partial-knowledge attacker, or on the contrary help him narrow his knowledge of the attribute. Consider for example record no. 1 in the original data, with a value of rank 2 for the third attribute. What the attacker gains as

---

[2] The two other attributes are not shown here due to space constraints but their reverse-mapped versions can be displayed in exactly the same way.

Table 2. Example of multiple reverse mapping on synthetic data sets

Original data set			Synthetic data set 1				Synthetic data set 2				Synthetic data set 3			
ID	X1	Rank of X1	X1	Rank of X1	Reverse-mapped X1	Small noises	X1	Rank of X1	Reverse-mapped X1	Small noises	X1	Rank of X1	Reverse-mapped X1	Small noises
1	38	3	46	9	51	−5	33	2	37	−4	38	4	39	−1
2	66	19	36	1	31	5	54	19	66	−12	46	14	57	−11
3	56	12	43	5	41	2	50	16	63	−13	42	8	50	−8
4	53	11	59	14	57	2	37	6	45	−8	41	6	45	−4
5	31	1	41	4	39	2	43	13	56	−13	49	16	63	−14
6	63	16	61	16	63	−2	45	15	61	−16	49	17	63	−14
7	39	4	44	7	49	−5	33	3	38	−5	56	20	70	−14
8	63	17	56	13	56	0	41	11	53	−12	42	9	51	−9
9	51	9	76	20	70	6	40	9	51	−11	45	12	56	−11
10	56	13	49	10	51	−2	37	5	41	−4	53	19	66	−13
11	70	20	65	17	63	2	37	4	39	−2	42	7	49	−7
12	61	15	59	15	61	−2	43	12	56	−13	35	3	38	−3
13	41	5	40	3	38	2	32	1	31	1	44	11	53	−9
14	49	7	43	6	45	−2	51	17	63	−12	47	15	61	−14
15	51	10	53	12	56	−3	58	20	70	−12	28	1	31	−3
16	64	18	51	11	53	−2	39	8	50	−11	50	18	64	−14
17	45	6	66	18	64	2	45	14	57	−12	33	2	37	−4
18	57	14	44	8	50	−6	39	7	49	−10	42	10	51	−9
19	37	2	72	19	66	6	41	10	51	−10	40	5	41	−1
20	50	8	39	2	37	2	53	18	64	−11	46	13	56	−10

information is wrong in each of the synthetic data sets, with a possible rank identified as ranging between 8 and 11. In fact, in that case, having multiple sets consistently orientates the partial-knowledge attacker in the wrong direction. The same is true for several records, e.g. nos. 7, 15, 17. For these individuals, it can be reasonably argued that synthetic data sets offer more privacy in the sense that they fool the attacker consistently across all sets released.

Now consider records nos. 2 and 18. Respectively the third and first synthetic data sets perfectly disclose the attribute values of these records. But because the other sets lead into another direction, the partial-knowledge attacker is again confused. As a result, synthetic data sets seem to provide here again better protection than non-synthetic approaches for these records. However, the partial-knowledge attacker can claim with reasonable confidence that the real value for record no. 2 is between ranks 4 and 18 of the original data and for record no. 18 between 3 and 7. That is, he can claim that the eighteenth individual has a value for the third attribute comprised between 2298 and 2428. Clearly, he has gained some information from the synthetic data sets.

The information can be also narrowed for records where no exact attribute disclosure occurs across the three synthetic data sets in the first place. Consider for example records nos. 4 and 20. For the former, the attacker can claim that the real value is comprised between 2336 and 2737; for the latter, he can claim it is between 2298 and 2704.

Table 3. Example of multiple rank-based record linkage: *third* attribute disclosure scenario

	Original data set		Multiple rank-based record linkage: ranks identified by the intruder for X3		
ID	X3	Rank of X3	Synthetic data set 1	Synthetic data set 2	Synthetic data set 3
1	2228	2	11	9	8
2	2299	4	12	18	4
3	2534	10	1	8	12,17
4	2526	9	5	17	11
5	2336	5	16	13	2
6	2598	13	19	19	3
7	2736	16	2	9	8
8	2557	11	12	3	10.9
9	2704	15	17,4,5	16	12,2
10	2513	8	5	17	13
11	2942	19	17	3	10
12	2737	17	18	7	3
13	2559	12	2	2	8
14	2809	18	8	16	16
15	2195	1	4.5	16	11
16	2655	14	6,19	11	4
17	2963	20	15	5	15
18	2298	3	3	7	7
19	2382	6	11	9	8
20	2428	7	15	15	3,14

Alternatively, assuming that the maximum-knowledge attacker now ignores his knowledge of the first attribute in the original data leads to the similar presence of edges in information (Table 4). For example, for records nos. 9 and 18 the knowledge of the first attribute is narrowed to a significant extent.

While these examples are meant to be illustrative, they however tend to suggest that synthetic data do not come always with low disclosure risk. Releasing multiple data sets can in fact be viewed as an additional privacy threat. Even if by definition no real individual is present in the synthetic data, some clones nonetheless are, and these clones can be re-identified to learn some information about some real individuals.

Originally, the proposal of releasing multiple data sets aimed at enhancing the quality of information offered by synthetic data. But, considering that such practice can be undoubtedly cumbersome for the users and that the quality of information can in some cases be made at least as good with a single data set [10], having multiple releases seem also to entail some previously uncharacterized privacy risks that render this practice questionable.

Table 4. Example of multiple rank-based record linkage: *first* attribute disclosure scenario

ID	X1	Rank of X1	Synthetic data set 1	Synthetic data set 2	Synthetic data set 3
			Multiple rank-based record linkage: ranks identified by the intruder for X1		
1	38	3	15,1	6	14
2	66	19	12	16,17	20
3	56	12	20	2,4	19,15
4	53	11	3,6	19	7,10
5	31	1	12,11	13	20
6	63	16	7	19,5	4,11
7	39	4	1,7,18	10	17
8	63	17	19	2,4	9
9	51	9	16	8	12
10	56	13	6	1,14	7,3
11	70	20	10	20	5
12	61	15	18	12,10	6
13	41	5	8,2	3	17,1
14	49	7	17	8	8
15	51	10	12	17	2
16	64	18	16	5	4
17	45	6	1	15	18
18	57	14	15	7	16
19	37	2	9	7	1,13
20	50	8	9,3,14	1,14	7,3,13

5 Conclusions and Future Work

Synthetic data is often perceived as having lower disclosure risk than other forms of SDC methods. In this paper, we show that this may not always be the case. Despite the fact that no real individuals are included in a data release, at least as far as fully synthetic data are concerned, synthetic and real individuals remain however linked by the information they convey. If an attacker is able to retrieve some information on real individuals that happens to be correct, it ultimately does not matter that this information is based on simulated data. Even if such a disclosure does not fall under the consideration of any legislation on privacy, it can nonetheless be viewed as unethical insofar as it affects real individuals.

The objective of this paper was thus to investigate the privacy guarantee of synthetic data. Using recent advances in the literature on the definition of an attacker in data anonymization, we confronted synthetic data to an attack by a maximum-knowledge intruder. While conservative in its stance, this model has the merit to establish a common benchmark to gauge the privacy guarantees of non-synthetic anonymization methods. It thus seems only fair to consider synthetic data in the same context. Actually, the maximum-knowledge attacker is the counterpart of the popular and widely used notion of known-plaintext attack in cryptography.

We first presented an extension of a reverse-mapping procedure that can be performed both by an attacker and a synthetic data releaser. Under a reasonable assumption on the size of the synthetic data sets to be released, this procedure shows that in fact any synthetic data set can always be expressed as a permutation of the original data, in a way similar to non-synthetic SDC techniques. This result offers applications beyond disclosure risk assessment. For one thing, it is always possible to release synthetic data sets with the same privacy properties but with an improved level of information, because the marginal distributions can be always preserved without increasing risk. On the privacy front, reverse mapping leads to the consequence that the distinction made in the literature between non-synthetic and synthetic data is not so clear-cut. Thus, both approaches must be evaluated against the same privacy challenges.

Next, we proposed an extension of the rank-based record linkage procedure that can also be performed both by the attacker and the synthetic data releaser. In particular, the latter can use it to assess the privacy guarantee of its synthetic data before release. This procedure shows that the practice of releasing several synthetic data sets for a single original data set entails privacy issues that do not arise in non-synthetic anonymization (where typically only one anonymized data set is released). Indeed, the multiple releases can lead to better privacy guarantees, by confusing the attacker, or facilitate attribute disclosure by helping the attacker narrow the range of the possible values that he is trying to retrieve. An empirical investigation in the last section illustrates those issues. We believe that this has interesting consequences for synthetic data releases that deserve further investigation.

The results presented in this paper are preliminary and illustrative. As future work, we plan to: (i) investigate the theoretical and empirical conditions under which multiple synthetic data sets can lead to more confusion than help for the attacker; (ii) assess the possibility of considering synthesizers as tools to generate different permutation patterns, which could offer some insights for non-synthetic anonymization techniques; (iii) enlarge the scope of the experimental work by using various synthetic data sets, and in particular assess the occurrence and the magnitude of range narrowing during an attack.

Acknowledgments and Disclaimer. The following funding sources are gratefully acknowledged by the third author: European Commission (project H2020-700540 "CANVAS"), Government of Catalonia (ICREA Acadèmia Prize) and Spanish Government (projects TIN2014-57364-C2-1-R "SmartGlacis" and TIN2015-70054-REDC). The views in this paper are the authors' own and do not necessarily reflect the views of UNESCO or any of the funders.

References

1. Domingo-Ferrer, J., Muralidhar, K.: New directions in anonymization: permutation paradigm, verifiability by subjects and intruders, transparency to users. Inf. Sci. **337**, 11–24 (2016)
2. Domingo-Ferrer, J., Ricci, S., Soria-Comas, J.: Disclosure risk assessment via record linkage by a maximum-knowledge attacker. In: 13th Annual International Conference on Privacy, Security and Trust-PST 2015, Izmir, Turkey, September 2015
3. Domingo-Ferrer, J., Sánchez, D., Rufian-Torrell, G.: Anonymization of nominal data based on semantic marginality. Inf. Sci. **242**, 35–48 (2013)
4. Drechsler, J.: Synthetic Datasets for Statistical Disclosure Control. Springer, New York (2011). https://doi.org/10.1007/978-1-4614-0326-5
5. Drechsler, J., Bender, S., Rässler, S.: Comparing fully and partially synthetic datasets for statistical disclosure control in the German IAB establishment panel. Trans. Data Priv. **1**, 105–130 (2008)
6. Hu, J., Reiter, J.P., Wang, Q.: Disclosure risk evaluation for fully synthetic categorical data. In: Domingo-Ferrer, J. (ed.) PSD 2014. LNCS, vol. 8744, pp. 185–199. Springer, Cham (2014). https://doi.org/10.1007/978-3-319-11257-2_15
7. Hundepool, A., et al.: Statistical Disclosure Control. Wiley, Hoboken (2012)
8. Muralidhar, K., Domingo-Ferrer, J.: Rank-based record linkage for re-identification risk assessment. In: Domingo-Ferrer, J., Pejić-Bach, M. (eds.) PSD 2016. LNCS, vol. 9867, pp. 225–236. Springer, Cham (2016). https://doi.org/10.1007/978-3-319-45381-1_17
9. Muralidhar, K., Domingo-Ferrer, J.: Microdata masking as permutation. In: UNECE/ EUROSTAT Work Session on Statistical Data Confidentiality, Helsinki, Finland, October 2015
10. Muralidhar, K., Sarathy, R.: A comparison of multiple imputation and data perturbation for masking numerical variables. J. Off. Stat. **22**, 507–524 (2006)
11. Muralidhar, K., Sarathy, R., Domingo-Ferrer, J.: Reverse mapping to preserve the marginal distributions of attributes in masked microdata. In: Domingo-Ferrer, J. (ed.) PSD 2014. LNCS, vol. 8744, pp. 105–116. Springer, Cham (2014). https://doi.org/10.1007/978-3-319-11257-2_9
12. Reiter, J.P., Wang, Q., Zhang, B.: Bayesian estimation of disclosure risks in multiply imputed, synthetic data. J. Priv. Confid. **6**(1), 17–33 (2014). Article no. 2
13. Reiter, J.P.: Satisfying disclosure restrictions with synthetic data sets. J. Off. Stat. **18**, 531–544 (2002)
14. Reiter, J.P.: Releasing multiply imputed, synthetic public use microdata: an illustration and empirical study. J. Roy. Stat. Soc. Ser. A **168**, 185–205 (2005)
15. Rubin, D.B.: Discussion: statistical disclosure control limitation. J. Off. Stat. **9**, 462–468 (1993)
16. Ruiz, N.: On some consequences of the permutation paradigm for data anonymization: centrality of permutation matrices, universal measures of disclosure risk and information loss, evaluation by dominance. Inf. Sci. **430–431**, 620–633 (2018)
17. Ruiz, N.: A general cipher for individual data anonymization. Inf. Sci. (2017, under review). (https://arxiv.org/abs/1712.02557)
18. Soria-Comas, J., Domingo-Ferrer, J.: A non-parametric model for accurate and provably private synthetic data sets. In: Proceedings of International Conference on Availability, Reliability and Security-ARES 2017, Article no. 3. ACM (2017)
19. Willenborg, L., De Waal, T.: Elements of Statistical Disclosure Control. Springer, New York (2001). https://doi.org/10.1007/978-1-4613-0121-9

The Quasi-Multinomial Synthesizer for Categorical Data

Jingchen Hu[1](✉) and Nobuaki Hoshino[2]

[1] Vassar College, Poughkeepsie, USA
jihu@vassar.edu
[2] Kanazawa University, Kanazawa, Japan

Abstract. We present a new synthesizer for categorical data based on the Quasi-Multinomial distribution. Characteristics of the Quasi-Multinomial distribution provide a tuning parameter, which allows a Quasi-Multinomial synthesizer to control the balance of the utility and the disclosure risks of synthetic data. We develop a Quasi-Multinomial synthesizer based on a popular categorical data synthesizer, the Dirichlet process mixtures of products of multinomial distributions. The general sampling methods and algorithm of the Quasi-Multinomial synthesizer are developed and presented. We illustrate its balance of the utility and the disclosure risks by synthesizing a sample from the American Community Survey.

Keywords: Bayesian · Dirichlet process · Microdata
Quasi-Multinomial · Synthetic

1 Introduction

The synthetic data approach to data confidentiality has gained attention and momentum in the past two decades. Based on the theory and applications of multiple imputation methodology for missing data problems (Rubin 1987); statistical models are first estimated from the original confidential data, and then multiply-imputed synthetic data is generated to provide high utility, low risks public microdata. Multiple synthetic datasets should be generated, and appropriate combining rules have been developed to provide accurate point estimates and variance estimates of parameters of interest. Refer to Reiter and Raghunathan (2007); Drechsler (2011) for details of the combining rules.

More recently, nonparametric Bayesian models have been further developed and turned into data synthesizers. Among them, the Dirichlet process mixtures of products of multinomials (DPMPM) synthesizer is worth particular attention. The DPMPM consists of a set of flexible Bayesian latent class models that have been developed to capture complex relationships among multivariate unordered categorical variables (Dunson and Xing 2009). Hu et al. (2014) implemented the DPMPM as a synthesizer on multivariate unordered categorical data and demonstrated its balance between data utility and disclosure risks.

© Springer Nature Switzerland AG 2018
J. Domingo-Ferrer and F. Montes (Eds.): PSD 2018, LNCS 11126, pp. 75–91, 2018.
https://doi.org/10.1007/978-3-319-99771-1_6

Drechsler and Hu (2017+) used the DPMPM synthesizer for generating partially synthetic data with geocoding information. Other work on some version of the DPMPM for synthesis include Manrique-Vallier and Reiter (2014), Hu *et al.* (2018), Manrique-Vallier and Hu (2018). The DPMPM has also been proposed as a multiple imputation engine for missing data problems when all variables are categorical (Si and Reiter 2013; Akande *et al.* 2017; Murray 2018+; Akande *et al.* 2017+).

While useful and promising, the characteristics of the utility and disclosure risks tradeoff of the DPMPM synthesizer have not yet been a research focus. The DPMPM synthesizer is based on the multinomial distribution, which has no parameter to control the tradeoff. On the other hand, the Quasi-Multinomial (QM) distribution of Consul and Mittal (1977) is a generalized multinomial distribution with an additional parameter, which can be effectively tuned to deliver a desired balance of utility and disclosure risks in the synthetic data products that statistical agencies would produce, if we construct a synthesizer based on the QM distribution.

In this paper, we focus on developing the QM-DPMPM synthesizer, and comparing it with the DPMPM synthesizer. Section 2 introduces the QM distribution, discusses the sampling methods and proposes an algorithm based on acceptance rejection sampling. The DPMPM and the QM-DPMPM synthesizers are introduced in Sect. 3. Section 4 presents an illustrative application comparing the two synthesizers, using a sample from the American Community Survey (ACS). Discussions and future work are given in Sect. 5.

2 The Quasi-Multinomial Distribution

2.1 Introducing the Quasi-Multinomial Distribution

The Quasi-Multinomial distribution (type 2) is a generalized multinomial distribution proposed by Consul and Mittal (1977). We define the QM distribution by the following probability mass function (pmf):

$$p(y_1, \ldots, y_F) = \frac{n!}{y_1! \ldots y_F!} \frac{1}{(1 + n\beta)^{n-1}} \prod_{f=1}^{F} \pi_f (\pi_f + y_f \beta)^{y_f - 1}, \qquad (1)$$

where $y_f, f = 1, 2, \ldots, F$, is a nonnegative integer, and $\sum_{f=1}^{F} y_f = n$. Similar to the multinomial distribution, y_f is regarded as the random frequency of the fth cell given a total frequency n. We denote this parameterization of the QM distribution in Eq. (1) by $\text{QM}(\pi_1, \ldots, \pi_F; n, \beta)$.

We consider the parameter β in Eq. (1) as a nonnegative real number. While this pmf is proper when $\beta > -\min_f \pi_f/n$, we disallow negative β to avoid the dependence of the parameter space on parameters themselves. This limitation is necessary for theorems in Sect. 2.2; it also guarantees that the QM will always increase the data protection.

When $\beta = 0$, Eq. (1) reduces to the pmf of the multinomial distribution with cell probabilities $(\pi_1, \ldots, \pi_F)$, which we denote as Multinomial$(\pi_1, \ldots, \pi_F; n)$. As β increases, the variance of any univariate marginal frequency increases (Consul and Mittal 1975).

Among the similarities of the QM distribution to the multinomial distribution, there are a few points worth mentioning. First, similar to the multinomial distribution, the parameters π_f's of Eq. (1) are nonnegative real number, and $\sum_{f=1}^{F} \pi_f = 1$. Hence π_f of the QM distribution is referred to by a cell probability. Second, the expectation of the fth marginal frequency is $n\pi_f$ regardless of β (Hoshino 2009). In other words, the fth sample relative frequency is the unbiased estimator of the fth cell probability for all β.

Then we note that since the variance of any univariate marginal frequency increases as β increases, the accuracy of the unbiased estimator of a cell probability can be limited by increasing β. This fact implies that replacing the multinomial distribution with the QM distribution in the generation of synthetic data facilitates to control the balance of the utility and the disclosure risk of synthetic data while the unbiasedness remains to hold. We thus regard β as a tuning parameter of the QM synthesizer determined by a statistical agency.

To sample from the QM distribution, we rely on a special case of it. When $F = 2$, the QM distribution becomes the Quasi-Binomial (QB) distribution (type 2), proposed by Consul and Mittal (1975). We denote the QB distribution by $QB(\pi; n, \beta)$, with its pmf:

$$p_{QB}(y) = \frac{n!}{y!(n-y)!} \frac{1}{(1+n\beta)^{n-1}} \pi(\pi+y\beta)^{y-1}(1-\pi)(1-\pi+(n-y)\beta)^{n-y-1}, \quad (2)$$

where $y = 0, 1, \ldots, n$, $0 \leq \pi \leq 1$, $\beta \geq 0$.

2.2   The Decomposability of the Quasi-Multinomial Distribution

Sampling from the QM distribution can be decomposed into simpler sampling of marginal variables. The characteristic nature of the QM distribution enables two general methods of such decomposition. The first one is the conditional distribution method (Devroye 1986). The second one is multi-stage sampling.

The first decomposition of sampling from the QM distribution exploits the following general relationship:

$$(Y_1, \ldots, Y_F) \stackrel{d}{=} (Y_1|Y_2, \ldots, Y_F) \ldots (Y_{F-2}|Y_{F-1}, Y_F)(Y_{F-1}|Y_F)Y_F, \quad (3)$$

where "$\stackrel{d}{=}$" denotes equality in distribution.

The right hand side of (3) is the product of the conditional distributions of univariate margins. An explicit formula of these distributions is given below on the QM distribution:

Theorem 1. *If* $(Y_1, \ldots, Y_F) \sim QM(\pi_1, \ldots, \pi_F; n, \beta)$ *then* $(Y_g|Y_{g+1} = y_{g+1}, \ldots, Y_F = y_F) \sim QB(\pi_g/(1 - \sum_{f=g+1}^{F} \pi_f); n - \sum_{f=g+1}^{F} y_f, \beta)$ *for* $g = 1, \ldots, F$.

Theorem 1 reads that $Y_F \sim QB(\pi_F; n, \beta)$. It is widely known that Theorem 1 holds for the case of $\beta = 0$ or the multinomial distribution. Theorem 1 follows from Theorem 2 below.

Combining Eq. (3) and Theorem 1, we observe that sampling from the QM distribution is accomplished by sequential sampling from the QB distribution. By symmetry, Theorem 1 holds even after exchanging the indices of variables. Therefore the resulting F dimensional sampling distribution of our procedure does not depend on the order of single margins to sample.

These single margins should be ordered in sampling so that corresponding cell probabilities are decreasing. This sequential sampling from larger cells is known to be efficient on the multinomial distribution (Ho *et al.* 1979).

The second decomposition exploits another property that the conditional QM distribution given the sum of partial frequencies is again QM:

Theorem 2. *If* $(Y_1, \ldots, Y_F) \sim QM(\pi_1, \ldots, \pi_F; n, \beta)$ *then for* $g = 1, \ldots, F$ *and* $m = 0, \ldots, n,$

$$(Y_1, \ldots, Y_g | \sum_{f=1}^{g} Y_f = m) \sim QM(\pi_1/(\sum_{f=1}^{g} \pi_f), \ldots, \pi_g/(\sum_{f=1}^{g} \pi_f); m, \beta). \quad (4)$$

Theorem 2 can be shown by the fact that the QM distribution is closed under the collapse of cells (Hoshino 2009). It is noteworthy that Theorem 2 holds after exchanging the indices of variables as Theorem 1 does.

Theorem 2 validates two-stage sampling from the QM distribution: The first stage generates the aggregated frequency of $m = Y_1 + \cdots + Y_g \sim$ QB $(\sum_{f=1}^{g} \pi_f; n, \beta)$; the second stage generates frequencies $Y_1, \ldots, Y_g$ given m, subject to Eq. (4). Then the resulting vector $(Y_1, \ldots, Y_F) \sim QM(\pi_1, \ldots, \pi_F; n, \beta)$. More generally a recursive argument validates multi-stage sampling from the QM distribution.

This type of multi-stage sampling has been used for the multinomial distribution to reduce computing time. For example, Malefaki and Iliopoulos (2007) provide an empirical support to import stages, which might seem redundant. On the QM distribution, we will see that multi-stage sampling can reduce computing time because it lowers the rejection rate of acceptance rejection sampling.

We have shown that sampling from the QM distribution can be expressed as the combination of sequential and multi-stage sampling from the QB distribution. Next, we discuss the method of sampling from the QB distribution.

2.3   Sampling from the Quasi-Binomial Distribution

To sample from the QB distribution, we prepare acceptance rejection sampling (also called the rejection method by Devroye (1986)). The key element of acceptance rejection sampling is a "proposal" distribution, which should be close to its "target" distribution and also easy to sample. We use the Beta-Binomial mixture (BB) distribution as our proposal distribution.

If $p \sim \text{Beta}(a_1, a_2)$ and $Y \mid p \sim \text{Binomial}(n, \pi)$ then Y is called $\text{BB}(a_1, a_2; n)$ distributed, with pmf

$$p_{BB}(y) = \frac{n!}{y!(n-y)!} \frac{\Gamma(a.)}{\Gamma(a.+n)} \frac{\Gamma(a_1+y)}{\Gamma(a_1)} \frac{\Gamma(a_2+n-y)}{\Gamma(a_2)}, \quad y = 0, \ldots, n. \quad (5)$$

The same as the QB distribution, the BB distribution belongs to the family of the Conditional Compound Poisson distributions (Hoshino 2009). Equalizing the canonical parameters of this family, the counterpart of $\text{BB}(a_1, a_2; n)$ is $\text{QB}(a_1/(a_1 + a_2); n, 1/(a_1 + a_2))$, whose pmf is

$$p_{QB}(y) = \frac{n!}{y!(n-y)!} \frac{1}{a.(a.+n)^{n-1}} a_1(a_1+y)^{y-1} a_2(a_2 + (n-y))^{n-y-1}, \quad (6)$$

where $a. := a_1 + a_2$, and y takes nonnegative integers from 0 to n.

We regard the case of $a_1 = a_2 = 0$ as improper, and this case is excluded from our argument. If $a_1 = 0$ and $a_2 > 0$ then $\text{QB}(a_1/(a_1 + a_2); n, 1/(a_1 + a_2))$ degenerates at 0, or it takes 0 with probability one. On the contrary if $a_2 = 0$ and $a_1 > 0$ then $\text{QB}(a_1/(a_1 + a_2); n, 1/(a_1 + a_2))$ degenerates at n. Hence these two cases do not need sampling, and they are also excluded from our argument henceforward.

The ratio of Eqs. (5) to (6) is

$$\frac{p_{BB}(y)}{p_{QB}(y)} = \frac{\Gamma(a.+1)\Gamma(a_1+y)\Gamma(a_2+n-y)(a.+n)^{n-1}}{\Gamma(a.+n)\Gamma(a_1+1)\Gamma(a_2+1)(a_1+y)^{y-1}(a_2+(n-y))^{n-y-1}}, \quad (7)$$

and we are interested in the minimum of Eq. (7) with respect to y, which equals the average acceptance rate of our acceptance rejection sampling. The proof of the following Theorem 3 is given in Appendix 1.

Theorem 3. *Suppose that n is a positive integer, and a_1, a_2 are positive real numbers. Denote the value of y that minimizes Eq. (7) by y^*. Then $y^* = n$ when $a_1 < a_2$, and $y^* = 0$ when $a_2 < a_1$. When $a_1 = a_2$, (7) is minimized at $y = 0$ and $y = n$.*

By symmetry we only consider the case of $a_1 < a_2$, where the average acceptance rate is assured by Theorem 3 to be

$$\min_y \frac{p_{BB}(y)}{p_{QB}(y)} = \frac{p_{BB}(n)}{p_{QB}(n)} = \frac{\Gamma(a.+1)\Gamma(a_1+n)}{\Gamma(a.+n)\Gamma(a_1+1)} \left(\frac{a.+n}{a_1+n}\right)^{n-1} =: r(a_1, a_2, n).$$

This rate converges to unity in the following sense:

Theorem 4. *Suppose that n is a positive integer, and $0 < \pi < 1$. Then*

$$\lim_{a \to \infty} r(\pi a, (1-\pi)a, n) = 1.$$

The fact that both $BB(a_1, a_2; n)$ and $QB(a_1/(a_1 + a_2); n, 1/(a_1 + a_2))$ are close to Binomial$(n, a_1/(a_1 + a_2))$ when $(a_1 + a_2)$ is large should suffice to prove Theorem 4. Consequently, our acceptance rejection sampling can be very efficient by taking β very close to 0 for fixed π and n.

On the other hand, one may be interested in the efficiency of our sampling for fixed π and β when n is large. Actually, $r(a_1, a_2, n) = O(n^{-a_2})$ as $n \to \infty$. Therefore large n may cause inefficient sampling, but multi-stage sampling can avoid this situation by repeating the division of samples into two groups of cells, where the sum of cell probabilities should be close to $1/2$ since it implies smaller a_2. Refer to Table 2 in Appendix 2 for the summary of average acceptance rates r of our QB sampler.

Next we derive the acceptance rate of our sampling for $a_1 < a_2$ depending on y:

$$\frac{p_{QB}(y)}{p_{BB}(y)} \frac{p_{BB}(n)}{p_{QB}(n)} = \frac{\Gamma(a_1 + n)\Gamma(a_2 + 1)}{\Gamma(a_1 + y)\Gamma(a_2 + n - y)} \frac{(a_1 + y)^{y-1}(a_2 + n - y)^{n-y-1}}{(a_1 + n)^{n-1}} =: \rho(y). \tag{8}$$

Consequently our QB sampler is summarized in the following (note that $U(0, 1)$ is for the standard uniform distribution):

Algorithm 1 (Acceptance rejection sampling from the QB distribution). *The following procedure generates a sample from $QB(a_1/(a_1 + a_2); n, 1/(a_1 + a_2))$ for a positive integer n.*

When $0 < a_1 < a_2$,

1. *Generate $p \sim Beta(a_1, a_2)$*
2. *Generate $y|p \sim Binomial(n, p)$*
3. *Generate $u \sim U(0, 1)$*
4. *If $u > \rho(y)$ then goto 1*
5. *Output y.*

When $0 < a_2 < a_1$,

1. *Swap a_2 and a_1.*
2. *Generate $p \sim Beta(a_1, a_2)$*
3. *Generate $y|p \sim Binomial(n, p)$*
4. *Generate $u \sim U(0, 1)$*
5. *If $u > \rho(y)$ then goto 2*
6. *Output $n - y$.*

3   The DPMPM and QM-DPMPM Synthesizers

3.1   The DPMPM Synthesizer

The DPMPM is a Bayesian version of latent class models. Consider a sample $\mathbf{X}$ consists of n records, and each record has p unordered categorical variables. The basic assumption of the DPMPM is that every record $\mathbf{X}_i = (X_{i1}, \cdots, X_{ip})$ belongs to one of F underlying unobserved/latent classes. Given the latent class assignment z_i of record i, as in Eq. (10), each variable X_{ij} independently follows a multinomial distribution, as in Eq. (9). Note that d_j is the number of categories of variable j, and $j = 1, \cdots, p$.

$$X_{ij} \mid z_i, \theta \overset{ind}{\sim} \text{Multinomial}(\theta_{z_i 1}^{(j)}, \ldots, \theta_{z_i d_j}^{(j)}; 1) \text{ for all } i, j \tag{9}$$

$$z_i \mid \pi \sim \text{Multinomial}(\pi_1, \ldots, \pi_F; 1) \text{ for all } i, \tag{10}$$

The DPMPM effectively clusters records with similar characteristics based on all p variables. Relationships among these p categorical variables are induced by integrating out the latent class assignment z_i. To empower the DPMPM to pick the effective number of occupied latent classes, the truncated stick-breaking representation (Sethuraman 1994) is used as in Eq. (11) through Eq. (14),

$$\pi_f = V_f \prod_{l<f}(1 - V_l) \text{ for } f = 1,\ldots,F \qquad (11)$$

$$V_f \overset{iid}{\sim} \text{Beta}(1,\alpha) \text{ for } f = 1,\ldots,F-1, \quad V_F = 1 \qquad (12)$$

$$\alpha \sim \text{Gamma}(a_\alpha, b_\alpha) \qquad (13)$$

$$\theta_f^{(j)} = (\theta_{f1}^{(j)},\ldots,\theta_{fd_j}^{(j)}) \sim \text{Dirichlet}(a_1^{(j)},\ldots,a_{d_j}^{(j)}). \qquad (14)$$

and a blocked Gibbs sampler is implemented for the Markov chain Monte Carlo sampling procedure (Ishwaran and James 2001; Si and Reiter 2013; Hu et al. 2014; Drechsler and Hu 2017+; Manrique-Vallier and Hu 2018; Hu et al. 2018).

Let p_0 be the number of variables to be synthesized. To generate one partially synthetic dataset $\mathbf{X}_{DPMPM}^*$ using the DPMPM synthesizer, we first generate sample values of (π, α, θ^s) from the posterior distribution (θ^s contains the sample values of variables to be synthesized). Through a multinomial draw with the samples of π, we can generate the vector of latent class assignments $\{z_i, i = 1,\cdots,n\}$, as in Eq. (10). Then through a multinomial draw with samples of θ^s, we can generate synthetic variable $\{X_{ij}^*, i = 1,\cdots,n, j = 1,\cdots,p_0\}$, as in Eq. (9).

The sampling process above can also be described as the process of distributing records over different values of $\mathbf{X}$. The probability of the ith record to take the values of $(x_{i1}, x_{i2}, \cdots, x_{ip_0})$ is expressed as $\prod_{j=1}^{p_0} \theta_{z_i x_{ij}}^{(j)}$. We note that any record in the same latent class f has the same probability of taking the values of $(x_1, x_2, \cdots, x_{p_0})$, which is $p(x_1, x_2, \cdots, x_{p_0}; f) = \prod_{j=1}^{p_0} \theta_{f x_j}^{(j)}$.

Regarding $(x_1, x_2, \cdots, x_{p_0})$ as the address of a cell in a p_0 dimensional contingency table, then $p(x_1, x_2, \cdots, x_{p_0})$ gives the cell probability of the corresponding cell. The total number of the cells is $\prod_{j=1}^{p_0} d_j =: D$, and the generation of one record in the fth latent class is equivalent to one multinomial draw from D cells with probabilities $\{p(x_1, x_2, \cdots, x_{p_0}; f), x_j = 1, 2, \cdots, d_j, j = 1, 2, \ldots, p_0\}$. Abbreviating these probabilities as $q_d, d = 1, 2, \cdots, D$, the DPMPM synthesizer actually distributes individuals over D cells as

$$(n_{f1}, n_{f2}, \ldots, n_{fD}) \sim \text{Multinomial}(q_1, q_2, \cdots, q_D; n_f), \qquad (15)$$

where n_{fd} denotes the number of records in the fth latent class taking the values of the dth cell, and n_f is the total number of records in the fth latent class.

It is worthy of note that

$$(n_1, n_2, \cdots, n_F) \sim \text{Multinomial}(\pi_1, \pi_2, \cdots, \pi_F; n). \qquad (16)$$

This additional view of the DPMPM leads to our QM-DPMPM synthesizer in the next subsection.

3.2   The QM-DPMPM Synthesizer

The QM-DPMPM synthesizer just replaces the multinomial draw in Eq. (15) of the DPMPM with

$$(n_{f1}, n_{f2}, \ldots, n_{fD}) \sim \mathrm{QM}(q_1, q_2, \cdots, q_D; n_f, \beta). \qquad (17)$$

Hence the QM-DPMPM obviously reduces to the DPMPM when $\beta = 0$. The parameter β is subjectively selected to take the balance of utility and disclosure risk of the synthetic data.

We provide a succinct overview of the QM-DPMPM synthesizer procedure. First we generate sample values of (π, α, θ^s) from the posterior distribution based on the DPMPM model. Through a multinomial draw with the samples of π, we can multinomially distribute n individuals over F latent classes with cell probabilities $\pi = (\pi_1, \ldots, \pi_F)$, as in Eq. (16). Finally, we quasi-multinomially distribute n_f individuals over D cells, as in Eq. (17). Note that n_{fd} is the number of individuals in cell d of class f.

As we can see, the QM-DPMPM synthesizer generates counts of combinations of synthesized variables. Once the count values of $\{n_{fd}, f = 1, \cdots, F, d = 1, \cdots, D\}$ are drawn, a partially synthetic dataset $\mathbf{X}^*_{QM-DPMPM,\beta}$ is obtained by duplicating the combinations of variables with $n_{fd} > 1$, and keeping the combinations of variables with $n_{fd} = 1$. Eventually, these synthesized combinations are attached to the un-synthesized variables to produce the partially synthetic dataset $\mathbf{X}^*_{QM-DPMPM,\beta}$.

3.3   Notes on Implementation

The `NPBayesImpute` R package is used for the DPMPM implementation. After the Markov chain Monte Carlo (MCMC) is converged, we generate $\mathbf{X}^*_{DPMPM}$ and $\mathbf{X}^*_{QM-DPMPM,\beta}$ within the Gibbs sampler and save these synthetic datasets. We repeat the above processes $m > 1$ times to obtain m synthetic datasets, using approximately independent draws of parameters obtained at MCMC iterations that are far apart.

4   Illustrative Application

We apply the DPMPM and QM-DPMPM synthesizers on a subset of a public available 2012 American Community Survey (ACS) sample. A similar dataset was used in Hu et al. (2014). We choose only $p = 10$ unordered categorical variables from the original $p = 14$ in Hu et al. (2014) because both the DPMPM synthesizer and the QM-DPMPM synthesizer have been developed for unordered categorical variables. To work with ordered categorical variables such as categorized age variables (levels: $1 = 18$–29, $2 = 30$–44, $3 = 45$–59, $4 = 60+$), methods such as probit models are needed. The multinomial-based synthesizers cannot incorporate the inherent order in those variables properly.

Table 1. Variables in the ACS sample, taken from the 2012 ACS public use microdata samples. PR stands for Puerto Rico. The Synthesized column records whether the variable is synthesized (yes) or not (no). The Known column records whether the variable is known to the intruder (yes) or not (no), for identification disclosure risks evaluation

Variable	Categories	Synthesized	Known
SEX	1 = male, 2 = female	Yes	Yes
RACE	1 = White alone, 2 = Black or African American alone, 3 = American Indian alone, 4 = other, 5 = two or more races, 6 = Asian alone	Yes	Yes
DIS	1 = has a disability, 2 = no disability	Yes	No
HICOV	1 = has health insurance coverage, 2 = no coverage	Yes	No
HISP	1 = not Spanish, Hispanic, or Latino, 2 = Spanish, Hispanic, or Latino	Yes	No
MAR	1 = married, 2 = widowed, 3 = divorced, 4 = separated, 5 = never married	No	Yes
MIG	1 = live in the same house (non movers), 2 = move to outside US and PR, 3 = move to different house in US or PR	No	Yes
LANX	1 = speaks another language, 2 = speaks only English	No	No
WAOB	born in: 1 = US state, 2 = PR and US island areas, oceania and at sea, 3 = Latin America, 4 = Asia, 5 = Europe, 6 = Africa, 7 = Northern America	No	No
SCH	1 = has not attended school in the last 3 months, 2 = in public school or college, 3 = in private school or college or home school	No	No

Our sample has $n = 10,000$ records and $p = 10$ unordered categorical variables. We synthesize 5 sensitive variables {SEX, RACE, DIS, HICOV, HISP}, and keep the remaining 5 variables un-synthesized {MAR, LANX, WAOB, MIG, SCH}. See Table 1 for the description and synthesis information of each variable.

We use the methods described in Sects. 3.1 and 3.2 to generate partially synthetic data from the DPMPM and QM-DPMPM synthesizers, respectively. For the QM-DPMPM synthesizer, we consider a sequence of 1000 values of β to assess its effect on the QM-DPMPM synthesizer ($\beta \in \{0.9991, 0.9981, \cdots, 0.0011, 0.0001\}$).

We generate $m = 20$ synthetic datasets from the DPMPM synthesizer, and $m = 20$ synthetic datasets from the QM-DPMPM synthesizer from one of the 1,000 β values ($\beta \in \{0.9991, \cdots, 0.0001\}$). We evaluate the utility and identification disclosure risks of each set of $m = 20$ synthetic datasets. We then evaluate and compare the utility and disclosure risks of $\mathbf{X}^*_{QM-DPMPM,\beta}$ for the range of β to those of $\mathbf{X}^*_{DPMPM}$.

4.1 Utility Evaluation

We first compare relative frequencies for various cross tabulations of all 10 variables in the original dataset and in the synthetic datasets. Specifically, we compute the relative frequencies for all one-way tables, two-way tables, and three-way tables. We then compare these relative frequencies in the original data to the synthetic data. Our approach is a modified version of that in Hu *et al.* (2014); Drechsler and Hu (2017+).

Figure 1 shows a clear trend of decreasing three-way deviation as β value decreases from 0.9991 to .0001. Plots of one-way and two-way deviation also show clear decreasing trends, and they are omitted for brevity. The shown trend is not surprising. Recall that the parameter β in the QM-DPMPM synthesizer effectively controls its similarity to the DPMPM synthesizer: larger β is associated with the larger variance of any univariate marginal frequency, resulting larger differences from the DPMPM synthesizer. Recall also that the QM-DPMPM synthesizer with $\beta = 0$ is the same as the DPMPM synthesizer. Figure 1 indicates that deviation-based utility of the QM-DPMPM synthesizer is higher when β decreases and approaches 0, and should be the highest (equal to that of the DPMPM synthesizer) when $\beta = 0$. The utility increases much more quickly when β drops under 0.25.

Analysts are often interested in regression analyses using synthetic data. To assess regression-based utility, we run logistic regression of disability status (DIS) on a number of predictors: health insurance coverage (HICOV), migration status (MIG), language use (LANX) and schooling (SCH). All predictors are treated as categorical. To deal with the separation problem in logistic regression, we use the `logistf` R package to fit a logistic regression model using Firth's bias reduction method (Firth 1993) to the original data, the DPMPM synthetic data, and the QM-DPMPM synthetic data.

We obtain the point estimates and 95% confidence intervals from the original and synthetic data and then make comparison to see how close the inferences are (closer to the original means higher utility). For the DPMPM synthetic data and the QM-DPMPM synthetic data, we obtain the point estimates and the 95% confidence intervals by using the combining rules for inference based on partial synthetic data (Reiter 2003; Drechsler 2011).

Using the results from the DPMPM synthetic data as a benchmark, we note that while some coefficients are preserved reasonably well (e.g. Intercept, MIG-3, LANX-2 and SCH-2), some differ to some degree (e.g. HICOV-2). However, every 95% confidence interval based on the DPMPM synthetic data includes the point estimate from the original data, indicating a reasonably high level of utility. Table 3 in Appendix 3 contains the detailed results.

Figure 2 plots inferences of the logistic regression coefficient of predictor SCH-2 based on the original data (the grey horizontal line), the DPMPM synthetic

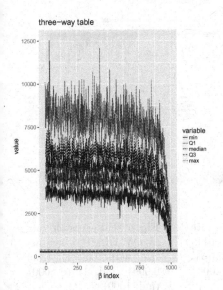

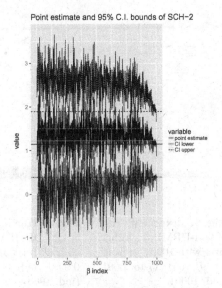

Fig. 1. Three-way table of utility of QM synthesizer with 1000 β's, from 0.9991 (ind. 1) to 0.0001 (ind. 1000). The minimum, Q1, median, Q3, and maximum of the utility of DPMPM synthesizer are 251.38, 284.95, 298.05, 318.66, and 390.78 respectively, and marked on the plot.

Fig. 2. Utility based on logistic regression coefficient estimates and 95% confidence intervals, for the SCH-2 predictor. Grey horizontal lines are values based on the original data; black horizontal lines are values based on the DPMPM synthetic data. β value ranges from 0.9991 (ind. 1) to 0.0001 (ind. 1000).

data (the black horizontal line), and the QM-DPMPM synthetic data (from 0.9991 (ind. 1) to .0001 (ind. 1000)). Plots on the remaining 6 coefficients show similar pattern, and are omitted for brevity. These results show that overall, the larger the β value, the greater the distance between the inferences based on the original data and those based on the QM-DPMPM synthetic data, indicating less accuracy in the inferences. Moreover, as β decreases, QM-DPMPM inference converges to DPMPM inference, showing the same trend of decreasing one-way, two-way, and three-way deviation as β decreases.

4.2   Identification Disclosure Risks Evaluation

For partially synthetic data, both identification disclosure and attribute disclosure risks possibly exist (Drechsler 2011; Hu 2018+). For illustrative purpose, we consider the identification disclosure risks in our application. That is, we evaluate the probability of identifying a record in the sample by matching with available external information.

Our evaluation approach is a Bayesian probabilistic matching procedure (Duncan and Lambert 1986, 1989; Lambert 1993; Fienberg *et al.* 1997; Reiter

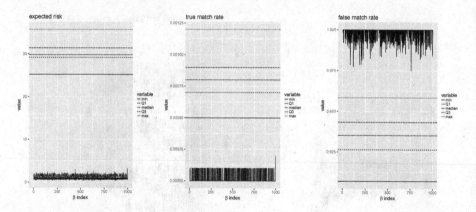

Fig. 3. Expected risk of QM-DPMPM synthesizer with 1000 β's, from 0.9991 (ind. 1) to .0001 (ind. 1000).

Fig. 4. True match rate of QM-DPMPM synthesizer with 1000 β's, from 0.9991 (ind. 1) to .0001 (ind. 1000).

Fig. 5. False match rate of QM-DPMPM synthesizer with 1000 β's, from 0.9991 (ind. 1) to .0001 (ind. 1000).

2005; Drechsler and Reiter 2008; Reiter and Mitra 2009; Drechsler and Reiter 2010; Drechsler and Hu 2017+). This general evaluation procedure considers the matching probability of a target vector **t** available to the intruder. This target record **t** contains some un-synthesized variables that are available through external files, denoted as $\mathbf{t}^{A_{us}}$, and some other synthesized and available variables, denoted as $\mathbf{t}^{A_s}$. Therefore $\mathbf{t} = (\mathbf{t}^{A_{us}}, \mathbf{t}^{A_s})$.

The identification disclosure risks evaluation aims at estimating the probability of the intruder being able to identify a record i with the available target vector **t**, by using the knowledge of un-synthesized variables in $\mathbf{t}^{A_{us}}$ and guessing the synthesized variables in $\mathbf{t}^{A_s}$. In the end, three summaries of identification disclosure probabilities are produced: (i) the expected match risk, an overall summary of all target records being the true match among all records with the highest match probability (similar to the measure proposed in Franconi and Polettini (2004)); (ii) the true match rate, the percentage of true unique matches among the target records; and (iii) the false match rate, the percentage of false matches among unique matches (Reiter and Mitra 2009; Drechsler and Reiter 2010; Drechsler 2011 Drechsler and Hu 2017+; Hu 2018+).

In our application, we assume the intruder knows the sex, race, martial status and migration status of all respondents through external files. Among these variables in **t**, sex and race are synthesized, while marital status and migration statues are un-synthesized. Therefore, $\mathbf{t}^{A_{us}} = \{\text{MAR, MIG}\}$ and $\mathbf{t}^{A_s} = \{\text{SEX, RACE}\}$. We treat all n = 10,000 records in the sample as target records. For each of $m = 20$ synthetic datasets we calculate risk measures, and the five number summary of these 20 risk values is plotted for each β in Figs. 3, 4 and 5. The five number summary is useful to evaluate the many scenarios of an intruder, who may combine the information of multiple synthetic datasets in various ways.

Figure 3 shows that the expected risk is stable against the change of β, which may seem opposite to our intuition. Nevertheless it reflects the fact that observed true matches are few. Let X be the number of matched sample records to a target record. The expected match risk increases by $1/X$ only when the target record is truly matched among the X records. Hence the expected match risk is close to zero when true matches are few. Then even the expectation of the expected match risk is increasing as $\beta \to 0$, it is hard to observe the increasing trend of the realized expected match risk.

The stable discrepancy of the true match rates in Fig. 4 between the QM-DPMPM and the DPMPM may look large, but this discrepancy is not large in a stochastic sense. To see this fact, let us focus on the number of true unique matches; in the case of DPMPM ($\beta = 0$) it ranges from 5 to 12, which are not very far from 0 to 2 of $\beta > 0$.

To confirm the convergence of the true match rate of the QM-DPMPM to that of the DPMPM, we need even smaller β, but we observe the clear convergence of unique match counts of the QM-DPMPM to that of the DPMPM (a plot is omitted for brevity).

5   Concluding Remarks

The properties of the QM distribution and its tuning parameter β have motivated us to develop a QM-DPMPM synthesizer based on the DPMPM synthesizer. We have seen that around $\beta = 0$, the utilities of the QM-DPMPM synthesizer measured in Sect. 4.1 are very close to those of the DPMPM. On the other hand, risk measures dealt with in Sect. 4.2 show various speeds of convergence as $\beta \to 0$. This difference implies that with only a slight loss of utility, a statistical agency may be able to generate a much safer data set by employing the QM-DPMPM synthesizer.

We believe the QM-DPMPM synthesizer is a promising method of generating synthetic categorical data that is worth further investigation. It would be interesting to experiment with smaller β values and evaluate the utility-risks tradeoff. Additionally, many other multinomial distribution based categorical data synthesizers can be turned into a QM distribution based synthesizer with desired utility-risks balance.

Appendix

Appendix 1

Proof of Theorem 3: By symmetry it suffices to show the case of $a_1 < a_2$. To simplify our argument, write $g(y; a) = \Gamma(a + y)/(\Gamma(a + 1)(a + y)^{y-1})$. Then $p_{BB}(y)/p_{QB}(y) = g(y; a_1)g(n - y; a_2)/g(n; a.)$. Hence $p_{BB}(y)/p_{QB}(y)$ is minimized when $g(y; a_1)g(n - y; a_2)$ is minimized. Also we note that $g(y; a)r(y; a) =$

$g(y+1; a)$, where $r(y; a) = (a+y)^y/(a+y+1)^y$. Since $r(0, a) = 1$ and $r(y, a) < 1$ for $y \geq 1$, $g(y; a)$ is monotonically decreasing when y increases. Therefore to minimize $g(y; a_1)g(n-y; a_2)$, we increase y_1 or y_2 one by one so that $g(y_1; a_1)g(y_2; a_2)$ is more reduced. Denote the ith step values by $(y_1, y_2)_i, i = 1, \ldots, n$. At each step y_1 increases by one if $r(y_1, a) \leq r(y_2, a)$ otherwise y_2 increases by one. Actually $r(0, a_1) = r(0, a_2) = 1$, but if $0 < a_1 < a_2$ then $r(1, a_1) < r(1, a_2)$. Hence we begin with $(y_1, y_2)_1 = (1, 0)$. Observing

$$\frac{d\log r(y,a)}{dy} = \log\left(1 - \frac{1}{a+y+1}\right) + \frac{y}{(a+y)(a+y+1)}$$

$$= -\frac{1}{a+y+1} - \frac{1}{2}\left(\frac{1}{a+y+1}\right)^2 - \cdots + \frac{y}{(a+y)(a+y+1)} < 0,$$

we note that $r(y, a) > r(y', a)$ when $y < y'$. Therefore $r(1, a_2) > r(1, a_1) > r(i, a_1)$ for $i \geq 2$, which leads to $(y_1, y_2)_n = (n, 0)$. □

Appendix 2

A summary of average acceptance rates r of our QB sampler proposed in Sect. 2.

Table 2. Average acceptance rates: $r(0.1/\beta, 0.9/\beta, n)$

n	β									
	2^{-1}	2^{-2}	2^{-3}	2^{-4}	2^{-5}	2^{-6}	2^{-7}	2^{-8}	2^{-9}	2^{-10}
10^1	0.13	0.06	0.03	0.02	0.03	0.06	0.14	0.29	0.50	0.69
10^2	0.00	0.00	0.01	0.07	0.26	0.51	0.71	0.84	0.92	0.96
10^3	0.12	0.34	0.59	0.77	0.87	0.94	0.97	0.98	0.99	1.00
10^4	0.81	0.90	0.95	0.97	0.99	0.99	1.00	1.00	1.00	1.00
10^5	0.98	0.99	0.99	1.00	1.00	1.00	1.00	1.00	1.00	1.00
10^6	1.00	1.00	1.00	1.00	1.00	1.00	1.00	1.00	1.00	1.00
10^7	1.00	1.00	1.00	1.00	1.00	1.00	1.00	1.00	1.00	1.00
10^8	0.84	0.84	0.84	0.84	0.84	0.84	0.84	0.84	0.84	0.84
10^9	0.00	0.00	0.00	0.00	0.00	0.00	0.00	0.00	0.00	0.00

Appendix 3

The table contains detailed results of regression-based utility of the DPMPM synthetic data in Sect. 4.

Table 3. 95% confidence intervals of logistic regression coefficients based on the original data and based on the m = 20 synthetic data generated by the DPMPM synthesizer.

Estimand	Original data		DPMPM (m = 20)	
	Estimate	95% CI	$\bar{q}_{20}$	95% CI
Intercept	2.12	[1.86, 2.39]	2.27	[1.53, 3.01]
HICOV - 2	0.60	[0.43, 0.77]	0.14	[−0.41, 0.69]
MIG - 2	0.39	[−0.61, 1.39]	0.06	[−1.31, 1.43]
MIG - 3	0.04	[−0.12, 0.19]	0.06	[−0.50, 0.63]
LANX - 2	-0.87	[−1.14, −0.60]	−0.96	[−1.71, −0.22]
SCH - 2	1.22	[0.94, 1.50]	1.14	[0.39, 1.90]
SCH - 3	1.63	[1.01, 2.25]	1.07	[0.07, 2.07]

Appendix 4

The Table 4 contains the minimum, first quartile, median, third quartile, and maximum of the identification disclosure risks of the DPMPM synthesizer. They are all marked as horizontal lines in Figs. 3, 4 and 5.

Table 4. Table of the minimum, first quartile (Q1), median, third quartile (Q3), and maximum of the identification disclosure risks of the DPMPM synthesizer.

Summary	Min	Q1	Median	Q3	Max
Expected risk	25.2011	29.1804	29.8434	31.4175	35.6882
True match rate	0.0005	0.0007	0.0008	0.0009	0.0012
False match rate	0.9070	0.9264	0.9352	0.9433	0.9583

References

Akande, O., Li, F., Reiter, J.P.: An empirical comparison of multiple imputation methods for categorical data. Am. Stat. **71**, 162–170 (2017)

Akande, O., Reiter, J. P., Barrientos, A. F.: Multiple imputation of missing values in household data with structural zeros (2017+). arXiv:1707.05916

Consul, P.C., Mittal, S.P.: A new urn model with predetermined strategy. Biometrische Zeitschrift **17**, 67–75 (1975)

Consul, P.C., Mittal, S.P.: Some discrete multinomial probability models with predetermined strategy. Biometrische Zeitschrift **19**, 161–173 (1977)

Devroye, L.: Non-uniform Random Variate Generation. Springer, New York (1986). https://doi.org/10.1007/978-1-4613-8643-8

Drechsler, J.: Synthetic Datasets for Statistical Disclosure Control. Springer, New York (2011). https://doi.org/10.1007/978-1-4614-0326-5

Drechsler, J., Hu, J.: Strategies to facilitate access to detailed geocoding information based on synthetic data (2017+). arXiv:1803.05874

Drechsler, J., Reiter, J.P.: Accounting for intruder uncertainty due to sampling when estimating identification disclosure risks in partially synthetic data. In: Domingo-Ferrer, J., Saygın, Y. (eds.) PSD 2008. LNCS, vol. 5262, pp. 227–238. Springer, Heidelberg (2008). https://doi.org/10.1007/978-3-540-87471-3_19

Drechsler, J., Reiter, J.P.: Sampling with synthesis: a new approach to releasing public use microdata samples of census data. J. Am. Stat. Assoc. **105**, 1347–1357 (2010)

Duncan, G.T., Lambert, D.: Disclosure-limited data dissemination. J. Am. Stat. Assoc. **10**, 10–28 (1986)

Duncan, G.T., Lambert, D.: The risk of disclosure for microdata. J. Bus. Econ. Stat. **7**, 207–217 (1989)

Dunson, D.B., Xing, C.: Nonparametric Bayes modeling of multivariate categorical data. J. Am. Stat. Assoc. **104**, 1042–1051 (2009)

Fienberg, S.E., Makov, U., Sanil, A.P.: A Bayesian approach to data disclosure: optimal intruder behavior for continuous data. J. Off. Stat. **13**, 75–89 (1997)

Firth, D.: Bias reduction of maximum likelihood estimates. Biometrika **80**, 27–38 (1993)

Franconi, L., Polettini, S.: Individual risk estimation in μ-argus: a review. In: Domingo-Ferrer, J., Torra, V. (eds.) PSD 2004. LNCS, vol. 3050, pp. 262–272. Springer, Heidelberg (2004). https://doi.org/10.1007/978-3-540-25955-8_20

Ho, F.C.M., Gentle, J.E., Kennedy, W.J.: Generation of random variates from the multinomial distribution. In: Proceedings of the American Statistical Association Statistical Computing Section (1979)

Hoshino, N.: The quasi-multinomial distribution as a tool for disclosure risk assessment. J. Off. Stat. **25**, 269–291 (2009)

Hu, J.: Bayesian estimation of attribute and identification disclosure risks in synthetic data (2018+). arXiv:1804.02784

Hu, J., Reiter, J.P., Wang, Q.: Disclosure risk evaluation for fully synthetic categorical data. In: Domingo-Ferrer, J. (ed.) PSD 2014. LNCS, vol. 8744, pp. 185–199. Springer, Cham (2014). https://doi.org/10.1007/978-3-319-11257-2_15

Hu, J., Reiter, J.P., Wang, Q.: Dirichlet process mixture models for modeling and generating synthetic versions of nested categorical data. Bayesian Anal. **13**, 183–200 (2018)

Ishwaran, H., James, L.F.: Gibbs sampling methods for stick-breaking priors. J. Am. Stat. Assoc. **96**, 161–173 (2001)

Lambert, D.: Measures of disclosure risk and harm. J. Off. Stat. **9**, 313–331 (1993)

Malefaki, S., Iliopoulos, G.: Simulating from a multinomial distribution with large number of categories. Comput. Stat. Data Anal. **51**, 5471–5476 (2007)

Manrique-Vallier, D., Hu, J.: Bayesian non-parametric generation of fully synthetic multivariate categorical data in the presence of structural zeros. J. Roy. Stat. Soc. Ser. A (2018, to appear)

Manrique-Vallier, D., Reiter, J.P.: Bayesian estimation of discrete multivariate latent structure models with structural zeros. J. Comput. Graph. Stat. **23**, 1061–1079 (2014)

Murray, J.S.: Multiple imputation: a review of practical and theoretical findings. Stat. Sci. (2018+)

Reiter, J.P.: Inference for partially synthetic, public use microdata sets. Surv. Methodol. **29**, 181–188 (2003)

Reiter, J.P.: Estimating risks of identification disclosure in microdata. J. Am. Stat. Assoc. **100**, 1103–1112 (2005)

Reiter, J.P., Mitra, R.: Estimating risks of identification disclosure in partially synthetic data. J. Priv. Confid. **1**, 99–110 (2009)

Reiter, J.P., Raghunathan, T.E.: The multiple adaptations of multiple imputation. J. Am. Stat. Assoc. **102**, 1462–1471 (2007)

Rubin, D.B.: Multiple Imputation for Nonresponse in Surveys. Wiley, New York (1987)

Sethuraman, J.: A constructive definition of Dirichlet priors. Statistica Sinica **4**, 639–650 (1994)

Si, Y., Reiter, J.P.: Nonparametric Bayesian multiple imputation for incomplete categorical variables in large-scale assessment surveys. J. Educ. Behav. Stat. **38**, 499–521 (2013)

Synthetic Data via Quantile Regression for Heavy-Tailed and Heteroskedastic Data

Michelle Pistner[1]([✉]), Aleksandra Slavković[1], and Lars Vilhuber[2][ID]

[1] Department of Statistics, The Pennsylvania State University,
University Park, PA, USA
pistner@psu.edu
[2] Economics Department, Cornell University, Ithaca, NY, USA

Abstract. Privacy protection of confidential data is a fundamental problem faced by many government organizations and research centers. It is further complicated when data have complex structures or variables with highly skewed distributions. The statistical community addresses general privacy concerns by introducing different techniques that aim to decrease disclosure risk in released data while retaining their statistical properties. However, methods for complex data structures have received insufficient attention. We propose producing synthetic data via quantile regression to address privacy protection of heavy-tailed and heteroskedastic data. We address some shortcomings of the previously proposed use of quantile regression as a synthesis method and extend the work into cases where data have heavy tails or heteroskedastic errors. Using a simulation study and two applications, we show that there are settings where quantile regression performs as well as or better than other commonly used synthesis methods on the basis of maintaining good data utility while simultaneously decreasing disclosure risk.

1 Introduction

Statistical disclosure control (SDC) encompasses a group of privacy-preserving techniques that introduce bias and variance in data in order to protect confidentiality of individual records. Many different methods have been proposed, and most involve some sort of perturbation. Traditional methods include the addition of random noise, rank-swapping, and top or bottom coding [17]. SDC methods are not equal in terms of the privacy protection or data utility they provide, and improperly protected data are vulnerable to re-identification attacks [9].

Utilizing ideas from multiple imputation, synthetic data was proposed as a privacy method to protect some or all variables in a data set [25,35]. Since its introduction, synthetic data methodology has been extensively studied [4, 7,13,31,33], and several synthetic data products have been released (e.g., [1, 6,20,27]). However, when the original data have heavy-tailed distributions or heteroskedastic errors, challenges remain from both the privacy-protection and modeling perspectives.

© Springer Nature Switzerland AG 2018
J. Domingo-Ferrer and F. Montes (Eds.): PSD 2018, LNCS 11126, pp. 92–108, 2018.
https://doi.org/10.1007/978-3-319-99771-1_7

In the general statistics literature, quantile regression has been proposed as a method that can alleviate some issues associated with heavy-tailed or censored data. First proposed by Koenker and Bassett [22], quantile regression has been extensively studied and applied to many scientific domains (e.g., ecology and economics [14,42]). It is a flexible modeling approach in the sense that the entire conditional distribution can be modeled and model assumptions are less restrictive when compared to least squares regression [10]. Quantile regression is an area of active research: algorithmic improvements for coefficient estimation have been developed [30], quantile-based techniques have been applied to other statistical models (e.g., random forests [28]), Bayesian approaches have been proposed [43], and strategies to alleviate monotonicity violations have been suggested [3,5,26].

Quantile regression has been also proposed as a synthetic data mechanism by Huckett and Larsen [15,16]. Using visual plots and basic measures, Huckett and Larsen [16] found that synthetic data generated via quantile regression preserves marginal and conditional distributions well when compared to the original data. However, they did not compare their synthesis with other commonly used methods or thoroughly study its performance in the context of heavy-tailed or heteroskedastic data. We revisit their methodology by addressing some open questions pertaining to quantile selection and estimation, proposing quantile regression as a privacy method for heavy-tailed and heteroskedastic applications, and comparing it to other commonly used synthesis methods such as classification and regression trees (CART).

The structure of the remainder of this paper is as follows. Section 2 discusses some fundamentals of quantile regression. Section 3 briefly discusses synthetic data methodology and defines the utility and risk measures we use. In Sect. 4, we describe our proposed algorithm and illustrate its workings by way of a simulation study. In Sect. 5, we present two applications to census and business data. Finally, we end with conclusions and future work in Sect. 6.

2   Quantile Regression

Quantile regression is a flexible modeling technique in the sense that it can model the entire conditional distribution of the response variable [22]. In contrast to linear regression which models the conditional mean of the response distribution, quantile regression models the τth conditional quantile of the response, where τ is any quantile ranging between zero and one.

More formally, quantile regression specifies a model of the form:

$$y = X^T \beta_\tau + \epsilon_\tau, \tag{1}$$

where $X \in \mathbb{R}^{n \times p}$ is a matrix of covariates, $\beta_\tau \in \mathbb{R}^{p \times 1}$ is a vector of coefficients, and ϵ_τ represents the corresponding error. Following the notation of [10], estimation of β_τ is solved by minimizing a weighted sum of absolute deviations:

$$\hat{\beta}_\tau = \operatorname{argmin}_\beta \sum_{i=1}^n w_\tau(y_i, \eta_{i\tau}) |y_i - \eta_{i\tau}|, \tag{2}$$

where $\eta_{i\tau} = x_i^T \beta_\tau$ and $w_\tau(y_i, \eta_{i\tau})$ equals τ if $y_i > \eta_{i\tau}$, $1 - \tau$ if $y_i < \eta_{i\tau}$, and 0 otherwise. This can be equivalently written using the so-called "check" function:

$$\hat{\beta}_\tau = \mathrm{argmin}_\beta \sum_{i=1}^{n} \rho_\tau(y_i - \eta_{i\tau}), \tag{3}$$

where $\rho_\tau(u) = u\{\tau - I(u < 0)\}$. Coefficient estimates can be solved using standard linear programming optimization techniques. Quantile regression is efficiently implemented using several different optimization routines in the R package *quantreg* [21]. For our simulations, we used the Frisch-Newton algorithm [30] from the *quantreg* package since it is computationally more efficient than the other algorithms, and thus more suitable for our syntheses.

As mentioned, quantile regression is a more flexible modeling tool than standard linear regression. It is a distribution-free method as it does not require a priori specification of a parametric distribution for the errors, for example, and it is well suited to address "extreme" observations in terms of covariates. There are no assumptions of homoskedasticity and common distribution of the errors [10]. Instead, the main assumptions are that τ^{th} quantile of the errors is zero and the errors are independent [10]. In addition, there is an implicit assumption of monotonicity. When estimating coefficients for multiple quantiles, the predicted estimates must be ordered for a given value of the independent variables, i.e., for a given set X, $X^T \beta_{0.25} \leq X^T \beta_{0.75}$. This is often violated in practice, especially when the number of estimated quantiles is large [3]. Several methods have been proposed to address this challenge [3,5,26].

Quantile regression has also been studied from a Bayesian perspective. In one characterization, Bayesian quantile regression assumes that errors arise from an asymmetric Laplace distribution (ALD), i.e., $\epsilon_{i\tau} \sim ALD(0, \sigma^2, \tau)$ [43]. If the prior distribution on β_τ is proportional to a constant, maximizing the resulting posterior is equivalent to minimizing the weighted absolute deviations of (2) [10]. To ease computations, Kozumi and Kobayashi [23] recommend to reparameterize the asymmetric Laplace distribution as a location-scale mixture of normal distributions. This method is implemented in the R package *bayesQR* [2]. Full implementation details are discussed in Sect. 4.

3   Synthetic Data Overview

3.1   General Methodology

Synthetic data methods use the original data to develop synthetic versions of sensitive values. Closely related to work in multiple imputation, synthetic data was proposed as a method to protect all or some observations that is variables in a data set [25,35]. Many different models have been proposed as synthesis methods, and definitions of utility and disclosure risk measures that can potentially be used within this setting have been formalized [7,17].

Let $X = (x_1, .., x_p)$ be an observed data set that has sensitive variables. Note that any combination of variables and rows can be synthesized, including all variables and all rows. These are sometimes referred to as *partial* and *full* syntheses, respectively [31].

Synthetic data aims to preserve and model the joint distribution $(f(x_1, \ldots, x_p))$ between these variables by modeling them as a series of conditional distributions, i.e.,

$$f(x_1, \ldots, x_p) = f(x_1)f(x_2|x_1) \ldots f(x_p|x_1, \ldots, x_{p-1}). \qquad (4)$$

The syntheses are typically conducted in a sequential manner. Any type of statistical model can be used to capture the conditional distributions in (4), including classification and regression trees (CART) [33], random forests [4], and kernel density estimators [41]. Choice of model depends on both the data structure and desired privacy levels. The R package *synthpop* contains functions to create synthetic data using several different statistical models [29].

In practice, agencies have begun using synthetic data products to release usable data sets to researchers with less stringent conditions or as public-use data tabulations. In particular, the U.S. Census Bureau has released partially-synthesized versions of the U.S. Census Bureau's Longitudinal Business Data (LBD) [20] and Survey of Income and Program Participation (SIPP) [1]. In general, for confidential micro data, researchers must apply for access and travel to one of a small number of secure data centers to use the data. In contrast, the synthetic data products are available via a remote desktop application to approved researchers. Synthetic data products are typically not used for final analyses but instead are used to write code and develop models (exception being *OnTheMap* [27] that releases synthetic data for direct analysis with no validation). Final analyses are conducted on the original data by agency employees, compared to the synthetic results, and the findings undergo a disclosure process before being released to the researcher. This type of process is sometimes referred to as a *gold-standard* analysis [37]. Other statistical agencies have developed synthetic data products such as the Scottish Longitudinal Survey [36] and the German IAB Establishment Data [6].

3.2 Utility Measures

Utility measures aim to quantify the degree of similarity between the original and synthetic data. They can quantify overall similarity or analysis-specific similarity between the original and synthetic data. Following [37], we refer to these metrics as *general* and *specific* utility, respectively.

Specific utility typically measures agreement across a specific analysis or group of analyses. Examples of specific utility measures include standardized distances between summary statistics calculated on the original and synthetic data or confidence interval overlap of estimated parameters [7,19].

General utility measures typically quantify distributional similarity (or divergence) between the original and synthetic data. Basic measures are based on empirical cumulative distribution functions (CDF) or Kullback-Leibler divergence [17]. Measures based on propensity scores have also been proposed

[19,37,40]. These measures are based on the idea that if the synthetic and original data are similar, data set membership should be indistinguishable between the two data sets. To quantify the general utility of synthetic data sets, we use a propensity score-based metric, *pMSE*, with CART as the classification algorithm (e.g., for more details see [40] and [37]). Since we assume that our synthetic data have the same sample size as the original data, *pMSE* produces a value between zero and 0.25. Ideal values are close to zero.

3.3   Risk Measures

In general, disclosure risk is categorized into two types [17]. *Attribute* disclosure refers to an intruder learning the value of a confidential variable for a given individual. *Identity* disclosure refers to an intruder matching a record with outside information to learn the entity corresponding to that record.

Traditional measures of disclosure risk are dependent on the underlying data structure and assumptions on intruder knowledge; in contrast to this is differential privacy (for a comprehensive review, see [8]) which we do not deal with in this paper. Intuitively, observations that are similar to other observations should have lower disclosure risk when compared to outlier observations. Separate measures have been developed for categorical, continuous, and combination data (e.g., for a review of these measures see [7,17]). Business data typically have many continuous variables with skewed distributions which warrant special risk measures [17].

Foschi [11] and Ichim [18] develop measures of disclosure specifically for heavily skewed business data. Foschi uses a robust finite Gaussian mixture model in conjunction with hypothesis tests to determine if a given observation is consistent with respect to one of the mixture distributions. Those observations unlikely to belong to any mixture component are classified as at-risk. Ichim relies on a density-based method relying on the local outlier factor (LOF). In this paper, we adopt Ichim's measure to compare disclosure risk across different synthetic data methods we implement, including our proposed quantile regression based synthesis.

4   Quantile Regression and Synthetic Data

Quantile regression as a synthetic data method has been previously introduced by Huckett and Larson as a component of a multimethod approach that also included hot deck imputation and rank-swapping [15,16]. They applied their method to a subset from the American Community Survey (ACS) and an Iowa tax database, and they explored the disclosure risk associated with quantile regression as a synthesis method [24]. They studied utility of synthetic data sets at a basic level by comparing empirical CDFs between the observed and synthetic data and by comparing basic results from several regressions. However, they failed to compare their method to other existing methods in the literature and did not thoroughly study quantile regression in the context of heavy-tailed or heteroskedastic data.

Following the basic algorithm of Huckett and Larson [16], we propose and implement an updated algorithm to produce a fully synthetic data as follows:

Algorithm 1. Data synthesis using quantile regression

Data: The original data set $X = (x_1, ..., x_p)$
Result: A synthetic data set $Z = (z_1, ..., z_p)$
for *each* $i \in 2, .., p$ **do**
  1a. Model $x_i | x_1, ..., x_{i-1}$ using quantile regression for a large number of quantiles;
  for *each* $j \in 1, ..., n$ **do**
    2a. Randomly select a quantile and choose the model corresponding to the quantile;
    2b. From this model, generate $z_{j,i}$ for each observation from $z_{j,1}, ..., z_{j,i-1}$ and return as synthetic data value for $x_{j,i}$;
  end
end

We introduce two novel ideas in Algorithm 1 to provide more motivation for quantile-based syntheses and ease of computations. First, in Step 2a, we follow Huckett and Larson's suggestion of sampling a random quantile, rather than all quantiles. Huckett and Larson provided little motivation for their choice of quantile. However, we suggest that the choice of random quantiles is closely linked to the theories underlying the inverse transform method for pseudo-random sampling (for a review, see [34]). This method generates random samples by inverting the cumulative distribution function at values chosen randomly from a uniform distribution. This link supports the choice of uniform random quantiles in Algorithm 1.

Even though the above approach addresses potential computational bottlenecks, there could still be a large number of quantiles that would require independent model fitting, that is a new estimated model for each chosen observation. Although computation time is linear in the number of quantiles, a large number of observations could make the random quantile-based approach infeasible in certain cases. To circumvent this, we also propose a *binned* approach as an alternative to Step 2a. Here, we propose to split the set of quantiles, i.e., (0,1), into a large number of bins, and compute the quantile regression estimates for the midpoint quantile in each bin. Then, a bin is chosen at random instead of a quantile in Step 2a. The model corresponding to that bin is used to obtain a synthetic value for a given observation.

More specifically, in our simulations and applications, we use 250 bins. A series of simulations showed that this number provided a good balance between data utility and disclosure risk. All of our simulations and applications used this binning technique. The set of quantiles is split into 250 bins—i.e. (0, 0.004], (0.004, 0.008], ..., (0.996, 1)—then coefficient estimates were calculated for the midpoint of each bin—i.e., 0.002, 0.006, ..., 0.998. Bins or, equivalently, quantiles are chosen for each observation, and corresponding coefficient estimates are used to generate synthetic data for that observation in Step 2b. Random bins were selected using the *runif* command in base R, and quantile regression coefficients were estimated using the Frisch-Newton algorithm from the *rq* function in the *quantreg* R package.

The algorithm for Bayesian quantile regression-based synthetic data is similar to Algorithm 1, where we use random bins but now we use the *bayesQR* R package to estimate model coefficients using the Gibbs sampling approach of [23]. We use a multivariate normal prior distribution with mean and variance equal to least-squares regression estimates. Following [43], we use an inverse Gamma distribution as the prior for variance.

We compare our quantile-based syntheses to CART-syntheses. This was motivated by previous research suggesting that CART is a flexible synthesis tool that results in syntheses with high utility [33]. We used the R package *synthpop* to synthesize data with CART [29]. Several considerations for generating synthetic data that are unique to each application were discovered. They will be discussed in more detail in Sects. 5.1 and 5.2, but first we consider a set of simulations.

4.1   Simulations

We used two simulation studies to compare quantile regression to CART as synthesis methods in terms of general utility and disclosure risk. We simulated data for two different data structures, one with heavy-tails and the other with heteroskedastic errors, both of $n = 5,000$. Our simulation scheme is detailed in Table 1.

Table 1. Simulation schemes for heavy-tailed and heteroskedastic data sets.

Heavy-tailed	Heteroskedastic errors
$X_1 \sim exp(0.1)$	$x_1 \sim N(0,1)$
$X_2\|X_1 = \alpha_0 + \beta_0 X_1 + \epsilon_i$	$X_2\|X_1 = \alpha_0 + \beta_0 X_1 + (\gamma_0 + \gamma_1 X_1)\epsilon_i$
$X_3\|X_1, X_2 = \alpha_1 + \beta_1 X_1 + \beta_2 X_2 + \gamma_i$	$X_3\|X_1, X_2 = \alpha_1 + \beta_1 X_1 + \beta_2 X_2 +$
	$(\gamma_2 + \gamma_3 X_1 + \gamma_4 X_2)\tau_i$
$\epsilon_i, \gamma_i \sim exp(0.1)$	$\epsilon_i, \tau_i \sim N(0,1)$
$(\alpha_0, \alpha_1, \beta_0 \beta_1, \beta_2) = (2, 5, 2, -2, 1)$	$(\alpha_0, \alpha_1) = (2, 5)$
	$(\beta_0, \beta_1, \beta_2) = (2, -2, 1)$
	$(\gamma_0, \gamma_1, \gamma_2, \gamma_3, \gamma_4) = (1.5, 5, 1.5, 5, -6)$

In total, 10 data sets were synthesized using our Algorithm 1 with both the frequentist and Bayesian quantile regression implementation, and CART via the *synthpop* package which relieves on the *party* and *rpart* packages for categorical and continuous variables, respectively [12,38]. 250 bins were used for the quantile regression syntheses. As suggested by [43], a multivariate normal prior was used for the coefficient estimates with mean and covariance equal standard linear regression estimates, and an inverse gamma prior was used for variance. For CART-based syntheses, the minimum number of observations per terminal node was set to five.

General utility and disclosure risk are calculated for each synthesis according to the *pMSE* measure and Ichim's method discussed in Sects. 3.2 and 3.3. Averages and standard errors of these measures are presented in Table 2. Under both data structures, we found that quantile regression outperforms CART on the basis of general utility, especially in the heteroskedastic setting. Bayesian quantile regression did not offer the same results, and, due to significantly longer computing time, we did not explore it further. In many additional simulations, not reported here, we show consistently similar results.

Table 2. Mean and standard errors over utility score and disclosure risk for the simulated data sets over 10 repetitions. This utility measure captures overall similarity between the original and synthetic data. Scores close to zero are optimal.

Method	Heavy-tailed utility	Heavy-tailed risk	Heteroskedastic utility	Heteroskedastic risk
QR	0.05 (0.015)	1.56 (0.20)	0.10 (0.003)	12.25 (0.46)
Bayes QR	0.07 (0.010)	1.67 (0.10)	0.11 (0.002)	12.4 (0.77)
CART	0.06 (0.015)	2.49 (0.31)	0.12 (0.002)	13.0 (0.70)

5 Applications

In this section, we consider two applications to census and business data in order to demonstrate the privacy-preserving and data utility potential of quantile-based regression, with a special focus on data with heavy tails and heteroskedastic errors[1].

5.1 1901 Census of Scotland

Data. The 1901 Census of Scotland collected information on household structures for all heads of households in the country. We use a subset of the original data ($n = 82{,}851$) [39]. For each head of household, individual data was collected, including age, race, geographic location, working status, and marital status. Additional variables were collected for total household size and the total number of household members in each individual category, including children, other relatives, servants, and boarders. These count variables had very heavy tails. Some variables such as total household size and rooms per person in each household are linear combinations of other variables (e.g., total household size was the sum across each individual categories) and were excluded from synthesis.

For our application, we use 11 out of the 26 total variables. Three individual-level attributes are used (sex, age, and marital status) and eight count-type variables (i.e., counts of household members by type and total number of rooms per house). See Appendix A for a summary of variables.

---

[1] All code is available on our Github repository at https://github.com/labordynamics institute/replication_qr_synthetic.

Results. A total of ten syntheses are produced from the quantile-based regression and using CART. General utility is computed for each synthesis according to the pMSE method discussed in Sect. 3.2. Quantile-based syntheses had a mean utility of 0.015 (SE: 0.0005) whereas CART-based syntheses had a mean utility of 0.002 (SE: 0.0003). In this setting, both methods lead to good utility based on the pMSE, but CART clearly outperforms our proposed quantile-based synthesis. This may be a result of CART treating the counts as categorical outcomes instead of continuous. However, as we will see further below, CART actually overfits the data leading to much higher disclosure risk.

We use several visual and empirical risk measures to assess the synthesis products for these two methods. Simple plots of the tails of the different count variables were created to compare the methods. An example plot in Fig. 1 shows original data for the number of servants, and data generated via CART and quantile regression. The graph is a snapshot of the right tail of the distribution (anything greater than or equal to 5 counts in order to see the results in the tails better). Figure 1 shows that data generated via CART closely matches the original data counts in right tail of the distributions, which potentially increases risk for these outlier observations. In terms of empirical risk measures, CART and quantile-based syntheses again show similar measures. However, for quantile-based syntheses, these risk measures are lower with a mean percentage of observations as risk of 12.8% (SE: 0.8%), whereas for CART-based syntheses, these risk measures were higher with a mean of 13.2% (SE: 0.4%).

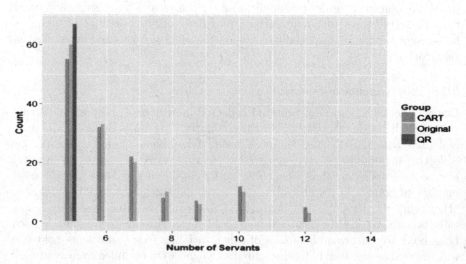

Fig. 1. Distribution of the tails of "Number of Servants" variable. Note that the synthetic data generated via CART closely matches the original data counts in the tails. This could increase disclosure risk.

5.2   Synthetic Longitudinal Business Database

Data. The Longitudinal Business Database (LBD) represents a census of all U.S.-based establishments. It contains payroll and employment information from 1976 to 2000 for over 21 million establishments. Since all establishments are included, there is no privacy protection offered via sampling. It is only accessible onsite at the secure Federal Statistical Research Data Centers in the United States.

The Synthetic Longitudinal Business Database (SynLBD) is a synthetic version of the LBD developed by researchers in conjunction with the U.S. Census Bureau [20]. Access to the data is available on servers housed at Cornell University via a remote desktop [32]. Users must apply for access through the U.S. Census Bureau. However, the process is much simpler when compared to access to the restricted data itself.

There are several key differences in variables between LBD and SynLBD. The LBD contains information on firm geography which was used to develop the SynLBD, but the original variables are removed from the synthetic data. Additionally, the Standard Industry Classification (SIC) code is available in full form in the LBD. SIC codes are four digits in length. For the SynLBD, these codes are truncated to the first three digits.

Several synthesis rules were used when constructing the SynLBD, as we discuss below. For a complete description of the synthesis process for the SynLBD, see the supplementary material to [20]. For this application, we treated the synthetic data as if it were the confidential data and created new synthetic data from it. No confidential data were used in this application.[2] For our synthesis, we used constraints similar to [20] when constructing quantile-based syntheses. We fit separate models for each SIC code, and within each SIC code, further distinguish models for single-unit versus multi-unit establishments. We generate synthetic versions for first year of unit establishment via a bootstrapped sample. Conditional on the first year, we synthesize the last year of unit establishment via quantile regression, i.e., Algorithm 1. Next, we synthesize the multi-unit status conditional on first year and last year of an establishment. Employment and payroll variables are then synthesized, starting with employment for 1976. All employment variables are synthesized sequentially through the last year of data availability (2000), based on the first and last years of an establishment, multi-unit status, and employment in the prior year. Finally, payroll data is synthesized, based on first and last years of an establishment, multi-unit status, contemporaneous employment and lagged payroll. Different models were fit for continuing units (current year greater than first year) and births (current year equal to birth year).[3] 100 bins were used in synthesizing the quantile-based data. For CART-based replicates, the minimum number of terminal points per node was set to five.

---

[2] We are discussing applying the methodology to the confidential data for the purpose of generating a new release of synthetic data.

[3] Note that neither geography nor the full SIC code were used in the quantile regression synthesis.

Results. Because the data are synthesized separately for each SIC code, we compute the general utility ($pMSE$) and risk measures for each of set of data by SIC code. A total of 5 synthetic data sets were created for each industry code using each method. Means and standard errors across the simulations are reported. Average utility difference between the two synthesis methods across all SIC codes was 0.02 (SE: 0.029). Risk measures are for the percentage of observations at risk according to the approach of [18]. For evaluation purposes, the SynLBD is treated as the original data.

A summary of results for selected SIC codes is presented in Table 3. Our results indicate that neither method strictly outperforms the other on the basis of either utility or risk. Rather, dependent on the SIC code, one method may outperform the other on the basis of utility, risk, or both, although the quantile-based regression performs better for most SIC codes. For example, for SIC industry *178*, quantile regression-based syntheses outperform CART-based syntheses in terms of both the utility and risk measures used. For SIC industry *829*, quantile-based syntheses have lower general utility scores, but it offers slightly lower risk compared to the CART-based syntheses. Overall, given our current analyses, there does not seem to be a consistent and obvious relationship as to which method would outperform the other. However, it is clear that quantile regression can yield synthetic data of higher quality when compared to CART based on the risk and utility measures used.

Table 3. General utility and disclosure risk values for selected industries.

SIC code	Sample size	Utility QR	Utility CART	Risk QR	Risk CART
178	10,975	0.067 (0.008)	0.095 (0.001)	5.73 (0.28)	6.55 (0.23)
239	27,396	0.052 (0.004)	0.053 (0.004)	4.91 (0.07)	6.20 (0.13)
328	3,069	0.079 (0.021)	0.112 (0.004)	7.08 (0.21)	8.21 (0.14)
354	27,411	0.050 (0.006)	0.065 (0.003)	6.08 (0.01)	7.82 (0.11)
473	20,193	0.061 (0.002)	0.064 (0.003)	6.12 (0.16)	7.08 (0.16)
511	49,370	0.074 (0.002)	0.037 (0.001)	4.06 (0.01)	4.98 (0.06)
542	39,670	0.049 (0.004)	0.042 (0.002)	3.83 (0.07)	4.87 (0.08)
703	19,624	0.076 (0.005)	0.076 (0.004)	7.11 (0.15)	8.86 (0.13)
829	45,709	0.051 (0.002)	0.028 (0.001)	3.89 (0.07)	4.84 (0.07)
865	11,691	0.050 (0.003)	0.062 (0.011)	7.88 (0.16)	11.66 (0.43)

Note: A total of 5 synthetic data sets were created for each industry code using each method. Means and standard errors (in parentheses) across the simulations are reported. Risk measures are for the percentage of observations at risk according to the approach of [18].

Since heavy-tailed variables were of particular interest, several graphical measures were used to examine the tails of selected distributions. Examples of such graphs can be found in Fig. 2, where we plot the tials of total payroll variable for year 2000 for two Standard Industry Codes (829-left plot and 865-right plot). These graphs indicate that CART has a tendency to closely match the original data in the tails of a distribution. This presents a clear potential for disclosure, especially for outlier observations. In comparison, quantile regression-based syntheses are much more random in the tails. For example, SIC code *829* represents "Schools and Educational Services, Not Classified Elsewhere." Examples of these industries include driving, cooking, or personal development schools. Large national chains are much more likely to be an outlier observation. In the case of CART-based syntheses, the tails of payroll variables closely match the original data. This could lead to higher risk for outlier observations. Quantile regression did not track the far extremes as carefully. However, further investigation is needed to see if this has an effect on any specific analyses.

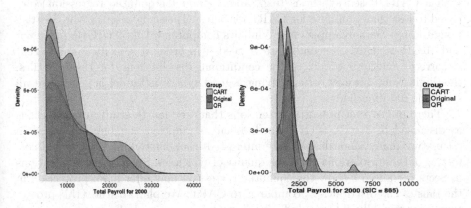

Fig. 2. Distribution of the tails of total payroll variable for 2000 for Standard Industry Codes 829 (left) and 865 (right).

6 Discussion

We explored quantile regression as a synthesis method for heavy-tailed and heteroskedastic data. The reason for this is twofold: quantile regression can be used to model the entire conditional distribution of a response variable and has less restrictive assumptions compared to standard linear regression. Further, we compared the utility and disclosure risk of the quantile-based method with a CART-based synthesis.

We applied our methods to a simulation study and two different data sets. Our simulation results indicated that quantile regression can outperform CART as a synthesis method on the basis of general utility when data have heavy tails or heteroskedastic errors. Our applications showed support for this claim; for certain data sets, quantile regression produced synthetic data with either higher utility, lower risk, or both. Furthermore, density plot analyses indicated that quantile regression methods may provide more privacy protection in the tails of skewed that is heteroskedastic distributions. Observations in the tails are often of high interest but are likely to have higher disclosure risk. More formal analyses are needed to justify this claim, but our work points to promising use of quantile regression for large-scale heavy-tailed data and those with heterosckedastic errors.

We note several potential limitations to the proposed quantile-based methods. First, quantile regression estimates can be expensive to compute in practice. With our analyses, quantile regression estimation was not found to be overly burdensome in its own right, but, in general, it lagged behind computation times with a CART-based synthesis (e.g., about 8 min for quantile regression compared to less than a minute for CART for one synthetic data set for the Scottish Longitudinal Data example on a Windows computer with a 2.60 GHz processor and 16 GB of RAM). Second, care is needed in order to specify a good (that is correct) synthesis model of the conditional distributions for the quantiles; this limitation, however, of specifying a good synthesis model is general to all synthetic data methods.

There are several potential extensions that we plan to study further. First, as discussed in Sect. 2, there is an implicit assumption of monotonicity when estimating many quantiles. We are interested in understanding how would correcting of these violations affect the quality of syntheses. Second, our applications in Sects. 5.1 and 5.2 show that quantile regression may offer more protection in the tails of the distributions compared to CART. We plan to study this protection more formally. This includes both studying potential effects on statistical inference and investigating different risk methods related to outlier observations. Third, we plan on developing synthetic versions of the original LBD data and comparing our findings to the original data.

Acknowledgments. The Synthetic LBD data were accessed through the Synthetic Data Server at Cornell University, which is funded through NSF Grant SES-1042181 and BCS-0941226 and a grant from the Alfred P. Sloan foundation. Access to the Synthetic LBD is described at https://www2.vrdc.cornell.edu/news/synthetic-data-server/step-1-requesting-access-to-sds/. Use of the Integrated Census Microdata Project at the University of Essex was facilitated by Gillian Raab.

A Supplementary Materials for 1901 Census of Scotland

There were a total of 82,851 observations in our extract. Of these observations, 20,303 were female and 62,548 were male. Additional statistics follow (Tables 4, 5 and Fig. 3).

Table 4. Summary statistics for continuous variables for extract of 1901 Census of Scotland. Note that the count variables had very heavy tails.

Variable	N	Mean	St. Dev.	Min	Max
Age	82,851	45.731	14.266	0.830	97.000
Number of Servants	82,851	0.048	0.369	0	12
Number of boarders	82,851	0.082	0.419	0	12
Number of lodgers	82,851	0.084	0.413	0	11
Number of Family	82,851	0.024	0.196	0	10
Number of relatives (age ≥ 15)	82,851	2.654	1.465	0	16
Number of relatives (age < 15)	82,851	1.385	1.750	0	19
Number relationship unknown	82,851	0.005	0.119	0	10

Table 5. Count of records by marital status.

Single	7,320
Widowed	15,821
Divorced	1
Married	55,288
Married - spouse absent	4,308
Not known	113

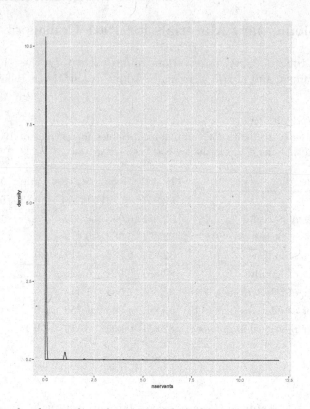

Fig. 3. Density for the number of servants. **Note:** Most observations have a value of zero.

References

1. Benedetto, G., Stinson, M., Abowd, J.M.: The creation and use of the SIPP Synthetic Beta. Mimeo, U.S. Census Bureau, April 2013. http://hdl.handle.net/1813/43924
2. Benoit, D.F., Van den Poel, D.: bayesQR: A Bayesian approach to quantile regression. J. Stat. Softw. **76**(7), 1–32 (2017)
3. Bondell, H.D., Reich, B.J., Wang, H.: Noncrossing quantile regression curve estimation. Biometrika **97**(4), 825–838 (2010)
4. Caiola, G., Reiter, J.P.: Random forests for generating partially synthetic, categorical data. Trans. Data Priv. **3**(1), 27–42 (2010)
5. Chernozhukov, V., Fernández-Val, I., Galichon, A.: Quantile and probability curves without crossing. Econometrica **78**(3), 1093–1125 (2010)
6. Drechsler, J.: Synthetic datasets for the German IAB establishment panel. Invited Paper WP.10, Joint UNECE/Eurostat work session on statistical data confidentiality (2009). http://www.unece.org/fileadmin/DAM/stats/documents/ece/ces/ge.46/2009/wp.10.e.pdf
7. Drechsler, J.: Synthetic Datasets for Statistical Disclosure Control: Theory and Implementation. LNS, vol. 201. Springer, New York (2011). https://doi.org/10.1007/978-1-4614-0326-5

8. Dwork, C., Roth, A.: The algorithmic foundations of differential privacy. Found. Trends Theor. Comput. Sci. **9**(3–4), 211–407 (2014)
9. Dwork, C., Smith, A., Steinke, T., Ullman, J.: Exposed! a survey of attacks on private data. Annu. Rev. Stat. Appl. **4**, 61–84 (2017)
10. Fahrmeir, L., Kneib, T., Lang, S., Marx, B.: Regression: Models, Methods and Applications. Springer, Heidelberg (2013). https://doi.org/10.1007/978-3-642-34333-9
11. Foschi, F.: Disclosure risk for high dimensional business microdata. In: Joint UNECE-Eurostat Work Session on Statistical Data Confidentiality, pp. 26–28 (2011). https://www.unece.org/fileadmin/DAM/stats/documents/ece/ces/ge.46/2011/03_Italy-Foschi.pdf
12. Hothorn, T., Hornik, K., Zeileis, A.: Unbiased recursive partitioning: a conditional inference framework. J. Comput. Graph. Stat. **15**(3), 651–674 (2006)
13. Hu, J., Reiter, J.P., Wang, Q.: Dirichlet process mixture models for modeling and generating synthetic versions of nested categorical data. Bayesian Anal. **13**(1), 183–200 (2018)
14. Huang, Q., Zhang, H., Chen, J., He, M.: Quantile regression models and their applications: a review. J. Biometr. Biostat. **8**, 354 (2017)
15. Huckett, J.C., Larsen, M.D.: Microdata simulation for confidentiality of tax returns using quantile regression and hot deck. In: Proceedings of the Third International Conference on Establishment Data. American Statistical Association (2007)
16. Huckett, J.C., Larsen, M.D.: Microdata simulation for confidentiality protection using regression quantiles and hot deck. In: Proceedings of the Survey Research Methods Section. American Statistical Association (2007)
17. Hundepool, A., et al.: Statistical Disclosure Control. Wiley, Hoboken (2012)
18. Ichim, D.: Disclosure control of business microdata: a density-based approach. Int. Stat. Rev. **77**(2), 196–211 (2009)
19. Karr, A., Oganian, A., Reiter, J., Woo, M.J.: New measures of data utility. In: Workshop Manuscripts of Data Confidentiality, A Working Group in National Defense and Homeland Security (2006). http://sisla06.samsi.info/ndhs/dc/Papers/NewDataUtility-01-10-06.pdf
20. Kinney, S.K., Reiter, J.P., Reznek, A.P., Miranda, J., Jarmin, R.S., Abowd, J.M.: Towards unrestricted public use business microdata: the synthetic Longitudinal Business Database. Int. Stat. Rev. **79**(3), 362–384 (2011)
21. Koenker, R.: quantreg: Quantile Regression (2017). R package version 5.34: https://CRAN.R-project.org/package=quantreg
22. Koenker, R., Bassett Jr., G.: Regression quantiles. Econometrica **46**, 33–50 (1978)
23. Kozumi, H., Kobayashi, G.: Gibbs sampling methods for Bayesian quantile regression. J. Stat. Comput. Simul. **81**(11), 1565–1578 (2011)
24. Larsen, M.D., Huckett, J.C.: Multimethod synthetic data generation for confidentiality and measurement of disclosure risk. Int. J. Inf. Priv. Secur. Integr. 2 **1**(2–3), 184–204 (2012)
25. Little, R.J.: Statistical analysis of masked data. J. Off. Stat. **9**(2), 407–426 (1993)
26. Liu, Y., Wu, Y.: Simultaneous multiple non-crossing quantile regression estimation using kernel constraints. J. Nonparametr. Stat. **23**(2), 415–437 (2011)
27. Machanavajjhala, A., Kifer, D., Abowd, J.M., Gehrke, J., Vilhuber, L.: Privacy: theory meets practice on the map. In: International Conference on Data Engineering (ICDE), pp. 277–286 (2008). https://doi.org/10.1109/ICDE.2008.4497436
28. Meinshausen, N.: Quantile regression forests. J. Mach. Learn. Res. **7**(Jun), 983–999 (2006)

29. Nowok, B., Raab, G.M., Dibben, C.: synthpop: Bespoke creation of synthetic data in R. J. Stat. Softw. **74**(11), 1–26 (2016)
30. Portnoy, S., Koenker, R.: The Gaussian hare and the Laplacian tortoise: computability of squared-error versus absolute-error estimators. Stat. Sci. **12**(4), 279–300 (1997)
31. Raab, G.M., Nowok, B., Dibben, C.: Practical data synthesis for large samples. J. Priv. Confid. **7**(3), 67–97 (2017)
32. RDC- Cornell University: Synthetic Data Server (2018). https://www2.vrdc.cornell.edu/news/synthetic-data-server/
33. Reiter, J.P.: Using CART to generate partially synthetic public use microdata. J. Off. Stat. **21**(3), 441–462 (2005)
34. Rizzo, M.L.: Statistical Computing with R. CRC Press, Boca Raton (2007)
35. Rubin, D.B.: Discussion: statistical disclosure limitation. J. Off. Stat. **9**(2), 461–468 (1993)
36. Scottish Longitudinal Study Development and Support Unit: Synthetic Data (2018). https://sls.lscs.ac.uk/guides-resources/synthetic-data/
37. Snoke, J., Raab, G.M., Nowok, B., Dibben, C., Slavkovic, A.: General and specific utility measures for synthetic data. J. Roy. Stat. Soc.: Ser. A (Stat. Soc.) **181**(3), 663–688 (2018)
38. Therneau, T., Atkinson, B., Ripley, B.: rpart: Recursive Partitioning and Regression Trees (2017). R package version 4.1-11: https://CRAN.R-project.org/package=rpart
39. University of Essex Department of History: I-CeM: Integrated Census Microdata Project (2018). https://www1.essex.ac.uk/history/research/icem/
40. Woo, M.J., Reiter, J.P., Oganian, A., Karr, A.F.: Global measures of data utility for microdata masked for disclosure limitation. J. Priv. Confid. **1**(1), 111–124 (2009)
41. Woodcock, S.D., Benedetto, G.: Distribution-preserving statistical disclosure limitation. Comput. Stat. Data Anal. **53**(12), 4228–4242 (2009)
42. Yu, K., Lu, Z., Stander, J.: Quantile regression: applications and current research areas. J. Roy. Stat. Soc.: Ser. D (Statistician) **52**(3), 331–350 (2003)
43. Yu, K., Moyeed, R.A.: Bayesian quantile regression. Stat. Probab. Lett. **54**(4), 437–447 (2001)

Some Clarifications Regarding Fully Synthetic Data

Jörg Drechsler[✉]

Institute for Employment Research,
Regensburger Str. 104, 90478 Nuremberg, Germany
joerg.drechsler@iab.de

Abstract. There has been some confusion in recent years in which circumstances datasets generated using the synthetic data approach should
be considered *fully synthetic* and which estimator to use for obtaining
valid variance estimates based on the synthetic data. This paper aims at
providing some guidance to overcome this confusion. It offers a review of
the different approaches for generating synthetic datasets and discusses
their similarities and differences. It also presents the different variance
estimators that have been proposed for analyzing the synthetic data.
Based on two simulation studies the advantages and limitations of the
different estimators are discussed. The paper concludes with some general recommendations how to judge which synthesis strategy and which
variance estimator is most suitable in which situation.

Keywords: Confidentiality · Multiple imputation · Fully synthetic
Variance estimation

1 Introduction

The synthetic data approach for data protection gained popularity over the last
decade as it offers a high level of data protection while ensuring good analytical properties in many circumstances. The approach replaces sensitive variables
and/or variables that pose a high risk of reidentification with synthetic values
drawn from a model fit to the original data. Two main approaches are generally
distinguished in the literature for generating synthetic data. The *fully synthetic*
approach, first proposed by Rubin [14] treats all units that did not participate
in the survey as missing data, multiply imputes those missing values to generate
synthetic populations and then disseminates synthetic samples from these populations. The *partially synthetic* approach proposed by Little [7] only synthesizes
sensitive records and/or records which have a risk of leading to identity disclosure (i.e. records which are sample uniques based on some key variables in the
data). See [2] for a detailed review of the two approaches.

The naming of the two approaches seems intuitive, since by definition all
values are synthetic for the imputed nonparticipating units based on the fully
synthetic approach and the partially synthetic approach will keep the original

© Springer Nature Switzerland AG 2018
J. Domingo-Ferrer and F. Montes (Eds.): PSD 2018, LNCS 11126, pp. 109–121, 2018.
https://doi.org/10.1007/978-3-319-99771-1_8

values for all records which do not pose any risk of disclosure. However as first pointed out by [1] there is nothing preventing us from using the partially synthetic approach to replace all values in the data by synthetic values, also obtaining a fully synthetic dataset. In fact, in several applications, in which researchers claimed to generate fully synthetic data, the synthesis strategy actually followed the approach proposed by Little, which is based on different methodology than the approach proposed by Rubin. As was first shown in [1] the variance estimator for fully synthetic data following Rubin's proposal is not strictly valid, if the data have been generated using the approach proposed by Little. On the other hand, [8] proposed a new variance estimator, which can be used if Little's approach has been used to synthesize the entire dataset. The major attractiveness of this estimator is that it can even be used, if only one synthetic data replicate is generated. However, as will be illustrated below, this variance estimator is not always appropriate, especially if Rubin's approach has been used to generate the synthetic data.

This paper aims to help overcome some of the confusion regarding fully synthetic datasets. I will illustrate the methodological differences of the two approaches and clarify, which variance estimator should be used in which situation.

The remainder of the paper is organized as follows: In Sect. 2 I review the methodological details of the two approaches to data synthesis and provide the formulae required to obtain valid inferences based on the two approaches. Section 3 contains simulation studies which illustrate which estimator will give valid inferences under which circumstances and highlight some of the advantages and limitations of the different variance estimators. The paper concludes with some general recommendations how to judge which synthesis strategy and which variance estimator is most suitable in which situation.

2   Two Approaches to Data Synthesis

In this section, I review the two common synthesis approaches: full synthesis as proposed by [7] and partial synthesis as proposed by [14]. I also present the formulae required to obtain valid variance estimates based on the different approaches. Note that the procedures for obtaining valid point estimates are identical for both approaches. Simply compute the estimate of interest on all synthetic datasets separately and get the final inference by averaging the different estimates.

2.1   The Fully Synthetic Approach According to Rubin

In Rubin's proposal the strong relationship to multiple imputation for missing data is most obvious. In fact, Rubin considered the problem as a missing data problem. He assumed that the data disseminating agency has information about some variables X for the whole population, for example from the sampling frame, and only the information from the survey respondents for the remaining variables

Y and that the goal is to release a protected version of the variables contained in Y. Let Y_{inc} contain the n records included in the survey and Y_{exc} be the $N - n$ records from the population which were not selected into the sample. Fully synthetic datasets are generated in two steps: First, construct m synthetic populations by imputing all the missing values in Y, i.e. by drawing Y_{exc} m times from the posterior predictive distribution $f(Y_{exc}|X, Y_{inc})$. Second, take simple random samples from these populations and release them to the public. In practice imputing the full population is not required. It suffices to take simple random samples from X and impute the missing Y information only for these cases.

Note that the term *fully synthetic data* is actually a misnomer for this approach since synthetic records are generated only for the $N - n$ records that were not part of the original sample. Thus, it can happen that the released samples from this population, comprising $N - n$ synthetic records and n original records, might still contain original values for Y. To avoid this problem, [9] suggest that "the whole population can be generated based on the posterior predictive distribution of "super" or "future" populations" ([9], p. 4), i.e. that synthetic values are also generated for those records that were already included in the original sample. However, as I will illustrate in Sect. 3, the inferential procedures developed by the authors are no longer strictly valid for finite population inference in this case.

In practice, it is not strictly required that some variables X are available for the full population. If X is empty, synthetic values could be generated by drawing from $f(Y_{exc}|Y_{inc})$ directly. But for the inferential procedures to be valid, it would still be necessary that only $N - n$ records are synthesized and that the released data contain random samples from the synthetic populations constructed by merging the synthetic records with the original sample.

Before presenting the variance estimator for fully synthetic data developed by [9], we need to introduce some additional notation: Let Q be the unknown scalar parameter of interest, such as the population mean or the regression coefficient in a linear regression. Inferences for this parameter derived from the original datasets usually are based on a point estimate q and an estimate for the variance of q, u. Let $q^{(i)}$ and $u^{(i)}$ for $i = 1, \ldots, m$ be the point and variance estimates for each of the m synthetic datasets. The following quantities are needed for inferences for scalar Q:

$$\bar{q}_m = \sum_{i=1}^{m} q^{(i)}/m, \tag{1}$$

$$b_m = \sum_{i=1}^{m} (q^{(i)} - \bar{q}_m)^2/(m-1), \tag{2}$$

$$\bar{u}_m = \sum_{i=1}^{m} u^{(i)}/m. \tag{3}$$

The analyst then can use $\bar{q}_m$ to estimate Q and its variance can be estimated using

$$T_f = (1 + m^{-1})b_m - \bar{u}_m. \tag{4}$$

Note, that T_f can be negative. For that reason, [10] suggests a slightly modified variance estimator that is always positive but conservative, $T_f^* = \max(0, T_f) + \delta(\frac{n_{syn}}{n}\bar{u}_m)$, where $\delta = 1$ if $T_f < 0$ and $\delta = 0$ otherwise. Here, n_{syn} is the number of observations in the released datasets sampled from the synthetic population. An alternative approach which uses posterior simulation to estimate the variance of the point estimates is discussed in [15]. While computationally more demanding, the approach offers the major advantage that the problem of negative variance estimates is avoided. Based on simulation studies [15] find that using posterior simulation is superior to using T_f coupled with the ad-hoc fix for negative variance estimates proposed in [10].

2.2   The Partially Synthetic Approach According to Little

With the approach proposed by Little only sensitive records and/or records that could be used for re-identification are typically replaced with synthetic values. Since some true values remain in the dataset the approach has been termed the partially synthetic data approach. The approach offers some flexibility over the fully synthetic data approach described above. The agency can decide which part of the data needs to be synthesized. The synthesis can range from synthesizing only some records for a single variable, for example all income values for individuals with an income above a given threshold, to synthesizing all variables, basically mimicking the fully synthetic data approach. Note, however that the general concept is different. In both approaches, the observed data are used to construct a synthesis models based on $f(Y|X)$, where X could potentially be empty. However, with partial synthesis, synthetic data are generated by simply drawing new values from the fitted model, potentially conditioning on the original values in X for the sampled cases. With full synthesis, a new sample X_{new} is drawn first and synthetic data are generated by drawing from $f(Y|X_{new})$. Additionally, the original Y values are kept for all those records in X_{new} which were already included in the original sample. Furthermore, as [13] showed, the first step of multiple imputation for nonresponse – drawing new values for the parameters of the imputation model from their posterior distribution – is not required for partial synthesis.

While $\bar{q}_m$ is still the appropriate estimator for Q, the formula for estimating the variance of $\bar{q}_m$ based on partially synthetic data differs from the one for fully synthetic data. As shown in [11] an unbiased variance estimator is given by:

$$T_p = \bar{u}_m + m^{-1}b_m. \tag{5}$$

This estimator is valid even if all records in the dataset are synthesized. Note that once we synthesize all records in the data, the restriction that the sample size of the original data and of the synthetic data need to be the same no longer applies (this restriction also does not apply if (sub)samples of the original data are disseminated as proposed in [4] or [5]). [1] suggested an adjusted version of the estimator if the sample sizes differ

$$T_{alt} = \frac{n_{syn}}{n_{org}} \bar{u}_m + b_m/m,$$ (6)

where n_{syn} is the sample size of the synthetic data and n_{org} is the sample size of the original data.

In a recent paper [8] showed that the variance estimator could be further simplified in this context:

$$T_s = (\frac{n_{syn}}{n_{org}} + 1/m)\bar{u}_m.$$ (7)

The authors also proposed a variant of the variance estimator which is valid even if new parameters are drawn from their posterior distribution. It is given by

$$T_{s(PPD)} = (\frac{n_{syn}}{n_{org}} + (1 + n_{syn}/n_{org})/m)\bar{u}_m.$$ (8)

This variance estimator (and its variant) is very attractive for two reasons: First, since the variability of $\bar{u}_m$ is less than the variability of b_m, T_s has less variability than T_{alt} (which itself has less variability than T_f). But more importantly, since the variance estimator only depends on $\bar{u}_m$, the variance can be estimated even if only one synthetic dataset is released. This is attractive, because the risk of disclosure generally increases with m [3,12] and thus some statistical agencies decided to release only one implicate of their synthetic data [6]. In this situation neither T_f nor T_p, nor T_{alt} could be used since b_m is not defined.

However, as one of the referees pointed out, neither T_{alt} nor T_s would be appropriate if the original data is a sample but the synthetic data cover the entire population. As illustrated in [9], $(1 + m^{-1})b_m$ would be an unbiased estimate for the variance in this setting. Given that $\bar{u}_m = 0$ if the data cover the entire population, only T_f would provide valid inferences; both T_{alt} and T_s would be biased. On the other hand, if both, the original and the synthetic data, cover the entire population, only T_p (and T_{alt}) would provide unbiased results, while T_f would overestimate and T_s would underestimate the true variance. But beyond these special cases it is important to note that T_s (and T_{alt}) are only valid for specific settings as I will illustrate in the next section.

3   Illustrative Simulation Studies

In this section, I present the results of two simulation studies. In the first simulation study X is empty, whereas the second simulation study assumes that some variables are available for the entire population. For both simulations, I use the same population consisting of $N = 100,000$ records comprising five variables, $Y_1, \ldots, Y_5$, drawn from $N(0, \Sigma)$, where Σ has variances equal to one and correlations ranging from 0.2 to 0.8. From this population, I repeatedly draw simple random samples using two different sampling rates $ss = \{1\%; 20\%\}$. These samples represent the cases for which survey data are collected. In the first simulation, I assume that all five variables are only available for the sampled cases. In the second simulation, I assume that $X = \{Y_1, Y_2\}$, that is Y_1 and Y_2 are available for the full population. In both simulations, I evaluate two synthesis strategies. With the first strategy, I follow Rubin's approach that is, I generate synthetic values only for those cases not included in the original sample. In the first simulation this is achieved by explicitly imputing all the $N - n$ cases in the population and then taking simple random samples from these synthetic populations. In the second simulation, I draw new simple random samples of X first and then synthesize $Y_3, \ldots, Y_5$ only for those records that are not included in the original sample. With the second strategy I follow Little's approach replacing all records by synthetic records. For each setup, I generate $m = \{5; 50\}$ synthetic datasets. Note that in the first simulation all five variables would be considered survey variables and hence all five variables would be released. In the second simulation setup Y_1 and Y_2 are treated as variables from the sampling frame which are neither synthesized nor released to the public (mimicking Rubin's original proposal).

To evaluate the impact of the different simulation designs on the different variance estimators, I look at several estimates of potential interest: the estimated variance for the means of Y_3 and Y_4, the conditional mean of Y_5 given $Y_3 > 0$, and all regression coefficients in the regressions of Y_5 on Y_3 and Y_4. All simulations are repeated $s = 5,000$ times and the size of the synthetic data is always twice the size of the original data. Posterior draws of the parameters are used in all synthesis models following Rubin's original proposal. I also ran the same simulations without posterior draws. Except for the expected undercoverage of T_f the general findings did not change and thus they are omitted for brevity.

3.1 Simulation Results if No Variables are Available for the Entire Population

Figures 1 and 2 contain the results for the simulation in which all variables are synthesized and no variables are available for the population. Each boxplot in each figure represents the ratio of estimated variance using the different variance estimators presented in the previous section divided by the true variance of the point estimate across the 5,000 simulation runs. If the variance estimator is unbiased, the boxplots should be centered around one. T_f and T_s represent the variance estimators described in the previous section (note that I always drop the subscript PPD for readability although the variance estimator appropriate for this synthesis setup is used). The superscripts indicate the synthesis strategy. R means that the synthetic data were generated following Rubin's proposal, i.e., only those cases that were not included in the original sample are synthesized. L means that the approach proposed by Little was used to generate the synthetic data, i.e., all records are synthesized. The numbers above the boxplots are the

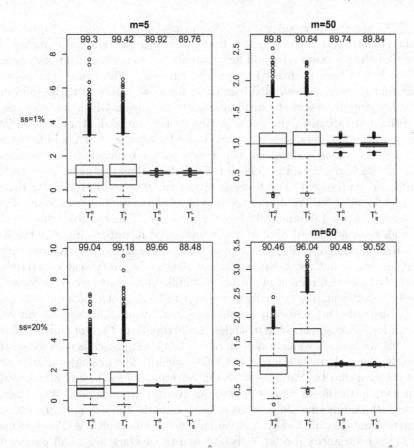

Fig. 1. Ratio of the estimated variance of $\hat{\bar{Y}}_4$ divided by the true variance of $\hat{\bar{Y}}_4$ across the 5,000 simulation runs for the two variance estimators for different synthesis designs. If the estimated variance is unbiased, the boxplots should be centered around one (indicated with a solid line).

true coverage rates (in percent) of 90% confidence intervals computed based on the synthetic data, i.e. they represent the percentage of times the 90% confidence intervals cover the true value from the population. For T_f t-distributions with the appropriate degrees of freedom according to [9] are used to compute the confidence intervals. For T_s the confidence intervals are based on normal approximations. Note that while the boxplots are based on the formula for T_f given above, the coverage rates for the fully synthetic variance estimate are based on the adjusted formula which only uses $\bar{u}_m$ whenever $T_f < 0$, since the confidence interval is not defined if $T_f < 0$.

Figure 1 contains the results for the mean of Y_4. Similar results were obtained for the mean of Y_3 and all the regression coefficients from the linear regression model. Thus, these results are excluded for brevity. Several points are noteworthy. First, T_s always has substantially less variability than T_f. Second, for

$m = 5$, T_f is negative in several simulation runs leading to coverage rates based on the adjusted variance estimator that far exceed the nominal coverage rate (note that the overcoverage could have potentially been avoided if posterior simulations would have been used as proposed in [15]). The figures also seem to imply that there is a downward bias in T_f for $m = 5$. However, the solid lines in the boxplots represent the median and not the mean. Given the skewness of the estimated variances, the mean is larger than the median and the average variance ratios for the different scenarios is 1.00 when $ss = 1\%$ and 0.98 when $ss = 20\%$ for Rubin's approach, i.e. the variance estimate is in fact unbiased. However, we find substantial bias for T_f if the sampling rate is large and all records are synthesized. The average of the variance ratio is 1.32 for $m = 5$ and 1.53 for $m = 50$. We do not see any bias if the sampling rate is small (the average ratios are 1.00 and 1.01 for this scenario). This confirms that synthesizing all records instead of only the non-sampled records to increase the level of protection offered by Rubin's approach as proposed in [9] is only valid if the sampling rate is not too large. Small sampling rates are typical for household surveys but the requirement might not be fulfilled for business surveys for which large establishments are typically sampled with large probabilities of selection.

We also note that T_s is almost unbiased in all simulation designs, can never be negative by design, and has much less variability than T_f. Only for $m = 5$ and $ss = 20\%$ we notice a small undercoverage if Little's approach is used for synthesis due to an underestimation of the true variance. The average variance ratio is 0.91. Given that the variance is also underestimated for $\bar{Y}_3$ and all regression coefficients in the linear regression with an average variance ratio of 0.94 across all parameters, this small bias seems to be systematic. Interestingly, the variance estimate is unbiased if the adjusted variance estimator as proposed in [1] is used. The average variance ratio for $\bar{Y}_4$ is 0.98 and the average across all parameters is 1.02. Thus, it seems that the problem arises if $\bar{u}_m$ is used as an estimator for the model uncertainty. However, the undercoverage is relatively small and only occurs in rather extreme scenarios with a small number of synthetic datasets and a large sampling rate. Thus, in general T_s should be preferred over T_f for the parameters considered here.

Next, we consider the results for the conditional mean of Y_5 given that $Y_3 > 0$. The results are presented in Fig. 2. The general findings for T_f are still the same as in Fig. 1. Many of the estimates are negative if $m = 5$ leading to overcoverage in the adjusted variance estimate. We also find that T_f is again substantially biased if the sampling rate is large and all records are synthesized. Results are always unbiased if the number of imputations is large enough and only those records are synthesized that were not included in the original sample.

The findings for T_s are different. The variance is always overestimated leading to coverage rates above the nominal coverage rate for all simulation setups. The overcoverage is always higher if Rubin's approach is used. The average variance ratio varies between 1.23 and 1.26 for the different simulation runs, except for the setup with $m = 5$ and a large sampling rate. For this setup the general

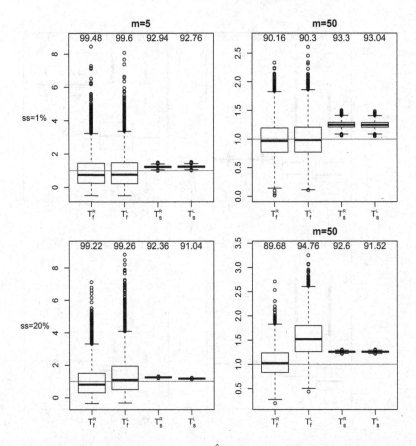

Fig. 2. Ratio of the estimated variance of $\hat{\bar{Y}}_5|Y_3 > 0$ divided by the true variance of $\hat{\bar{Y}}_5|Y_3 > 0$ across the 5,000 simulation runs for the two variance estimators for different synthesis designs. If the estimated variance is unbiased, the boxplots should be centered around one (indicated with a solid line).

underestimation of the variance found in Fig. 1 has a counter balancing effect and thus the average variance ratio is 1.16.

The results seem to imply that if estimates based on subsets of the data are considered, T_f should be preferred over T_s (unless the sampling rate is large and Little's approach was used to generate the data in which case none of the variance estimators is valid).

3.2 Simulation Results if Some Variables are Available for the Entire Population

Figure 3 contains the results for $\bar{Y}_4$ for the simulation in which Y_1 and Y_2 are available for the entire population and are used as explanatory variables when synthesizing Y_3 to Y_5, but only Y_3, Y_4, and Y_5 are released to the public. The

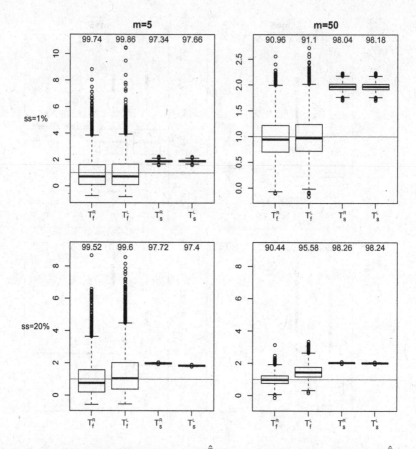

Fig. 3. Ratio of the estimated variance of $\hat{\bar{Y}}_4$ divided by the true variance of $\hat{\bar{Y}}_4$ across the 5,000 simulation runs for the two variance estimators for different synthesis designs. If the estimated variance is unbiased, the boxplots should be centered around one (indicated with a solid line).

results were similar for all other estimates considered in this simulation study and thus are omitted for brevity. Note that this simulation design exactly matches the setup in [14]. In this simulation the cases from the original sample are used to estimate a model for $f(Y_3, Y_4, Y_5 | Y_1, Y_2)$ using the sequential regression approach. The fitted models are used to generate synthetic data by drawing from $f(Y_3, Y_4, Y_5 | Y_1^{new}, Y_2^{new})$ where Y_1^{new}, Y_2^{new} are the records contained in the new sample drawn from the population. Again, only those cases not included in the original sample are synthesized for Rubin's approach, whereas all records are synthesized if Little's approach is used.

The results for T_f are essentially the same as in the setup in which no variables are available for the full population. For T_s the findings change. Now, the estimate is always biased even for those estimates that are based on all records such as $\bar{Y}_4$ considered here. The average estimated variance is almost twice the variance of the true variance. The overestimation is not really surprising. The synthetic sample size is twice the sample size of the original data and unlike in the previous simulation the additional records contain additional information since they consist of the original values for Y_1 and Y_2. This is not reflected in T_s but it is reflected in T_f.

Thus, if some information is available for the full population and this information is exploited when generating the synthetic data, it is mandatory to use T_f and not T_s.

It might seem reasonable to abandon Rubin's proposal in this situation and only use the sampled cases of $Y_1, \ldots, Y_5$ mimicking the approach of the previous simulation study when generating synthetic data. After all, as discussed previously, T_s has some nice properties: it can never be negative, has less variability than T_f, and can also be used if only one replicate is released.

However, we would throw away valuable information in this case, since the observed values in Y_1 and Y_2 can be used to reduce the uncertainty in the point estimates. For example, the true variance of $\hat{\bar{Y}}_4$ across the 5,000 simulation runs with $m = 50$ and $ss = 1\%$ is $1.03 * 10^{-3}$ if the design of Sect. 3.1 is used, that is, if only in the information from the sample is exploited for Y_1 and Y_2. It reduces to $5.14 * 10^{-4}$ if true values are used for the n_{syn} records for these two variables. Thus, there are efficiency gains to be expected from using Rubin's proposal if variables are available for the full population. Obviously, the efficiency increases, the larger the sample size of the synthetic sample and the higher the correlation between the variables which are available for the full population and the variables of interest. With Rubin's approach it is possible in theory that the estimate based on the synthetic data is more efficient, that is, has less variability, than the estimate based on the original data. With the approach proposed by Little this will never be possible no matter how large the size of synthetic data is. Note that the adjustment factor n_{syn}/n in T_s always ensures that the adjusted $\bar{u}_m$ term is approximately equal to the variance based on the original data. Thus T_s always has to be larger than the variance in the original data. This is to be expected since we can only use the information from the sample to generate our synthetic data. With the approach of Rubin we can use additional information beyond the sampled cases to improve the information contained in the estimates of interest.

4 Conclusions

There has been some confusion lately, in which cases a synthetic dataset should be labeled fully synthetic data and which variance estimator should be preferred. [8] even suggested to introduce new terminology to overcome this confusion. They talk about *complete* and *incomplete* synthesis to distinguish whether all records in Y_{obs} are synthesized or not. They also show that their proposed variance estimator T_s is valid even if some unsynthesized variables X are available as long as the analyst always conditions on these X variables in her inferences. However, this is a very restrictive assumption. On the one hand, the X variables from the sampling frame might not even be released, as they might themselves be sensitive or deemed irrelevant for the study. On the other hand, even if they would be released, it seems unlikely that analysts would only be interested in inferences conditional on these X variables. For example, if the users would be interested in the marginal distribution of some of the variables or the unconditional relationship between some of the synthesized variables, the variance estimator would give biased results as illustrated in the second simulation study.

Thus, I recommend that users always rely on the variance estimator T_p for partially synthetic data whenever some records in the released data are not synthesized. If all records are synthesized, agencies releasing the synthetic data should additionally inform the users, whether the data were generated using additional variables available for the full population or not. If yes, only T_f will give valid results. If not, analysts might consider using T_s if inferences are based on all records in the released sample. As pointed out previously using T_s offers several advantages in this case: T_s will never be negative and will have less variability than T_f. It will also be unbiased even if all records are synthesized and the sampling rate is large. Furthermore, it will also be valid, if synthetic data have been generated without drawing new parameters from their posterior predictive distributions, which reduces the variance in the synthetic data and also simplifies the synthesis. Finally, it can also be computed even if only one replicate of the synthetic data is released.

However, T_s can only be used if the synthetic data comprise a sample and not the entire population and the results from the simulation study indicate that users should be careful when analyzing only subsets of the data. While T_f still gave valid results for the conditional mean of Y_5 (at least for sufficiently large m), T_s showed a substantial positive bias. Whether such bias arises in all situations in which only subsets of the data are analyzed and understanding the reasons for the bias are interesting areas of future research.

References

1. Drechsler, J.: Improved variance estimation for fully synthetic datasets. In: Proceedings of the Joint UNECE/EUROSTAT Work Session on Statistical Data Confidentiality (2011)
2. Drechsler, J.: Synthetic Datasets for Statistical Disclosure Control: Theory and Implementation. LNS, vol. 201. Springer, New York (2011). https://doi.org/10.1007/978-1-4614-0326-5
3. Drechsler, J., Reiter, J.P.: Disclosure risk and data utility for partially synthetic data: an empirical study using the German IAB establishment survey. J. Off. Stat. **25**, 589–603 (2009)
4. Drechsler, J., Reiter, J.P.: Sampling with synthesis: a new approach for releasing public use census microdata. J. Am. Stat. Assoc. **105**(492), 1347–1357 (2010)
5. Drechsler, J., Reiter, J.P.: Combining synthetic data with subsampling to create public use microdata files for large scale surveys. Surv. Methodol. **38**, 73–79 (2012)
6. Kinney, S.K., Reiter, J.P., Reznek, A.P., Miranda, J., Jarmin, R.S., Abowd, J.M.: Towards unrestricted public use business microdata: the synthetic longitudinal business database. Int. Stat. Rev. **79**, 362–384 (2011)

7. Little, R.J.A.: Statistical analysis of masked data. J. Off. Stat. **9**, 407–426 (1993)
8. Raab, G.M., Nowok, B., Dibben, C.: Practical data synthesis for large samples. J. Priv. Confid. **7**(3), 4 (2017)
9. Raghunathan, T.E., Reiter, J.P., Rubin, D.B.: Multiple imputation for statistical disclosure limitation. J. Off. Stat. **19**, 1–16 (2003)
10. Reiter, J.P.: Satisfying disclosure restrictions with synthetic data sets. J. Off. Stat. **18**, 531–544 (2002)
11. Reiter, J.P.: Inference for partially synthetic, public use microdata sets. Surv. Methodol. **29**, 181–189 (2003)
12. Reiter, J.P., Drechsler, J.: Releasing multiply-imputed, synthetic data generated in two stages to protect confidentiality. Stat. Sin. **20**, 405–421 (2010)
13. Reiter, J.P., Kinney, S.K.: Inferentially valid, partially synthetic data: generating from posterior predictive distributions not necessary. J. Off. Stat. **28**(4), 583–590 (2012)
14. Rubin, D.B.: Discussion: statistical disclosure limitation. J. Off. Stat. **9**, 462–468 (1993)
15. Si, Y., Reiter, J.P.: A comparison of posterior simulation and inference by combining rules for multiple imputation. J. Stat. Theory Pract. **5**(2), 335–347 (2011)

Differential Correct Attribution Probability for Synthetic Data: An Exploration

Jennifer Taub[✉], Mark Elliot, Maria Pampaka, and Duncan Smith

The University of Manchester, Manchester M13 9PL, UK
jennifer.taub@postgrad.manchester.ac.uk,
{mark.elliot,maria.pampaka,duncan.smith}@manchester.ac.uk

Abstract. Synthetic data generation has been proposed as a flexible alternative to more traditional statistical disclosure control (SDC) methods for limiting disclosure risk. Synthetic data generation is functionally distinct from standard SDC methods in that it breaks the link between the data subjects and the data such that reidentification is no longer meaningful. Therefore orthodox measures of disclosure risk assessment - which are based on reidentification - are not applicable. Research into developing disclosure assessment measures specifically for synthetic data has been relatively limited. In this paper, we develop a method called Differential Correct Attribution Probability (DCAP). Using DCAP, we explore the effect of multiple imputation on the disclosure risk of synthetic data.

Keywords: Synthetic data · Disclosure risk · CART

1 Introduction

With the increasing centrality of data in our lives, societies and economies, and the drive for greater government transparency and release of open data, there has been a concomitant increase in demand for public release microdata. However, many of the traditional SDC techniques still are subject to disclosure risks. For example, Dinur and Nissim (2003) showed that additive noise (a common SDC method) is not protective against certain kinds of attacks/adversaries and Elliot et al. (2016) demonstrate that standard SDC controlled datasets are vulnerable to reidentification attacks using a jigsaw identification if released as open data.

An alternative to traditional SDC techniques is the use of synthetic data. The idea of synthetic data was first introduced by Rubin (1993), who proposed treating each observed data point as if it were missing and imputing it conditional on the other observed data points to produce a fully synthetic dataset. As an alternative, Little (1993) introduced a method that would replace only the sensitive variables in the observed data, to produce what is referred to as

© Springer Nature Switzerland AG 2018
J. Domingo-Ferrer and F. Montes (Eds.): PSD 2018, LNCS 11126, pp. 122–137, 2018.
https://doi.org/10.1007/978-3-319-99771-1_9

partially synthetic data. Since fully synthetic data does not contain any original data, the disclosure of information from the synthetic data is less likely to occur. Likewise for partially synthetic data, the sensitive values are synthesised, and thus the risk of disclosure of sensitive information is lessened compared to the original data.

Rubin's initial proposal for producing synthetic data was based on multiple imputation (MI) techniques using parametric modelling. Rubin originally developed multiple imputation in the 1970s as a solution to deal with missing data by replacing missing values with multiple values, to account for the uncertainty of the imputed values. Alternatively, synthetic data can be created using single imputation (SI) wherein the uncertainty of a missing value is accounted for by a different set of rules for variance estimation (Raab et al. 2016). Recent research has examined non-parametric methods - including machine learning techniques - which are better at capturing non-linear relationships (Drechsler and Reiter 2011). These methods include classification and regression trees (CART) (Reiter 2005), random forests (Caiola and Reiter 2010), bagging (Drechsler and Reiter 2011), support vector machines (Drechsler 2010), and genetic algorithms (Chen et al. 2016). In this paper we will be using synthetic data generated from both traditional parametric modelling and from CART.

Synthetic data generation is still a relatively new method of data protection and there has been a relative dearth of research into assessment of residual disclosure risk. The way that it operates is functionally distinct from standard SDC methods in that it breaks the link between the data subjects and the output data in a way that SDC methods do not. Thus a common sense view might posit that synthetic data are without disclosure risk since the data do not represent real individuals. However, synthetic data may pose a disclosure risk through the attributions or inferences that it allows. Therefore, orthodox measures for disclosure risk which are based on the notion of re-identification are not applicable for fully synthetic data. Research into measuring risk for synthetic data has been limited to a few papers (notably Reiter and Mitra (2009); Reiter et al. (2014)).

In this paper we will be developing and using a measure introduced procedurally by Elliot (2014) to capture the disclosure risk of multiply imputed parametric and CART synthetic data. The remainder of this article is structured as follows: Sect. 2 presents an overview of disclosure risk with a focus on disclosure risk for synthetic data. Section 3 discusses the dataset being used, how the synthetic files are generated, and how we use DCAP. Section 4 presents the results of our analyses and discusses the relationship between disclosure risk and number of imputations, leading to our conclusion in Sect. 5.

2   Disclosure Risk

Data disclosure can occur in different forms. The most important forms are re-identification and attribution. Re-identification occurs when an identity is attached to a data unit, while attribution is when some attribute can be associated with a population unit (eg. Elliot et al. 2016). For example, attribution might occur if some microdata reveal that all men age 65+ in a particular geographical area have prostate cancer. Therefore, a data intruder who lives in that area would learn that their 65+ male neighbour has prostate cancer, which is information that the survey was clearly not meant to disclose. Re-identification and attribution frequently occur together but can occur separately. Some authors also distinguish another type, inference attacks, wherein an attacker can infer information at a high degree of confidence[1]. However, we would argue that the distinction is somewhat arbitrary since all attributions involve some degree of uncertainty. We instead define attributions as inferences that a population unit has a certain characteristic and define a subclass of those attributions as *correct attributions*, wherein an intruder correctly identifies an attribute for a given respondent. It follows that a well formed measure of attribution risk would capture the proportion of attributions that are correct.

With synthetic data, as Drechsler et al. (2008) write, "the intruder faces the problem that he never knows (i) if the imputed values are anywhere near the true values and (ii) if the target record is included in one of the different synthetic samples[2]" (p. 1018). Following this reasoning, it is widely understood that thinking of risk within synthetic data in terms of re-identification, which is how many other SDC methods approach disclosure risk, is not meaningful and therefore we must develop measures of attribution risk (see for example Reiter and Mitra (2009)).

2.1   Disclosure Risk Measures for Synthetic Data

Unfortunately, the bulk of the SDC literature (for example: Winkler 2005; Skinner and Elliot 2002; Elliot et al. 2002; Yancey et al. 2002; Fienberg and Makov 1998; Kim and Winkler 1995, etc.) has focused on re-identification in the form of record linkage, which is not meaningful for fully synthetic data. By focusing their efforts on re-identification, they do not address disclosure risk that occurs in the form of attribution without re-identification, such as that addressed by Smith and Elliot (2008) and Machanavajjhala et al. (2007).

Previous methods for synthetic data risk estimation include Reiter et al.'s (2014) Bayesian estimation approach and Reiter and Mitra's (2009) matching probability of partially synthetic data. Reiter and Mitra compare perceived match risk, expected match risk, and true match risk. Reiter et al. assume that an intruder seeks a Bayesian posterior distribution. In Reiter et al.'s framework, the intruder is assumed to know all of the records but one. This scenario is very unlikely to occur in the real world, and even the authors noted that it is a conservative estimate. Essentially this is an approach that overestimates disclosure risk.

Differential Correct Attribution Probability (DCAP). Elliot (2014) introduced an approach that combines the notion that one should measure attribution risk as the probability of an attribution being correct and that one can

---

[1] The level of confidence which is regarded as disclosive is a subjective judgement.
[2] A synthetic dataset often contains multiple synthetic samples (m).

then compare that probability to those obtained on the original data and against some baseline. Here we develop this approach more formally and will refer to it as the *Differential Correct Attribution Probability* (DCAP).

The underlying measure of DCAP is the correct attribution probability and this has some similarity to Reiter and Mitra's method. Both methods employ a matching mechanism with the assumption that the intruder knows the true values of a key consisting of non-target variables. However, the framing of the two approaches is different. Given that Reiter and Mitra are matching using partially synthetic data, when a match occurs amongst the statistical uniques[3], the intruder can be certain that the match occurring refers to the same record. However, what is uncertain in the Reiter and Mitra method, is whether the synthetic version of a target record is the same as that of the real target. While in DCAP, since the matching is occurring with fully synthetic data, the matching has considerably less certainty since the keys do not directly map onto one another. In Reiter and Mitra's method they are comparing assumed risk against actual matches. While in DCAP the correct attribution rate for synthetic data is compared to that of the original dataset and a baseline univariate distribution. In the DCAP method one record can have multiple matches even with the original dataset and therefore, it is merely a probability of attributing the correct target variable to the key, since it is not only concerned with uniques and nor does it assume verifiable matches, as in Reiter and Mitra's scenario. In essence Reiter and Mitra's method is best for partially synthetic data, while DCAP is better suited for fully synthetic data.

DCAP works on the assumption that the intruder knows the values of a set of key variables for a given unit and is seeking to learn the specific value of a target variable. Where the target variable is categorical[4] the method works as follows: We define d_o as the original data and K_o and T_o as vectors for the key and target information

$$d_o = \{K_o, T_o\} \tag{1}$$

Likewise, d_s is the synthetic dataset.

$$d_s = \{K_s, T_s\} \tag{2}$$

The Correct Attribution Probability (CAP) for the record indexed j is the empirical probability of its target variables given its key variables,

$$CAP_{o,j} = Pr(T_{o,j}|K_{o,j}) = \frac{\sum_{i=1}^{n}[T_{o,i} = T_{o,j}, K_{o,i} = K_{o,j}]}{\sum_{i=1}^{n}[K_{o,i} = K_{o,j}]} \tag{3}$$

where the square brackets are Iverson brackets and n is the number of records.

---

[3] A statistically unique record is a record in the dataset, in which no other record in the dataset has that particular combination of characteristics.

[4] Elliot (2014) presents a variant where the target is continuous but we do not consider that here.

The CAP for record j based on a corresponding synthetic dataset d_s is the same empirical, conditional probability but derived from d_s,

$$CAP_{s,j} = Pr(T_{o,j}|K_{o,j})_s = \frac{\sum_{i=1}^{n}[T_{s,i} = T_{o,j}, K_{s,i} = K_{o,j}]}{\sum_{i=1}^{n}[K_{s,i} = K_{o,j}]} \qquad (4)$$

For any record in the original dataset for which there is no corresponding record in the synthetic dataset with the same key variable values, the denominator in Eq. 4 will be zero and the CAP is therefore undefined. In Sect. 3.3 we describe two methods for dealing with this.

The baseline CAP for record j is the marginal probability of its target variables estimated from the original dataset,

$$CAP_{b,j} = Pr(T_{o,j}) = \frac{1}{n}\sum_{i=1}^{n}[T_{o,i} = T_{o,j}] \qquad (5)$$

In principle, the baseline could be set to any level. However, the choice of the univariate baseline in Eq. 5 as the default for the approach, is based on the pragmatic assumption that the intruder will routinely know the univariate distribution of a target variable for the population. Another way to look at this is that a univariate distribution is often considered releasable into the public domain; univariate distributions are frequently published as summary statistics, and so a synthetic CAP score equal to or lower than the baseline CAP is likely to be considered as a sufficiently low risk in most conceivable situations.[5] On the other hand if the synthetic CAP score is equal to the CAP score for the original data that would imply that the synthetic data is as disclosive as the original data and, therefore, the synthetic data generator is not sufficiently protecting the data. The original data and baseline CAP scores therefore represent effective operating bounds in which the risk associated with synthetic data is likely to fall.

We propose addressing multiple imputations for DCAP by pooling the matches from the multiple synthetic samples:

$$CAP_m = Pr(T_1 + ... + T_m = T_o|K_1 + ... + K_m = K_s) \qquad (6)$$

where m is indicative of the imputation. This is preferable to averaging because we believe that an actual intruder would be viewing a synthetic dataset in its entirety. This paper will address the following research questions:

1. How does using Multiple Imputation affect the CAP score?
2. How do statistical uniques affect the CAP score?
3. Do parametrically and CART generated data have similar CAP scores?
4. Is synthetic data differentially confidential?

---

[5] It is worth noting that if the mean CAP score of the whole synthetic dataset is at the baseline, that effectively means that the target is independent of the key which may be indicative that the data have a utility issue.

3   Empirical Demonstation of the DCAP Approach

3.1   Data Sources

Following Elliot (2014) we used the Living Costs and Food Survey (LCF) Office for National Statistics (2016) as one of our test datasets. The LCF has many characteristics that make it a good candidate as a test dataset. First, it has a small size, which will allow for the dataset to be quickly synthesised. Additionally, the LCF has detailed information, making it vulnerable to disclosure. Furthermore, since the LCF was used by Elliot (2014) this allows us to use his keys and targets. We used the 2014 LCF, which consists of 5133 records. We used the following variables: Government office region (GOR), household size, output area classifier, tenure, economic position of reference person, dwelling type, number of workers, number of cars, and, internet in household (synthesised in that order). We utilised three different versions of the LCF: (1) the original 2014 LCF; (2) a CART generated 2014 LCF $m = 10$, and (3) a parametrically generated 2014 LCF $m = 10$. Our second test dataset is the British Social Attitudes Survey (BSA) NatCen Social Research (2014). The 2014 BSA consists of 2878 records. We used the following variables: GOR, higher education qualification, marital status, age category, gender, social class, and household income. We will also use three different versions of the BSA: (1) the original 2014 BSA; (2) a CART generated 2014 BSA $m = 10$, and (3) a parametrically generated 2014 $m = 10$.

3.2   Creation of Synthetic Data Files

The synthetic data are generated using the r-package, synthpop version 1.3-0 (Nowok et al. 2016). Synthpop was used to generate both the parametric and CART synthetic datasets. The parametric synthetic dataset was generated using logistic regression and polytomous logistic regression, since all variables being used are categorical. The missing data from the original LCF and BSA are left unchanged in the synthesis process.

3.3   Parameters for the Experiments

Our key variables for the LCF are as follows: GOR, Output area classifier, tenure, dwelling type, internet in household, and household size. The target variable is economic position of reference person. The first four variables of the key and the target variable are the same as in Elliot (2014). For the BSA the key variables are: GOR, education qualification, marital status, age, gender, social class and household income. The target variable is banded income. (Different key sizes for the LCF are in Appendix A).

Treatment of Non-matches in the CAP Score. DCAP, like many previous disclosure risks measures, works on the basis of matching on key variables. However, here we are not primarily concerned with the status of those matches but whether they lead to correct or incorrect attributions. The CAP score is the proportion of matches leading to correct attributions out of total matches. However, when measuring the CAP score for synthetic data, not every combination of keys from the original dataset will be present in the synthetic dataset. Elliot presented two different resolutions for this issue: recording the CAP values for the non-matches as zero or treating a non-match on the key as undefined, whereby the record is discounted and does not count towards n. The basis for assigning a zero is that a non-match has a zero probability of yielding a correct attribution. However, the logic behind recording non-matches as undefined is that an actual intruder is more likely to stop their attempt with a non-match. Elliot (2014) found that treating non-matches as undefined leads to higher CAP scores. In this paper we will be exploring CAP scores for both when non-matches are recorded as zero and coded as undefined.

Different Intruder Scenarios. When originally proposed, the CAP score was intended to be averaged across the whole dataset, however there is nothing intrinsic to DCAP that requires the use of the entire dataset. DCAP can be used for a variety of different intruder scenarios. However, in all scenarios, it is assumed that the intruder knows the information from the key for the original dataset and that the target variable is unknown to the intruder.

We will be exploring DCAP in three different scenarios, one where the entire dataset is in use, a second where the intruder is only interested in respondents who are statistically unique for the key (this would make it equivalent to the Reiter and Mitra method introduced earlier), and a third where only the special uniques are considered. Informally, a statistical unique can occur by either chance (random unique) or it can occur because of a rare combination of traits (special unique). Special uniques are deemed more risky than random uniques; Elliot et al. (2002) define an algorithm for scoring statistical uniques according to how special they are.

To identify statistical uniques and special uniques, we used the Special Uniques Detection Algorithm (SUDA) software (Elliot et al. 2002). For each statistical unique, SUDA generated a score using the Data Intrusion Simulation (DIS) method, which estimates the intruder confidence in a match leading to correct inference (see Elliot et al. 2002, for more details). We used the records in the top decile of scores generated by SUDA so as to examine the most risky of records (the special uniques). For the LCF there were 1867 statistically unique records and 251 special unique records, while for the BSA there were 2120 statistically unique records and 235 special unique records.

4   Results and Discussion

The CAP scores for the LCF dataset are shown in Table 1[6]. It shows that for the synthetic datasets all CAP scores are smaller than the CAP score for the

---

[6] The different imputation levels (m) are nested, rather than independent synthetic datasets.

Table 1. Mean CAP scores for the original and synthetic LCF datasets for two methods handling non-matches, two synthesis methods, three different intruder scenarios; full set, statistical uniques, and special uniques and ten levels of multiple imputation.

	Non-matches as zero			Non-matches as undefined		
Scenario	Full set	Statistical uniques	Special uniques	Full set	Statistical uniques	Special uniques
Original	0.750	1	1	0.750	1	1
Baseline	0.266	0.255	0.226	0.266	0.255	0.226
CART						
$m = 1$	0.334	0.180	0.074	0.498	0.548	0.549
2	0.393	0.273	0.110	0.503	0.554	0.530
3	0.416	0.324	0.154	0.501	0.549	0.568
4	0.435	0.361	0.162	0.505	0.545	0.535
5	0.443	0.380	0.176	0.502	0.537	0.525
6	0.448	0.388	0.192	0.501	0.532	0.525
7	0.453	0.397	0.212	0.500	0.524	0.522
8	0.459	0.411	0.218	0.502	0.529	0.507
9	0.463	0.421	0.242	0.502	0.529	0.523
10	0.465	0.427	0.242	0.502	0.528	0.519
Parametric						
$m = 1$	0.296	0.138	0.0418	0.459	0.433	0.525
2	0.346	0.208	0.0531	0.460	0.433	0.430
3	0.364	0.251	0.0774	0.452	0.435	0.485
4	0.378	0.277	0.0817	0.450	0.434	0.437
5	0.388	0.295	0.0920	0.449	0.431	0.436
6	0.393	0.304	0.101	0.447	0.426	0.408
7	0.397	0.315	0.0987	0.445	0.427	0.393
8	0.403	0.324	0.108	0.447	0.428	0.394
9	0.406	0.328	0.115	0.446	0.424	0.384
10	0.420	0.355	0.142	0.455	0.437	0.410

original data. When the non-matches are coded as undefined all datasets have a CAP that is larger than the baseline CAP, there is no substantial effect of the number of imputations. The differences between the CAP scores for the three scenarios (full set, statistical uniques, and special uniques) are inconsistent but not large.

On the other-hand, when the non-matches are coded as zero, the full set, statistical uniques, and special uniques have different CAP sizes, with the CAP size becoming smaller as the records become riskier. Additionally, as m increases the CAP score increases. We found for the full set that the CAP scores were larger than the baseline CAP score. However, for the special uniques, the synthetic CAP score was similar to or smaller than the baseline CAP scores, but as m increases the synthetic CAP score becomes larger than the baseline CAP score.

This relationship between the synthetic CAP scores and the baseline CAP score is different than the findings of Elliot (2014), who found that the synthetic and baseline had similar scores. This difference most likely stems from two factors: (1) Elliot was using a smaller key size. Appendix A shows that when smaller keys are observed the difference between the synthetic CAP score, the original CAP score, and baseline CAP score is less dramatic. (2) Elliot was only using single imputation. As seen in Table 1 as m increases so does the CAP score, hence a dataset that is only $m = 1$, would be less disclosive and therefore closer to the baseline CAP score.

Additionally, when the non-matches are coded as zero, Table 1 shows that the statistical unique CAP scores tend to be smaller than the full set CAP scores, and the special unique scores are smaller than the statistical unique scores. This trend is not so when the non-matches are undefined. The statistical uniques and special uniques are actually a bit larger than the full set. This indicates that while riskier records, as designated by the statistical uniques and special uniques, are less likely to have a match, if they do have a match it is just as likely to be correct as any other match.

Table 1 also shows that for the synthetic CAP scores the parametrically generated synthetic dataset had smaller CAP scores than the CART generated synthetic dataset. This shows that - in this experiment at least - the parametrically generated synthetic data has less risk than the CART generated synthetic data.

4.1   CAP Scores Regressed on Number of Imputations

To explore the relationship between the CAP scores and the number of imputations (m), we put the CAP scores into a simple linear regression analysis where y is the CAP score and m is a continuous variable, shown in Table 2. Table 2 shows that when the non-matches are coded as zero (Models 1–3) the number of imputations (m) has a significant and positive effect. When the non-matches are coded as undefined, the regression models show a different relationship. Table 2 shows that when the non-matches are coded as undefined, for the full set (Model 4), the relationship between m and the CAP score is not significant for CART, but has a significant, if small, negative coefficient for the parametric synthetic data. For model 5 when the data is parametric the relationship is not significant, but has a small negative coefficient for the CART data. For the special uniques (Model 6) there is not significant relationship between m and the CAP score. The m coefficient for Models 4, 5 and 6, is, even when significant, considerably smaller than the m coefficient for Models 1, 2, or 3.

When the non-matches are coded as CAP=0, m has a significant positive effect on the CAP score, when non-matches are coded as undefined for the CAP score there is essentially no effect. The increase when non-matches are zero, is an artefact of the lower number of non-matches. With more synthetic samples, non-matches are less likely to occur and this increases the CAP score. However, when the non-matches are undefined there is no reason for the CAP score to

Table 2. Simple linear regression of CAP score on the number of imputations - LCF

	Term	Non-matches as zero			Non-matches as undefined		
		Model 1	Model 2	Model 3	Model 4	Model 5	Model 6
		Full set	Statistical uniques	Special uniques	Full set	Statistical uniques	Special uniques
CART	Intercept	0.367***	0.221***	0.0809***	0.501***	0.548***	0.552***
	m	0.0117***	0.0234***	0.0177***	5.64e-05	−0.00283*	−0.00393
Parametric	Intercept	0.367***	0.169***	0.0431***	0.458***	0.438***	0.489***
	m	0.0117***	0.0186***	0.00838***	−0.00141*	−0.00165	−0.0114

*$p < 0.05$ **$p < 0.01$ ***$p < 0.001$

change in either direction, hence m has either a very small coefficient and mostly non-significant coefficient.

4.2 Comparing the LCF CAP Scores to the BSA CAP Scores

The results for the BSA dataset mostly confirm the results from the exploratory regression for the LCF. (The average CAP scores for the BSA can be found in Appendix B). Table 3 is an exploratory regression analysis of the BSA CAP scores. Like the LCF, when the non-matches are coded as zero (Models 1, 2, and 3) m has a significant positive relationship to the CAP score, showing that as m increases the likelihood of a match increases. However, for the CAP scores when the non-matches are excluded the BSA is different than the LCF. For the full set (Model 4) for the parametric synthetic data m has a significant negative coefficient. While for Model 5 the statistical uniques have significant m coefficients Model 4 probably has a significant coefficient since the BSA has a larger proportion of statistical uniques than the LCF (see Sect. 3.3) and therefore Model 4 will look more similar to Model 5. Additionally, Appendix B shows that like the LCF, for the BSA the CART CAP scores are larger than the parametric CAP scores.

Table 3. Simple linear regression of CAP score on the number of imputations - BSA

	Term	Non-matches as zero			Non-matches as undefined		
		Model 1	Model 2	Model 3	Model 4	Model 5	Model 6
		Full set	Statistical uniques	Special uniques	Full set	Statistical Uniques	Special uniques
CART	Intercept	0.143***	0.114***	0.0109	0.302**	0.114***	0.337***
	m	0.0137***	0.0157***	0.0201***	0.000517	0.0157***	0.00908
Parametric	Intercept	0.0615***	0.0420***	0.00335	0.177***	0.169***	0.220**
	m	0.00641***	0.00677***	0.00445***	−0.00226**	−0.00219*	0.00539

*$p < 0.05$ **$p < 0.01$ ***$p < 0.001$

4.3   Is Synthetic Data Differentially Confidential?

Here we introduce the notion of *differential confidentiality* as an alternative way to using the CAP score to the DCAP method described in Sect. 2.1. We determine that a dataset is differentially confidential in respect of a given target and key if on average there is no difference in the CAP score for a record whether the record is in the original dataset or not.

To demonstrate this concept for synthetic data we partitioned the data into two equal sized datasets and synthesised one (A) but not the other(B). We then calculated the synthetic CAP score for both A and B. Table 4 shows that synthetic data is not inherently differentially confidential, since if a record is included in the synthesis model it has a higher average CAP score than were it not included in the synthesis model. It is noteworthy that the differences between the two sets are larger for the CART synthetic data than it is for the parametric synthetic data. Also of note, for the t-tests [7], while all t-statistics for the CART synthetic data were significant at the $p < 0.001$ level, for the parametric synthetic data, the t-statistics were significant at the $p < 0.05$ level if they were significant at all, which some were not. Furthermore, as m increases the CART synthetic data becomes less differentially confidential, while that is not the case for the parametrically generated synthetic data.

The reader may have noted a superficial similarity of this concept with *differential privacy*. However, this is a post hoc test where as the former is a mechanism for achieving a standard[8]. It is possible that a dataset could be differentially private but not differentially confidential and vice versa.

4.4   Summary

With respect to DCAP the greater number of imputations, the more likely a match is to occur. However the likelihood of said match leading to a correct inference is not affected by the number of imputations, this picture can only fully be seen by looking at both the CAP scores when non-matches are coded as zero and undefined. In all instances the synthetic CAP score is lower than that of the original, showing that the synthetic data does decrease the risk of disclosure. While the disclosure risk is less than that of the original, it does not satisfy differential confidentiality.

That being said, synthetic data does particularly decrease the disclosure risk of special uniques, which are the most risky records in any microdata. The special uniques had CAP scores at the same level or lower than the baseline CAP, showing that the risk to special uniques from synthetic data is less than releasing a univariate distribution. In all scenarios, the CART synthetic data had a higher CAP score than the parametric synthetic data. This all suggests that parametrically synthetic data has less disclosure risk than CART synthetic data.

---

[7] We used Welch's T-Test DF $= 5,131$.

[8] See for example Abowd and Vilhuber, 2008; Charest 2010 for uses of differential privacy in the synthesizing mechanism.

Table 4. Results of the differential confidentiality test for the LCF synthetic dataset-with non-matches as zero.

	Included in synthesis	Not included in synthesis	Difference	T-test
CART				
$m = 1$	0.295	0.207	0.0875	8.0987***
2	0.370	0.252	0.1181	10.835***
3	0.404	0.269	0.134	12.514***
4	0.426	0.282	0.144	13.553***
5	0.442	0.294	0.148	14.13***
6	0.455	0.300	0.155	14.85***
7	0.462	0.303	0.160	15.503***
8	0.473	0.312	0.161	15.691***
9	0.476	0.315	0.162	15.841***
10	0.481	0.319	0.161	15.871***
Parametric				
$m = 1$	0.253	0.231	0.0215	2.0214*
2	0.318	0.289	0.0295	2.6972**
3	0.341	0.319	0.0216	1.9972*
4	0.361	0.339	0.0213	1.9825*
5	0.372	0.347	0.0248	2.3498*
6	0.376	0.350	0.0258	2.4819*
7	0.384	0.361	0.0234	2.271*
8	0.386	0.368	0.0186	1.8221
9	0.391	0.375	0.0156	1.5404
10	0.397	0.378	0.0185	1.8377

5   Conclusion

In this paper we have developed the methods introduced by Elliot (2014) for measuring attribute disclosure risk in synthetic data. The CAP score appears to be a simple but robust measure of attribute disclosure risk and the two approaches to using that measure - DCAP and differential confidentiality - seem to provide some traction on the difficult problem of measuring attribute disclosure. Indeed, we note that although these methods have been developed with a view to measuring the disclosure risk for synthetic data, they could be used on any structured datasets. Indeed, they might be useful for the calculation of the relative residual risk of different disclosure control and privacy protection methods. In future work we hope to develop such a general comparative methodology.

A   An exploration into the CAP scores when smaller sized keys are used or the LCF

Key 6: GOR, Output area classifier, tenure, dwelling type, internet in hh, household size
Key 5: GOR, Output area classifier, tenure, dwelling type, internet in hh
Key 4: GOR, Output area classifier, tenure, dwelling type
Key 3: GOR, Output area classifier, tenure (Table 5).

Table 5. Mean CAP scores for the original and synthetic LCF datasets for for two methods of handling non-matches, two synthesis methods, three different key sizes, three different intruder scenarios;and ten levels of multiple imputation.

Scenario	Full set			Statistical uniques			Special uniques		
File	Key 5	Key 4	Key 3	Key 5	Key 4	Key 3	Key 5	Key 4	Key 3
Original	0.610	0.560	0.466	1	1	1	1	1	1
Baseline	0.266	0.266	0.266	0.244	0.248	0.239	0.242	0.246	0.239
Non-matches as zero									
CART									
$m = 1$	0.378	0.381	0.395	0.200	0.196	0.234	0.109	0.159	0.234
2	0.411	0.401	0.402	0.289	0.275	0.292	0.147	0.191	0.292
3	0.422	0.411	0.404	0.327	0.324	0.338	0.169	0.233	0.338
4	0.428	0.414	0.406	0.350	0.333	0.343	0.186	0.229	0.343
5	0.431	0.416	0.406	0.367	0.352	0.345	0.191	0.239	0.345
6	0.433	0.418	0.406	0.373	0.363	0.351	0.202	0.239	0.351
7	0.434	0.419	0.406	0.375	0.367	0.355	0.202	0.243	0.355
8	0.437	0.421	0.407	0.381	0.371	0.351	0.216	0.265	0.351
9	0.437	0.421	0.407	0.380	0.373	0.352	0.240	0.279	0.352
10	0.439	0.422	0.408	0.385	0.376	0.358	0.249	0.289	0.358
Parametric									
$m = 1$	0.360	0.362	0.388	0.153	0.152	0.197	0.0727	0.0685	0.197
2	0.387	0.382	0.396	0.217	0.203	0.231	0.0720	0.0929	0.231
3	0.395	0.383	0.393	0.242	0.221	0.282	0.0988	0.147	0.282
4	0.400	0.388	0.395	0.263	0.238	0.288	0.0874	0.137	0.288
5	0.404	0.390	0.396	0.279	0.255	0.297	0.0908	0.151	0.297
6	0.405	0.391	0.395	0.290	0.267	0.307	0.124	0.162	0.307
7	0.406	0.392	0.395	0.297	0.279	0.302	0.146	0.193	0.302
8	0.406	0.392	0.395	0.296	0.274	0.289	0.166	0.206	0.289
9	0.408	0.392	0.394	0.309	0.278	0.289	0.188	0.234	0.289
10	0.415	0.398	0.396	0.328	0.306	0.308	0.214	0.278	0.308
Non-matches as undefined									
CART									
$m = 1$	0.444	0.421	0.404	0.466	0.441	0.432	0.632	0.702	0.432
2	0.448	0.422	0.405	0.478	0.437	0.358	0.523	0.518	0.358
3	0.449	0.425	0.406	0.471	0.44	0.376	0.476	0.502	0.376
4	0.450	0.427	0.406	0.469	0.441	0.365	0.445	0.458	0.365
5	0.449	0.427	0.407	0.472	0.445	0.364	0.428	0.455	0.364
6	0.449	0.427	0.407	0.467	0.443	0.370	0.436	0.436	0.370
7	0.449	0.427	0.407	0.460	0.440	0.371	0.419	0.425	0.371
8	0.450	0.428	0.407	0.457	0.431	0.360	0.417	0.427	0.360
9	0.449	0.427	0.407	0.449	0.425	0.358	0.433	0.427	0.358
10	0.449	0.428	0.408	0.448	0.425	0.361	0.441	0.434	0.361

(continued)

Table 5. (*continued*)

Scenario File	Full set			Statistical uniques			Special uniques		
	Key 5	Key 4	Key 3	Key 5	Key 4	Key 3	Key 5	Key 4	Key 3
Original	0.610	0.560	0.466	1	1	1	1	1	1
Baseline	0.266	0.266	0.266	0.244	0.248	0.239	0.242	0.246	0.239
Parametric									
$m = 1$	0.428	0.405	0.397	0.373	0.324	0.318	0.471	0.338	0.318
2	0.429	0.406	0.400	0.373	0.320	0.287	0.377	0.325	0.287
3	0.425	0.400	0.395	0.372	0.309	0.320	0.402	0.398	0.320
4	0.424	0.401	0.396	0.374	0.312	0.314	0.332	0.319	0.314
5	0.425	0.400	0.396	0.376	0.315	0.310	0.322	0.334	0.310
6	0.423	0.399	0.396	0.376	0.318	0.318	0.367	0.325	0.318
7	0.422	0.399	0.395	0.375	0.323	0.310	0.382	0.330	0.310
8	0.421	0.398	0.395	0.365	0.312	0.293	0.381	0.327	0.293
9	0.421	0.398	0.395	0.370	0.314	0.294	0.414	0.352	0.294
10	0.425	0.402	0.396	0.382	0.333	0.308	0.429	0.390	0.308

B   The average CAP scores for the BSA

(See Table 6).

Table 6. Mean CAP scores for the original and synthetic BSA datasets for two methods handling non-matches, two synthesis methods, three different intruder scenarios; full set, statistical uniques, and special uniques and ten levels of multiple imputation.

Scenario	Non-matches as zero			Non-matches as undefined		
	Full set	Statistical uniques	Special uniques	Full set	Statistical uniques	Special uniques
Original	0.876	1	1	0.876	1	1
Baseline	0.0853	0.0851	0.0869	0.0853	0.0851	0.0869
CART						
$m = 1$	0.115	0.0871	0.0127	0.291	0.304	0.273
2	0.173	0.146	0.0503	0.311	0.329	0.370
3	0.196	0.172	0.0631	0.302	0.313	0.309
4	0.218	0.197	0.0971	0.306	0.316	0.362
5	0.230	0.211	0.130	0.307	0.316	0.424
6	0.238	0.221	0.149	0.304	0.312	0.417
7	0.248	0.233	0.160	0.308	0.316	0.409
8	0.251	0.238	0.173	0.306	0.314	0.419
9	0.257	0.246	0.187	0.307	0.315	0.414
10	0.259	0.248	0.190	0.305	0.312	0.407
Parametric						
$m = 1$	0.0474	0.0309	0.00426	0.168	0.154	0.200
2	0.0716	0.0517	0.0128	0.168	0.159	0.273
3	0.0896	0.0701	0.0213	0.176	0.171	0.263
4	0.0998	0.0812	0.0241	0.173	0.170	0.258
5	0.101	0.0829	0.0241	0.161	0.156	0.227
6	0.106	0.0875	0.0281	0.161	0.153	0.220
7	0.111	0.0943	0.0338	0.163	0.156	0.240
8	0.113	0.0959	0.0380	0.159	0.151	0.255
9	0.114	0.0974	0.0416	0.155	0.148	0.271
10	0.115	0.100	0.0506	0.153	0.147	0.297

References

Caiola, G., Reiter, J.: Random forests for generating partially synthetic, categorical data. Trans. Data Priv. **3**, 27–42 (2010)

Charest, A.: How can we analyze differentially-private synthetic datasets? J. Priv. Confid. **2**(2), 21–33 (2010)

Chen, Y., Elliot, M., Sakshaug, J.: A genetic algorithm approach to synthetic data production. In: PrAISe 2016 Proceedings of the 1st International Workshop on AI for Privacy and Security, Hague, Netherlands. ACM (2016)

Dinur, I., Nissim, K.: Revealing information while preserving privacy. In: Principles of Database Systems, pp. 202–210 (2003)

Drechsler, J.: Using support vector machines for generating synthetic datasets. In: Domingo-Ferrer, J., Magkos, E. (eds.) PSD 2010. LNCS, vol. 6344, pp. 148–161. Springer, Heidelberg (2010). https://doi.org/10.1007/978-3-642-15838-4_14

Drechsler, J., Bender, S., Rässler, S.: Comparing fully and partially synthetic datasets for statistical disclosure control in the German IAB establishment panel. Trans. Data Priv. **1**, 105–130 (2008)

Drechsler, J., Reiter, J.: An empirical evaluation of easily implemented, non-parametric methods for generating synthetic data. Comput. Stat. Data Anal. **55**, 3232–3243 (2011)

Elliot, M.: Final Report on the Disclosure Risk Associated with the Synthetic Data Produced by the SYLLS Team. CMIST (2014).http://hummedia.manchester.ac.uk/institutes/cmist/archive-publications/reports/. Accessed 17 Mar 2017

Elliot, M., Mackey, E., O'Hara, K., Tudor, C.: The Anonymisation Decision-Making Framework, 1st edn. UKAN, Manchester (2016)

Elliot, M., Mackey, E., O'Shea, S., Tudor, C., Spicer, K.: End user licence to open government data? A simulated penetration attack on two social survey datasets. J. Off. Stat. **32**(2), 329–348 (2016)

Elliot, M., Manning, A., Ford, R.: A computational algorithm for handling the special uniques problem. Int. J. Uncerta. Fuzziness Knowl.-Based Syst. **10**(5), 493–509 (2002)

Fienberg, S., Makov, U.: Confidentiality, uniqueness, and disclosure limitation for categorical data. J. Off. Stat. **14**(4), 385–397 (1998)

Kim, J., Winkler, W.: Masking microdata files. In: Proceedings of the Survey Research Methods Section, pp. 114–119. American Statistical Association (1995)

Little, R.: Statistical analysis of masked data. J. Off. Stat. **9**(2), 407–426 (1993)

Machanavajjhala, A., Kifer, D., Gehrke, J., Venkitasubramaniam, M.: L-diversity: privacy beyond k-anonymity. ACM Trans. Knowl. Discov. Data **1**(1), 1–52 (2007)

NatCen Social Research: British Social Attitudes Survey, 2014, [data collection], UK Data Service, 2nd edn (2016). Accessed 30 Apr 2018. SN: 7809. https://doi.org/10.5255/UKDA-SN-7809-2

Nowok, B., Raab, G., Dibben, C.: synthpop: Bespoke creation of synthetic data in R. J. Stat. Softw. **74**(11), 1–26 (2016)

Office for National Statistics: Department for Environment, Food and Rural Affairs, Living Costs and Food Survey, 2014, [data collection], UK Data Service, 2nd edn (2016). Accessed 08 Mar 2018. SN: 7992. https://doi.org/10.5255/UKDA-SN-7992-3

Raab, G., Nowok, B., Dibben, C.: Practical data synthesis for large samples. J. Priv. Confid. **7**(3), 67–97 (2016)

Reiter, J.: Using CART to generate partially synthetic, public use microdata. J. Off. Stat. **21**, 441–462 (2005)

Reiter, J., Mitra, R.: Estimating risks of identification disclosure in partially synthetic data. J. Priv. Confid. **1**(1), 99–110 (2009)

Reiter, J., Wang, Q., Zhang, B.: Bayesian estimation of disclosure risks for multiply imputed, synthetic data. J. Priv. Confid. **6**(1), 17–33 (2014)

Rubin, D.B.: Statistical disclosure limitation. J. Off. Stat. **9**(2), 461–468 (1993)

Skinner, C., Elliot, M.: A measure of disclosure risk for microdata. J. R. Stat. Soc. Ser. B **64**(4), 855–867 (2002)

Smith, D., Elliot, M.: A measure of disclosure risk for tables of counts. Trans. Data Priv. **1**(1), 34–52 (2008)

Winkler, W.: Re-identification methods for evaluating the confidentiality of analytically valid microdata. Research Report Series, 9 (2005)

Yancey, W., Winkler, W., Creecy, R.: Disclosure risk assessment in perturbative microdata protection. Research Report Series, 1 (2002)

pMSE Mechanism: Differentially Private Synthetic Data with Maximal Distributional Similarity

Joshua Snoke$^{(\boxtimes)}$ and Aleksandra Slavković

Department of Statistics, Pennsylvania State University,
University Park, PA 16802, USA
{snoke,sesa}@psu.edu

Abstract. We propose a method for the release of differentially private synthetic datasets. In many contexts, data contain sensitive values which cannot be released in their original form in order to protect individuals' privacy. Synthetic data is a protection method that releases alternative values in place of the original ones, and differential privacy (DP) is a formal guarantee for quantifying the privacy loss. We propose a method that maximizes the distributional similarity of the synthetic data relative to the original data using a measure known as the *pMSE*, while guaranteeing ϵ-DP. We relax common DP assumptions concerning the distribution and boundedness of the original data. We prove theoretical results for the privacy guarantee and provide simulations for the empirical failure rate of the theoretical results under typical computational limitations. We give simulations for the accuracy of linear regression coefficients generated from the synthetic data compared with the accuracy of non-DP synthetic data and other DP methods. Additionally, our theoretical results extend a prior result for the sensitivity of the Gini Index to include continuous predictors.

Keywords: Differential privacy · Synthetic data · Classification trees

1 Introduction

In many contexts, researchers wish to gain access to data which are restricted due to privacy concerns. While there are many proposed methods for allowing researchers to fit models or receive query responses from the data, there are other cases where either due to methodological familiarity or modeling flexibility, researchers desire to have an entire dataset rather than a set of specific queries. This paper proposes a method for releasing synthetic datasets under the framework of ϵ-*differential privacy*, which formally quantifies and guarantees the privacy loss from these releases.

Differential privacy (DP), originally proposed by Dwork et al. (2006), is a formal method of quantifying the privacy loss related to any release of information based on private data; for a more in-depth review see Dwork and Roth (2014),

© Springer Nature Switzerland AG 2018
J. Domingo-Ferrer and F. Montes (Eds.): PSD 2018, LNCS 11126, pp. 138–159, 2018.
https://doi.org/10.1007/978-3-319-99771-1_10

and for a non-technical primer see Nissim et al. (2017). Since its inception, DP has spawned a large literature in computer science and some in statistics. It has been explored in numerous contexts such as machine learning algorithms (e.g., Blum et al. (2005); Kasiviswanathan et al. (2011); Kifer et al. (2012), categorical data (e.g., Barak et al. (2007); Li et al. (2010), dimension reduction (e.g., Chaudhuri et al. (2012); Kapralov and Talwar (2013), performing statistical analysis (e.g., Wasserman and Zhou (2010); Karwa and Slavković (2016), and streaming data (e.g., Dwork et al. (2010), to name a few.

While DP is a rigorous risk measure, it has lacked flexible methods for modeling and generating synthetic data. Non-differentially private synthetic data methods (e.g., see Raghunathan et al. (2003); Reiter (2002, 2005); Drechsler (2011); Raab et al. (2017)) while not offering provable privacy, provide good tools for approximating accurate generative models reflecting the original data. Our proposed method maintains the flexible modeling approach of synthetic data methodology, and in addition maximizes a metric of distributional similarity, the *pMSE*, between the released synthetic data and the original data, subject to satisfying ϵ-DP. We also do *not* require one of the most common DP assumptions concerning the input data, namely that it is bounded, and we do not limit ourselves to only categorical or discrete data. We find that our method produces good results in simulations, and it provides a new avenue for releasing DP datasets for potentially a wide-range of applications.

Our specific contributions are: (1) the combination of the flexible synthetic data modeling framework with the guarantee of ϵ-DP, (2) the relaxation of DP assumptions concerning boundedness or discreteness of the input data, (3) the embedding of a metric within our mechanism guaranteeing maximal distributional similarity between the synthetic and original data, and (4) a proof for the sensitivity bound of the Gini Index for CART models in the presence of continuous predictors (not just discrete).

The rest of the paper is organized as follows. Section 2 gives a review of important results from differential privacy that we rely on and provides a review of related methods to ours which we use for comparison in our simulation study. Section 3 details our proposed methodology for sampling differentially private data with maximal distributional similarity. Section 4 provides theoretical results for the privacy guarantees of our algorithm. Section 5 shows simulations that support our theoretical findings and provide an empirical estimate of the privacy loss under standard computational practices. Section 6 provides simulation results for the comparison of the accuracy of linear regression coefficients calculated from our method and other DP methods. Section 7 details conclusions and future considerations.

2   Differential Privacy Preliminaries

Differentially privacy is a formal framework for quantifying the disclosure risk associated with the release of statistics or raw data derived from private input data. The general concept relies on defining a randomized algorithm which has similar output probabilities regardless of the presence or absence of any given record, as formalized in Definition 1. We replicate the definitions and theorems here using notation assuming $X \in \mathbb{R}^{n \times q}$ and $\theta \in \mathbb{R}^k$. X is an original data matrix, and we wish to release instead a private version, X^s, with the same dimension. θ is a vector of parameters corresponding to a chosen parameteric model, which can be used to generate synthetic data that reflects the generative distribution of X. Further restrictions may be placed on θ depending on the parametric model.

Definition 1 (Dwork et al. (2006)). *A randomized algorithm, $\mathcal{M}$, is ϵ-Differentially Private if for all $\mathcal{S} \subseteq Range(\mathcal{M})$ and for all X, X' such that $\delta(X, X') = 1$:*

$$\frac{P(\mathcal{M}(X') \in \mathcal{S})}{P(\mathcal{M}(X) \in \mathcal{S})} \leq exp(\epsilon).$$

The privacy is controlled by the ϵ parameter, with values closer to zero offering stronger privacy. Relaxations of ϵ-DP have been proposed to reframe the privacy definition or to improve the statistical utility. A few examples are approximate differential privacy (also known as (ϵ, δ)-DP, see Dwork et al. (2006)), probabilistic differential privacy (Machanavajjhala et al. (2008)), on-average K-L privacy (Wang et al. (2016)), or concentrated privacy (Dwork and Rothblum (2016); Bun and Steinke (2016)). We do not cover these relaxtions further, since our work relies on the stronger ϵ-DP.

A general example of an ϵ-DP mechanism, which produces private outputs, is the Exponential Mechanism defined by McSherry and Talwar (2007); see Definition 2. For a given θ that we wish to release, this mechanism defines a distribution from which private samples, $\tilde{\theta}_i$, can be made and released in place of the original vector.

Definition 2 (McSherry and Talwar (2007)). *The mechanism that releases values with probability proportional to*

$$exp\left(\frac{-\epsilon \, u(X, \theta)}{2 \, \Delta u}\right),$$

where $u(X, \theta)$ is a quality function that assigns values to each possible output, θ, satisfies ϵ-DP.[1]

Δu is the global sensitivity, and it is defined as the greatest possible change in the u function for any two inputs, differing in one row. Note that some definitions of DP use the addition or deletion of a row, but here we assume X and X' have the same dimension. More formally:

---

[1] We put the minus sign before the u function because our quality function decreases for more desirable outputs.

Definition 3 (Dwork et al. (2006)). *For all X, X' such that $\delta(X, X') = 1$, the Global Sensitivity (GS) of a function $u : \mathbb{R}^{n \times q} \to \mathbb{R}_{\geq 0}$ is defined as:*

$$\Delta u = \sup_{\theta} \sup_{\delta(X, X') = 1} |u(X, \theta) - u(X', \theta)|$$

We also rely on two theorems, known as post-processing and sequential composition. The first, stated in Proposition 1, says that any function applied to the output from a differentially private algorithm is also differentially private. We use this to generate synthetic data based on private parameters, rather than directly generating differentially private data.

Proposition 1 (Dwork et al. (2006); Nissim et al. (2007)). *Let $\mathcal{M}$ be any randomized algorithm, such that $\mathcal{M}(X)$ satisfies ϵ-differential privacy, and let g be any function. $g(\mathcal{M}(X))$ also satisfies ϵ-differential privacy.*

Sequential composition, stated in Theorem 1, says that for multiple outputs from a differentially private algorithm, one must compose the ϵ values for each output to produce the overall privacy loss of the process. We need to compose the privacy if we make multiple draws of private parameters from which we produce multiple private synthetic datasets. We may want to release multiple synthetic datasets for better accuracy of estimates based on the data. Estimates based on multiple datasets are calculated using appropriate combining rules; see Raab et al. [2017] for details.

Theorem 1 (McSherry (2009)). *Suppose a randomized algorithm, $\mathcal{M}_j$, satisfies ϵ_j-differential privacy for $j = 1, ..., q$. The sequence $\mathcal{M}_j(X)$ carried out on the same X provides $(\Sigma_j \epsilon_j)$-differential privacy.*

These theorems and definitions lay the groundwork for our method. Next, we give a brief overview of some related methods to ours which we use for comparison in our simulations in Sect. 6.

2.1   Review of Related Methods

A number of different methods have been proposed for releasing differentially private synthetic datasets, although only a few are focused on real-valued, $n \times q$ matrices. Bowen and Liu (2016) proposed drawing data from a noisy Bayesian Posterior Predictive Distribution (BPPD), and Wasserman and Zhou (2010) generate data from smooth histograms. Bowen and Liu (2016) provides a fairly comprehensive list of DP synthetic data methods. Many of these methods are limited to specific data types, such as categorical data (e.g., Abowd and Vilhuber (2008); Charest (2011)), or network data (e.g., Karwa et al. (2016, 2017)), or they are computational infeasibile for data with a reasonable number of dimensions, such as the Empirical CDF (Wasserman and Zhou (2010)), NoiseFirst/StructureFirst (Xu et al. (2013)), or the PrivBayes (Zhang et al. (2017)), though some recent work has proposed ways to reduce the computation time and improve the accuracy (Li et al. (2018)).

The noisy BPPD method from Bowen and Liu (2016) is similar to ours in the sense that focuses on drawing generative model parameters from a noisy distribution and then using these private parameters generates private synthetic data according to post-processing. In this case private parameters are drawn from posterior distribution $f(\theta|s^*)$ where s is the Bayesian sufficient statistic and s^* is the statistic perturbed according to the Laplace mechanism (e.g., see Dwork et al. (2006)). Bowen and Liu (2016) recommend drawing multiple sets of private parameters and producing a synthetic dataset for each one, which requires composing ϵ.

The smooth histogram method from Wasserman and Zhou (2010) works non-parametrically by binning the data, using these bins to estimate a consistent density function, and applying smoothing to the function which guarantees DP before drawing new samples. The DP smooth histogram is defined as:

$$\hat{f}_K^*(\mathbf{x}) = (1 - \lambda)\hat{f}_K(\mathbf{x}) + \lambda\Omega \tag{1}$$

where K is the total number of bins, $\Omega = \left(\prod_{j=1}^{p}(c_{j1} - cj0) \right)^{-1}$, $\lambda \geq \frac{K}{K+n(e^{\epsilon/n}-1)}$, and $\hat{f}_K(\mathbf{x})$ is a mean-squared consistent density histogram estimator. This method does not need to generate multiple datasets since it is not redrawing model parameters, and accordingly does not need to split ϵ across multiple synthetic datasets.

However, both of these methods require bounded data. This can be seen explicitly in the smooth histogram formulation where we assume the j_{th} variable has bounds $[c_{j0}, c_{j1}]$. We also need to assume bounds in order to create (and sample from) a finite K bins. The boundedness assumption is less explicit in the noisy BPPD method, but it comes into play when calculating the sensitivity of the statistics. If the data were unbounded, the sensitivity could be infinite, which would mean we have to sample the noise from a Laplace distribution with infinite variance.

We want to avoid assuming bounds as it is problematic when it comes to approximating the underlying generative distribution. Continuous data may be naturally unbounded, and at best in many real data scenarios we do not know what the bounds should be. If our bounds are too loose, we introduce more noise than necessary through the privacy process, limiting our accuracy. On the other hand, we introduce bias if we set the bounds too low because that truncates the generative distribution below its true range. We further explore the effect of these assumed bounds using a simulation study in Sect. 6.

Our method avoids the bounding problem by sampling from a distribution that shrinks in probability as we move to the tails. We do not bound the sample space, but we have low probability of sampling values which are far from the truth. This allows us to produce private data that accurately reflects the natural range of the data. We describe this further in Sect. 3.1.

Furthermore, the smooth histogram method suffers from computational limitations as the number of variables increases, since it divides the data matrix into bins, the number of which grows $O(p^p)$. Our method has computational limitations too, though of a different nature, which we discuss further in Sect. 3. One nice aspect of the noisy BPPD method is that it is computationally fast.

Finally, our method improves over these methods by incorporating a measure of distributional similarity on the resulting synthetic data. The noisy BPPD and smooth histogram add noise to the generative model for the synthetic data. These mechanisms concern only minimizing the noise added to the parameters, but they guarantee nothing concerning the data generated using these parameters. Our method on the other hand, finds the private parameters which can be used to generate synthetic data that will maximize the distributional similarity with respect to the original data. Regardless of the dataset, our method finds the best private parameters for an assumed synthesizing model. Sections 3 and 4 give a detailed explaination and theoretical results for our method.

3   Sampling Differentially Private Synthetic Data via the *pMSE* Mechanism

We propose to release a DP synthetic data matrix, X^s, in place of the original data X and with the same dimension, $n \times q$. In practice it is infeasible to sample a matrix of this size using the Exponential Mechanism. Instead, we sample private model parameters and then generate synthetic data based on these noisy parameters. We know from the results on post-processing (Proposition 1) that data generated based on these DP parameters are also DP. Based on the exponential mechanism we draw DP samples using

$$f(\theta) \propto exp\left(\frac{-\epsilon\, u(X, \theta)}{2\Delta u} \right), \tag{2}$$

where ϵ is our privacy parameter, $u(X, \theta)$ is our quality (or utility) function, and Δu is the sensitivity of the quality function. In practice we use a Markov chain Monte Carlo (MCMC) with the Metropolis algorithm to generate samples from this unnormalized density, since we do not know the value of the u function a priori. Next we define our quality function and derive a bound on its sensitivity.

3.1   Defining the Quality Function Using the *pMSE*

We base our quality function on the *pMSE* statistic developed in Woo et al. (2009) and Snoke et al. (2018):

$$pMSE = \frac{1}{N} \Sigma_{i=1}^{N} (\hat{p}_i - 0.5)^2,$$

where $\hat{p}_i$ are predicted probabilities (i.e., propensity scores) from a chosen classification algorithm. Algorithm 1 gives the steps for calculating the *pMSE* statistic. The *pMSE* is simply the mean-squared error of the predicted probabilities from this classification task, and it is a metric to assess how well we are able to discern between datasets based on a classifier. If we are unable to discern, the two datasets have high distributional similarity. A *pMSE* = 0 means every $\hat{p}_i = 0.5$, and it implies the highest utility. There has been much work dedicated to tuning models for out-of-sample prediction, but for our purposes we only use the classifier to get estimates of the in-sample predicted probabilities.

Algorithm 1. General Method for Calculating the *pMSE*

1: stack the n rows of original data X and the n rows of masked data X^s to create X^{comb} with $N = 2n$ rows
2: add an indicator variable, I, to X^{comb} s.t. $I = \{1 : x_i^{comb} \in X^s\}$
3: fit a model to predict I using predictors $Z = f(X^{comb})$
4: find predicted probabilities of class 1, $\hat{p}_i$, for each row of X^{comb}
5: obtain the $pMSE = \frac{1}{N} \Sigma_{i=1}^{N} (\hat{p}_i - 0.5)^2$

To make our quality function a function of θ, the vector of parameters we wish to sample, we use the expected value of the *pMSE* given θ, i.e.,

$$u(X, \theta) = E[pMSE(X, X_\theta^s)|X, \theta], \tag{3}$$

where $X^s \sim g(\theta)$. In practice we approximate this by generating l datasets for a given set of parameters and calculate the average *pMSE* across each data set. This approximation does not affect the privacy guarantee (as shown in the proof for Theorem 2), but for accuracy l should be large enough to give satisfactory results for estimating the expected value of $u(X, \theta)$.

As we mentioned before, this quality function makes no assumptions concerning whether the original data are categorical, discrete, or continuous. Secondly, because the *pMSE* is a function of the predicted probabilities, $\hat{p}$, which are bounded $\in [0, 1]$, the *pMSE* is bounded $\in [0, 0.25]$. This is true regardless of the range of the data, X, so we do not need to assume any kinds of bounds on the data.

We refer to our method as the *pMSE Mechanism*, since we rely on the *pMSE* for our quality function in the exponential mechanism. Algorithm 2 outlines the steps of the *pMSE* mechanism. The main assumption we need is that a reasonable generative model for the data, $g(\theta)$, exists.

Algorithm 2. Sampling DP Synthetic Data via the *pMSE* Mechanism

---

Input: Original dataset: X, chosen value: ϵ, synthesis model: $g(\theta)$, quality function: $u(X, \theta)$
1: Sample l vectors $\{\tilde{\theta}_1, ..., \tilde{\theta}_l\}$ from a density proportional to equation 2
2: For each $\tilde{\theta}_i$ generate synthetic data set $X_i^s \sim g(\tilde{\theta}_i)$, giving l total synthetic datasets $\{X_1^s, ..., X_l^s\}$ each with the same dimension as X
3: Releasing $\{X_1^s, ..., X_l^s\}$ satisfies $(l\epsilon)$-DP

3.2 Estimating the *pMSE* using Classification and Regression Tree (CART) Models

A key component to defining our quality function is the classification model used to estimate the predicted probabilities, $\hat{p}$, used in computing the *pMSE*. We choose the classification trees (Breiman et al. (1984)) fit using the Gini Index, for two primary reasons. First, we need a tight bound on the sensitivity of $u(X, \theta)$. While other machine learning models have been shown to outperform CART in many applications, we would have a far weaker bound on the sensitivity and would need to add much more noise. Secondly, as was shown in Snoke et al. (2018), CART models exhibit at least satisfactory performance in determining the distributional similarity. Future work may prove desirable bounds on the sensitivity of the *pMSE* when using stronger classifiers, in which case those models should certainly be adopted.

We use the impurity function known as Gini Index from Breiman et al. (1984), defined as:

$$GI = argmin \ \Sigma_{i=1}^{D+1} a_i \left(1 - \frac{a_i}{m_i}\right), \tag{4}$$

where m_i are the total number of observations in each node, a_i are the number of observations labeled 1 in each node, and D is the total number of nodes. In practice these models are fit in a greedy manner for computational purposes. The process is to make the first optimal split that minimizes the impurity for two nodes, and then to make proceeding splits and adding additional nodes if doing so continues to minimize the impurity according to a chosen cost function. If computation is not a concern, it would also be possible for any fixed D to do a full grid search to determine the optimal D splits that minimize the impurity over $D + 1$ nodes. The difference between globally optimal and greedy trees is important for our theoretical results. In our theoretical results in Sect. 4 we prove the sensitivity bound when trees are fit based on the globally optimal Gini Index, and in our simulations in Sect. 5 we perform an empirical examination of how frequently the greedy fitting violates our theoretical results.

4 Theoretical Results for the Sensitivity Bound

In order to sample from the exponential mechanism, we need a bound on the sensitivity of the quality function. The *pMSE* function is naturally bounded, but in Theorem 2 we prove a much tighter bound.

Theorem 2. *Given* $u(X, \theta) = E_\theta[pMSE(X, X_\theta^s)|X, \theta]$ *where* $pMSE = \Sigma_{i=1}^{2n} \frac{(\hat{p}_i - 0.5)^2}{2n}$ *with* $\hat{p}_i$ *estimated from a classification tree with optimal splits found using the Gini Index. Then*

$$\Delta u = \sup_\theta \ \sup_{\delta(X,X')=1} |u(X, \theta) - u(X', \theta)| \leq \frac{1}{n},$$

where $X, X_\theta^s \in \mathbb{R}^{n \times q}$.

The proof, given in the Appendix, intuitively follows from the fact that we can relate the *pMSE* to the Gini Index. We can then bound the change in Gini Index given a change in one row of the input data due to the fact that we are finding the global optimum. We will not suddenly do much better or much worse. In fact, we can quantify exactly how much better or worse we can do, which leads to the bound.

This bound is nice because it decreases with n, meaning the noise added decreases as the number of observations increase. This bound matches the results derived for the sensitivity of the Gini Index when assuming discrete predictors from Friedman and Schuster (2010). Our proof shows that this bound remains the same when using continuous predictors as well. The result in Friedman and Schuster (2010) was used for performing classification under differential privacy, rather than producing synthetic data, and we see our extension of the proof to include continuous predictors as a useful side result of this paper.

It is important to note that this proof is for the theoretical case when we can find the optimal partitioning for any number of nodes. The greedy method can violate the bound because we can no longer control how much the Gini Index can change after changing one row. If we violate the sensitivity bound, the method will not guarantee exact ϵ-DP. While it would be possible to use our method with a full grid search, computationally it is a poor idea. An alternative could be to use adaptive composition, i.e., fit the CART models greedily but in a way that satisfies DP. We could then compose the privacy between fitting the CART model and sampling from the exponential distribution, which we explore in future work.

5   Empirical Failure Rate of the Sensitivity Bound

These simulations show the empirical rate for which the greedy fitting violates the bound. We can also view the maximum simulated value as an empirical estimate of the sensitivity for this particular dataset, but we are more interested in the failure rate. We generated datasets, X, with $q = 2$ and $n = 5000$. $X_1 \sim N(2, 10)$ and $X_2 \sim N(-2.5 + 0.5x_1, 3)$, and we produced X' by taking X and adding random Gaussian noise, $N(0, 25)$, to each variable for one observation. We then drew a synthetic dataset X^s with $X_1^s \sim N(\theta_1, \theta_2)$ and $X_2^s \sim N(\theta_3 + \theta_4 x_1^s, \theta_5)$ where $\theta_i \sim N(0, 10)$. We estimated the *pMSE* with respect to X^s for both X and X' and calculated the difference. Recall the theoretical sensitivity bound is $1/n = 0.0002$, and any values larger than this violate the bound. We repeated this process 1,000,000 times each using CARTs of depths 1, 2, 5, and unlimited for the *pMSE* model. For all trees we included a complexity parameter (*cp*) requiring a certain percentage improvement in order to make an additional split. This parameter is necessary in order to not produce trees that are fully saturated (one terminal node per observation) when there is no depth limitation.

Figure 1 shows a sample of the simulated empirical sensitivity results. There are four groupings, for the trees with different depths, and darker points denote those violating the theoretical bound due to the greedy fitting algorithm. Table 1

Table 1. Empirical failure rates of 1,000,000 simulations for the sensitivity bound when using the greedy CART fitting algorithm for different tree sizes and different complexity parameters.

Tree depth	cp	Percentage violating bound
Depth 1	0.01	0.0%
Depth 2	0.01	0.3%
Depth 5	0.01	0.6%
Depth unlimited	0.01	0.7%
Depth 1	0.001	0.0%
Depth 2	0.001	0.5%
Depth 5	0.001	2.0%
Depth unlimited	0.001	2.5%

gives the full result. Recall that violations of the bound are more likely to occur when the greedy fitting makes globally non-optimal cuts. The percentage of simulations which violate the bound increases with tree depth size, and in the unlimited case for $cp = 0.01$ the empirical failure rate is $\leq 1\%$. As expected, there are no results which violate the bound when only one split is made. This confirms our theoretical results because with only one split, greedy is equivalent to optimal, so the bound is never violated in simulation. The empirical sensitivity also depends on the cp, so we ran simulations for two different values. A lower cp will lead to larger trees (subject to depth constraints), which means we are making more greedy splits and increasing the chance of violating the bound.

6   Empirical Evaluation of Differentially Private Linear Regression

In order to assess the practical statistical utility of our method, we ran simulations testing the accuracy of an estimated linear regression model. Our method guarantees maximal distributional similarity of the synthetic data based on the *pMSE* metric, but many researchers may be interested in more specific comparisons such as regression outputs.

We simulate datasets, X, in the same way as in Sects. 5 and 7. Using this data, we regress X_2 on X_1 and get ordinary least squares (OLS) estimates of the intercept ($\hat{\beta}_0$) and slope ($\hat{\beta}_1$) coefficients. We calculate the absolute difference between these estimates and the corresponding estimates we get by fitting the same model with the differentially private methods, i.e., $|\hat{\beta} - \hat{\beta}^{priv}|$. Our comparison methods are the noisy Bayesian method from Bowen and Liu (2016) and the smooth histogram from Wasserman and Zhou (2010), both described in Sect. 2.1. We also compare with methods that do not produce synthetic data but produce DP regression estimates such as the Functional Mechanism of Zhang et al. (2012) and Awan and Slavkovic (2018), considering L_1 and L_∞

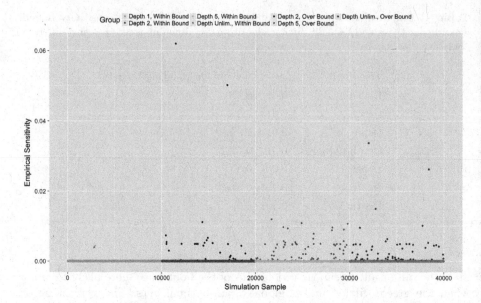

Fig. 1. Random sample of 10,000 simulations for each tree depth. Values shown are differences between $u(X, \theta)$ calculated with X and X'. Darker points violate the theoretical bound. $Cp = 1\%$.

mechanisms, respectively. Note that these methods require the data to be bounded in the same way as the noisy Bayes and smooth histogram methods. Finally we compare with estimates from non-DP synthetic data, sampled from the unperturbed BPDD.

For the *pMSE* mechanism, we carry out the simulations using trees of depths 1, 2, 5, and unlimited with a $cp = 0.01$. We see that the utility significantly improves as we move from depth 1 to 2 and from 2 to 5, but there is little change from 5 to unlimited. This is likely because trees of depth 5 are large enough to evaluate this dataset. The tree size is a potential tuning parameter for future work using this method. Astute readers may have noticed that the unnormalized distribution we propose for the *pMSE* mechanism does not necessarily exist, since the probability in the tails remains very flat. To fix this, we add a very flat prior, $N(0, 100, 000)$, to each of our parameters when sampling. The flatness does not affect the utility, but by adding it we ensure the probability in the tails eventually goes to zero.

We run the mechanisms with values $\epsilon \in \{0.25, 0.5, 1\}$. For the *pMSE* mechanism and the noisy BPDD we generate $l = 10$ private datasets each satisfying (ϵ/l)-DP, and for the smooth histogram and functional mechanisms we produce only one output satisfying ϵ-DP. This ensures all mechanisms satisfy the same level of privacy. The non-DP synthetic data method does not guarantee any privacy.

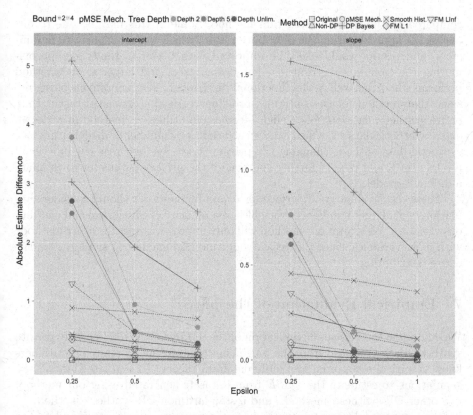

Fig. 2. Lineplots showing the mean simulation results. X-axis indicates different values of ϵ. Lines are also subdivided within methods by the tree depth and the bound.

For the noisy Bayes, smooth histogram, and functional mechanism methods, we ran the simulations truncating the data at different assumed bounds. For both variables, we set these bounds at two, four, five, or ten times the standard deviation. Four or five can be thought of as roughly the appropriate bounds, since this is Gaussian data and most observations will fall into those ranges. Two was chosen to be a range that is too narrow and excludes part of the true distribution, and ten was chosen for a looser bound. We see from the results that the smaller bound achieves better results on the regression, even when it is more narrow even than the truth. This is an artifact of the model we chose, and if an even tighter bound was chosen it may have greater adverse effects. The loose bound (10 times) performs quite poorly, since we must add much more noise. Figure 2 visualizes the better performing results (limited only for readability). Full results for all tree sizes and bounds are shown in Fig. 4 in the Appendix.

Overall our method outperforms the other two synthesis methods when using trees of depth 5 or unlimited, regardless of the bounds chosen. Even trees of depth 2 perform roughly the same or better than the other methods. At lower values of $\epsilon = 0.25$ the performance starts to become noticeably worse, especially when compared to the smooth histogram method. Some of this may be due to our method of approximating the distribution from which we sample private parameters, and results would likely be improved with better methods for approximating the sampling distribution. For deeper trees and larger ϵ, our method performs almost as well as the functional mechanism. This is quite encouraging, since that method focuses only on providing regression estimates rather than entire synthetic datasets. For a slight decrease in utility our method provides an entire synthetic dataset, which can be used to fit any number of models using our synthetic data without changing the privacy guarantee, whereas the functional mechanism would require further splitting of the privacy parameter to estimate a different model.

These results show good performance, and further work should consider simulations with larger numbers of variables or a mixture of categorical and continuous variables. We expect our method will only improve against the other methods with more variables, since theoretically our method maximizes similarity on the entire distribution.

7   Empirical Evaluation of the *pMSE*

We guarantee theoretical maximization of the *pMSE* for the differentially private synthetic data produced from the *pMSE* mechanism, but as many practitioners know empirical tests often look different from theory. To evaluate this, we ran simulations to estimate the *pMSE* from datasets generated using our method, two other DP synthesis methods, and a standard non-DP synthesis method.

We again simulate datasets, X, in the same way as in Sects. 5 and 6. For each X we generated synthetic datasets X^s and then calculated the *pMSE* using X and X^s. Our four synthesis methods were the *pMSE* mechanism, the noisy Bayesian method from Bowen and Liu (2016), the smooth histogram from Wasserman and Zhou (2010), and sampling from the non-differentially private BPPD using fully conditional sequential models. We ran 2,500 simulations each for five different values of $\epsilon \in \{0.25, 0.5, 1, 2, 4\}$. For our method we used CART trees with unlimited depth, and for the other two DP methods we assumed a bound on the data of four times the standard deviation, which is roughly the correct bound given that it is Gaussian data. Figure 3 shows the results for the mean simulations results and Table 2 shows the full mean and variance of the results.

As expected, the *pMSE* mechanism offers either the best or one of the best values of the *pMSE* among the methods guaranteeing DP. The smooth histogram method is fairly good as well offering comparable values at $\epsilon = 0.25$ or $\epsilon = 4$. The noisy BPPD method on the other hand is bad, even at high levels of ϵ, so should be used with caution.

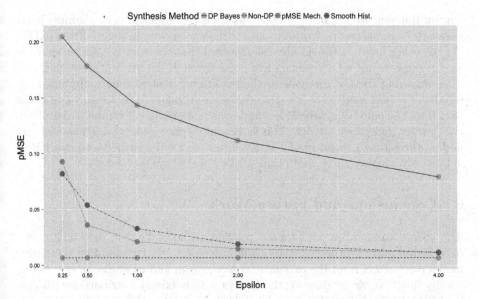

Fig. 3. Simulations results showing the mean *pMSE* calculated using synthetic producing according to four different methods. *pMSE* is calculated from comparison with original data, with values closer to 0 implying higher utility.

Table 2. Simulation results giving the mean and variance of the *pMSE* values calculated using four different synthesis methods and five different levels of ε.

ε	Simulated values	Non-DP	*pMSE Mech.*	DP Bayes	Smooth hist
0.25	*pMSE* Mean	0.00660	0.09281	0.20509	0.08206
	pMSE Var	8.681e−07	1.347e−03	3.079e−03	2.826e−05
0.5	*pMSE* Mean	0.00663	0.03610	0.17876	0.05398
	pMSE Var	8.929e−07	8.020e−05	5.914e−03	1.809e−05
1	*pMSE* Mean	0.00661	0.02107	0.14372	0.03278
	pMSE Var	8.342e−07	1.648e−05	8.579e−03	1.069e−05
2	*pMSE* Mean	0.00660	0.01459	0.11177	0.01892
	pMSE Var	8.342e−07	1.648e−05	8.579e−03	1.069e−05
4	*pMSE* Mean	0.00660	0.01161	0.07919	0.01129
	pMSE Var	8.671e−07	3.329e−06	9.087e−03	5.964e−06

We also see that the variance in the estimated *pMSE* values changes quite a bit depending on the method and level of ε. Both the *pMSE* mechanism and the smooth histogram show higher variances for either low (0.25) or high (4) values of ε, while the noisy BPPD method increases in variance as ε increases. This variance is something to keep in mind both in choosing a protection method and in developing the practical implementation. We could likely improve our current implementation of the *pMSE* mechanism in order to better sample noisy parameters and generate synthetic data with less variance in the resulting *pMSE*. On the other hand, it should also be expected that the variance decreases some as ϵ grows because we are adding less noise through the privacy mechanism.

Comparing the DP methods to the traditional synthetic data approach, we see that the best method at $\epsilon = 4$ produces an average *pMSE* roughly two times that from the non-DP synthesis method, which is producing synthetic data from the correct generative model. This is actually quite good considering we are adding the strong guarantee of DP. Even for $\epsilon = 1$ our method produces *pMSE* values only roughly three times that of the non-DP synthetic data.

8 Conclusions and Future Work

The *pMSE* mechanism we propose provides a novel flexible method to produce high-quality synthetic datasets guaranteeing ϵ-DP. By sampling generative model parameters from the exponential mechanism and using the *pMSE* as our quality function, we produce synthetic data with maximal distributional similarity to the original data. By using the *pMSE*, we ensure the sensitivity depends neither on the dimension nor the range of the data, and the bound decreases as we increase the sample size. This allows us to use this mechanism for continuous data, and the amount of noise we add will not grow with the dimension (apart from sampling from a more complex distribution).

Our simulations in Sects. 6 and 7 confirm that the *pMSE* mechanism generally performs as well or better than the other standard DP synthesis methods. In the case of linear regression the *pMSE* mechanism even performs roughly as well as methods that produce estimates of regression coefficients only rather than entire synthetic datasets. In the case of the empirical *pMSE*, as expected our method performs worse than non-DP synthetic data, but the utility cost seems reasonable for the privacy gain.

The *pMSE* mechanism relies on defining an appropriate form for the generative distribution from which to draw synthetic values. It is possible to misspecify this model, which would lead to poor utility. This is one drawback of the synthetic approach as opposed to simply adding noise. Fortunately, this aspect has been addressed in great detail in the synthetic data literature, so we feel that finding an appropriate model is possible without too much difficulty.

Our primary limitation is the computational feasibility to ensure the theoretical sensitivity bound. From the empirical simulations we saw that the bound does not always hold when using typical greedy fitting algorithms. Fitting the models using the global optimum would ensure the theoretical bound and guarantee ϵ-DP. Proposals have been made to carry out machine learning using global optimums, such as Bertsimas and Dunn (2017), so methods may exist to aid the computation.

An alternative implementation of our method would be to consider fitting the CART models in a way that satisfies ϵ-DP and then composing this ϵ with that from sampling from the *pMSE* mechanism. This is similar to the approach of Li et al. (2018). This is desirable because we could use any standard CART software to implement the method. Other future work could consider using different impurity measures than the Gini Index, deriving measures of choosing the best tree size, or best practices for sampling from the unnormalized density we get through the exponential mechanism.

Acknowledgments. This work is supported by the U.S. Census Bureau and NSF grants BCS-0941553 and SES-1534433 to the Department of Statistics at the Pennsylvania State University. Thanks to Bharath Sriperumbudur for special aid in deriving the final form of the theoretical proof.

10   Appendix: Proof of Theorem 4.1

Proof. We first show that using the expected value, and approximating it, can be bounded above by the supremum across all possible datasets X^s generated using θ.

$$\Delta u = \sup_{\theta} \sup_{\delta(X,X')=1} |u(X,\theta) - u(X',\theta)| \tag{5}$$

can be rewritten as

$$\Delta u = \sup_{\theta} \sup_{\delta(X,X')=1} |E_\theta[pMSE(X,X_\theta^s)|X,\theta] - E_\theta[pMSE(X',X_\theta^s)|X,\theta]| \tag{6}$$

where $u(X,\theta) = E_\theta[pMSE(X,X_\theta^s)|X,\theta]$. Since the absolute value is a convex function, we can apply Jensen's inequality and get

$$\leq \sup_{\theta} \sup_{\delta(X,X')=1} E_\theta[|pMSE(X,X_\theta^s) - pMSE(X',X_\theta^s)||X,\theta]. \tag{7}$$

Then by taking the supremum over any data set X_θ^s, we obtain

$$\leq \sup_{X_\theta^s} \sup_{\delta(X,X')=1} |pMSE(X,X_\theta^s) - pMSE(X',X_\theta^s)|. \tag{8}$$

This also bounds our approximation of the expected value that we propose to use in practice, since the supremum is also greater than or equal to the sample mean.

Now writing this explicitly in terms of the CART model, we get

$$\sup_{a_i,\, m_i,\, a_i',\, m_i'} \frac{1}{2n} \left| \Sigma_{i=1}^{D+1} m_i \left(\frac{a_i}{m_i} - 0.5 \right)^2 - m_i' \left(\frac{a_i'}{m_i'} - 0.5 \right)^2 \right| \tag{9}$$

where a_i, m_i, and D are defined as before, and a_i' and m_i' are the corresponding values for the model fit using X'. Expanding this we get

$$\sup_{a_i,\, m_i,\, a_i',\, m_i'} \frac{1}{2n} \left| \Sigma_{i=1}^{D+1} \left(\frac{a_i^2}{m_i} - a_i - 0.25 m_i \right) - \left(\frac{a_i'^2}{m_i'} - a_i' - 0.25 m_i' \right) \right| \quad (10)$$

and we can cancel the third terms because $\Sigma_{i=1}^{D+1} m_i = \Sigma_{i=1}^{D+1} m_i'$. When we multiple by $2n$, the remaining inside term is equivalent to the sensitivity of the impurity, i.e.,

$$\sup_{a_i,\, m_i,\, a_i',\, m_i'} \left| GI(X, X^s, D) - GI(X', X^s, D) \right| = \Delta GI \quad (11)$$

By bounding the impurity, we bound the *pMSE*. We can rewrite the above as

$$\left| \min_D GI(X, X^s, D) - \min_D GI(X', X^s, D) \right| \quad (12)$$

since the optimal CART model finds the minimum impurity across any D. The greatest possible difference then is the difference between these two optimums. And we can bound this above by

$$\leq \sup_D \left| GI(X, X^s, D) - GI(X', X^s, D) \right|. \quad (13)$$

Let X^{comb} and X'^{comb} be the combined data matrices as described in Algorithm 1, including the $0, 1$ outcome variable. Recall that only one record has changed between X^{comb} and X'^{comb} (total number of records staying fixed), and it is labeled 0. We know that for a given D optimal split points producing $D + 1$ nodes on X^{comb}, there are a_i records labeled 1 and $\tilde{m}_i$ total records in each bin, such that $\exists\, j \neq k \neq l_1 \neq ... \neq l_{D-1}$ s.t. $\tilde{m}_j - m_j = m_k - \tilde{m}_k = 1$, $\tilde{m}_{l_v} = m_{l_v}$ for $v = \{1, ..., D-1\}$. In the same way, for a given D optimal split points producing $D + 1$ nodes on X'^{comb}, there are a_i' records labeled 1 and $\tilde{m}_i'$ total records in each bin, such that $\exists\, j' \neq k' \neq l_1' \neq ... \neq l_{D-1}'$ s.t. $\tilde{m}_{j'}' - m_{j'}' = m_{k'}' - \tilde{m}_{k'}' = 1$, $\tilde{m}_{l_v}' = m_{l_v}'$ for $v = \{1, ..., D-1\}$. What this simply means is that after changing one record, the discrete counts in the nodes change by at most one in two of the nodes and does not change in the other $D - 1$ nodes.

Due to the fact that the CART model produces the D splits that minimize the impurity, we know both that

$$\Sigma_{i=1}^{D+1} a_i' \left(1 - \frac{a_i'}{m_i'} \right) \leq \Sigma_{i=1}^{D+1} a_i \left(1 - \frac{a_i}{\tilde{m}_i} \right) \quad (14)$$

and

$$\Sigma_{i=1}^{D+1} a_i \left(1 - \frac{a_i}{m_i}\right) \leq \Sigma_{i=1}^{D+1} a_i' \left(1 - \frac{a_i'}{\tilde{m}_i'}\right). \tag{15}$$

The inequality (14) implies that after changing one record, if new split points are chosen, the impurity must be equivalent or better than simply keeping the previous splits and changing the counts. The inequality (15) implies that the first split points chosen must be equivalent or better than using the new splits with the changed counts. If this were not the case, the first split points would have never been made in the first place. These lead to the final step.

Because we have an absolute value, we consider two cases.

Case 1: $GI(X, X^s, D) \geq GI(X', X^s, D)$

$$\sup_{D} \left| \Sigma_{i=1}^{D+1} a_i \left(1 - \frac{a_i}{m_i}\right) - \Sigma_{i=1}^{D+1} a_i' \left(1 - \frac{a_i'}{m_i'}\right) \right| \leq$$

$$\sup_{D} \left| \Sigma_{i=1}^{D+1} a_i' \left(1 - \frac{a_i'}{\tilde{m}_i'}\right) - \Sigma_{i=1}^{D+1} a_i' \left(1 - \frac{a_i'}{m_i'}\right) \right| =$$

$$\left| a_{j'}' \left(1 - \frac{a_{j'}'}{\tilde{m}_{j'}'}\right) - a_{j'}' \left(1 - \frac{a_{j'}'}{m_{j'}'}\right) + a_{k'}' \left(1 - \frac{a_{k'}'}{\tilde{m}_{k'}'}\right) - a_{k'}' \left(1 - \frac{a_{k'}'}{m_{k'}'}\right) \right| =$$

$$\left| \frac{a_{j'}'^2 (\tilde{m}_{j'}' - m_{j'}')}{\tilde{m}_{j'}' m_{j'}'} + \frac{a_{k'}'^2 (\tilde{m}_{k'}' - m_{k'}')}{\tilde{m}_{k'}' m_{k'}'} \right| = \left| \frac{a_{j'}'^2}{\tilde{m}_{j'}' m_{j'}'} - \frac{a_{k'}'^2}{\tilde{m}_{k'}' m_{k'}'} \right| \leq 2 \tag{16}$$

The last step we know because $a_i \leq m_i$, and $\frac{n^2}{n(n-1)} \leq 2$.

Case 2: $GI(X', X^s, D) \geq GI(X, X^s, D)$

$$\sup_{D} \left| \Sigma_{i=1}^{D+1} a_i \left(1 - \frac{a_i}{m_i}\right) - \Sigma_{i=1}^{D+1} a_i' \left(1 - \frac{a_i'}{m_i'}\right) \right| \leq$$

$$\sup_{D} \left| \Sigma_{i=1}^{D+1} a_i \left(1 - \frac{a_i}{\tilde{m}_i}\right) - \Sigma_{i=1}^{D+1} a_i \left(1 - \frac{a_i}{m_i}\right) \right| =$$

$$\left| a_j \left(1 - \frac{a_j}{\tilde{m}_j}\right) - a_j \left(1 - \frac{a_j}{m_j}\right) + a_k \left(1 - \frac{a_k}{\tilde{m}_k}\right) - a_k \left(1 - \frac{a_k}{m_k}\right) \right| =$$

$$\left| \frac{a_j'^2 (\tilde{m}_j - m_j)}{\tilde{m}_j m_j} + \frac{a_k'^2 (\tilde{m}_k - m_k)}{\tilde{m}_k m_k} \right| = \left| \frac{a_j'^2}{\tilde{m}_j m_j} - \frac{a_k'^2}{\tilde{m}_k m_k} \right| \leq 2 \tag{17}$$

Finally, this gives us $\Delta GI \leq 2 \implies \frac{\Delta GI}{2n} = \Delta u \leq \frac{1}{n}$.

11   Appendix: Full Simulation Results

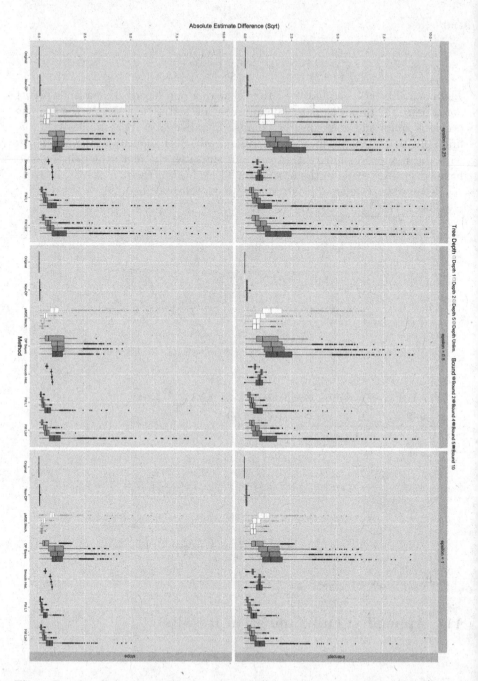

Fig. 4. Boxplots showing simulation results. The rows indicate the different coefficients, and the columns indicate different values of ϵ. Boxplots are also subdivided within methods by the tree depth (for the *pMSE* mechanism method) and the bound (for others).

References

Abowd, J.M., Vilhuber, L.: How protective are synthetic data? In: Domingo-Ferrer, J., Saygın, Y. (eds.) PSD 2008. LNCS, vol. 5262, pp. 239–246. Springer, Heidelberg (2008). https://doi.org/10.1007/978-3-540-87471-3_20

Awan, J., Slavkovic, A.: Structure and sensitivity in differential privacy: comparing k-norm mechanisms. arXiv preprint arXiv:1801.09236 (2018)

Barak, B., Chaudhuri, K., Dwork, C., Kale, S., McSherry, F., Talwar, K.: Privacy, accuracy, and consistency too: a holistic solution to contingency table release. In: Proceedings of the Twenty-Sixth ACM SIGMOD-SIGACT-SIGART Symposium on Principles of Database Systems, pp. 273–282. ACM (2007)

Bertsimas, D., Dunn, J.: Optimal classification trees. Mach. Learn. **106**(7), 1039–1082 (2017)

Blum, A., Dwork, C., McSherry, F., Nissim, K.: Practical privacy: the SuLQ framework. In: Proceedings of the Twenty-Fourth ACM SIGMOD-SIGACT-SIGART Symposium on Principles of Database Systems, pp. 128–138. ACM (2005)

Bowen, C.M., Liu, F.: Comparative study of differentially private data synthesis methods. arXiv preprint arXiv:1602.01063 (2016)

Breiman, L., Friedman, J.H., Olshen, R.A., Stone, C.J.: Classification and Regression Trees. Wadsworth, Belmont (1984)

Bun, M., Steinke, T.: Concentrated differential privacy: simplifications, extensions, and lower bounds. In: Hirt, M., Smith, A. (eds.) TCC 2016 Part I. LNCS, vol. 9985, pp. 635–658. Springer, Heidelberg (2016). https://doi.org/10.1007/978-3-662-53641-4_24

Charest, A.-S.: How can we analyze differentially-private synthetic datasets? J. Priv. Confid. **2**(2), 3 (2011)

Chaudhuri, K., Sarwate, A., Sinha, K.: Near-optimal differentially private principal components. In: Advances in Neural Information Processing Systems, pp. 989–997 (2012)

Drechsler, J.: Synthetic Datasets for Statistical Disclosure Control: Theory and Implementation, vol. 201. Springer, New York (2011). https://doi.org/10.1007/978-1-4614-0326-5

Dwork, C., Kenthapadi, K., McSherry, F., Mironov, I., Naor, M.: Our data, ourselves: privacy via distributed noise generation. In: Vaudenay, S. (ed.) EUROCRYPT 2006. LNCS, vol. 4004, pp. 486–503. Springer, Heidelberg (2006). https://doi.org/10.1007/11761679_29

Dwork, C., McSherry, F., Nissim, K., Smith, A.: Calibrating noise to sensitivity in private data analysis. In: Halevi, S., Rabin, T. (eds.) TCC 2006. LNCS, vol. 3876, pp. 265–284. Springer, Heidelberg (2006). https://doi.org/10.1007/11681878_14

Dwork, C., Naor, M., Pitassi, T., Rothblum, G.N., Yekhanin, S.: Pan-private streaming algorithms. In: ICS, pp. 66–80 (2010)

Dwork, C., Roth, A.: The algorithmic foundations of differential privacy. Found. Trends®Theor. Comput. Sci. **9**(3–4), 211–407 (2014)

Dwork, C., Rothblum, G.N.: Concentrated differential privacy. arXiv preprint arXiv:1603.01887 (2016)

Friedman, A., Schuster, A.: Data mining with differential privacy. In Proceedings of the 16th ACM SIGKDD International Conference on Knowledge Discovery and Data Mining, pp. 493–502. ACM (2010)

Kapralov, M., Talwar, K.: On differentially private low rank approximation. In: Proceedings of the Twenty-Fourth Annual ACM-SIAM Symposium on Discrete Algorithms, pp. 1395–1414. SIAM (2013).

Karwa, V., Krivitsky, P.N., Slavković, A.B.: Sharing social network data: differentially private estimation of exponential family random-graph models. J. R. Stat. Soc.: Ser. C (Appl. Stat.) **66**(3), 481–500 (2017)

Karwa, V., Slavković, A.: Inference using noisy degrees: differentially private β-model and synthetic graphs. Ann. Stat. **44**(1), 87–112 (2016)

Kasiviswanathan, S.P., Lee, H.K., Nissim, K., Raskhodnikova, S., Smith, A.: What can we learn privately? SIAM J. Comput. **40**(3), 793–826 (2011)

Kifer, D., Smith, A., Thakurta, A.: Private convex empirical risk minimization and high dimensional regression. In: Conference on Learning Theory, p. 25-1 (2012)

Li, B., Karwa, V., Slavković, A., Steorts, B.: Release of differentially private high dimensional histograms (2018, Pre-print)

Li, C., Hay, M., Rastogi, V., Miklau, G., McGregor, A.: Optimizing linear counting queries under differential privacy. In: Proceedings of the Twenty-Ninth ACM SIGMOD-SIGACT-SIGART Symposium on Principles of Database Systems, pp. 123–134. ACM (2010)

Machanavajjhala, A., Kifer, D., Abowd, J., Gehrke, J., Vilhuber, L.: Privacy: theory meets practice on the map. In: IEEE 24th International Conference on Data Engineering, ICDE 2008, pp. 277–286. IEEE (2008)

McSherry, F., Talwar, K.: Mechanism design via differential privacy. In: 48th Annual IEEE Symposium on Foundations of Computer Science, FOCS 2007, pp. 94–103. IEEE (2007)

McSherry, F.D.: Privacy integrated queries: an extensible platform for privacy-preserving data analysis. In: Proceedings of the 2009 ACM SIGMOD International Conference on Management of data, pp. 19–30. ACM (2009)

Nissim, K., Raskhodnikova, S., Smith, A.: Smooth sensitivity and sampling in private data analysis. In: Proceedings of the Thirty-Ninth Annual ACM Symposium on Theory of Computing, pp. 75–84. ACM (2007)

Nissim, K., et al.: Differential privacy: a primer for a non-technical audience (2017). https://www.ftc.gov/system/files/documents/public_comments/2017/11/00023-141742.pdf

Raab, G.M., Nowok, B., Dibben, C.: Practical data synthesis for large samples. J. Priv. Confid. **7**(3), 4 (2017)

Raghunathan, T.E., Reiter, J.P., Rubin, D.B.: Multiple imputation for statistical disclosure limitation. J. Off. Stat. **19**(1), 1–17 (2003)

Reiter, J.P.: Satisfying disclosure restrictions with synthetic data sets. J. Off. Stat. **18**(4), 531–544 (2002)

Reiter, J.P.: Using cart to generate partially synthetic, public use microdata. J. Off. Stat. **21**(3), 441–462 (2005)

Snoke, J., Raab, G.M., Nowok, B., Dibben, C., Slavkovic, A.: General and specific utility measures for synthetic data. J. R. Stat. Soc.: Ser. A (Stat. Soc.) **181**(3), 663–688 (2018)

Wang, Y.-X., Lei, J., Fienberg, S.E.: On-average KL-privacy and its equivalence to generalization for max-entropy mechanisms. In: Domingo-Ferrer, J., Pejić-Bach, M. (eds.) PSD 2016. LNCS, vol. 9867, pp. 121–134. Springer, Cham (2016). https://doi.org/10.1007/978-3-319-45381-1_10

Wasserman, L., Zhou, S.: A statistical framework for differential privacy. J. Am. Stat. Assoc. **105**(489), 375–389 (2010)

Woo, M.-J., Reiter, J.P., Oganian, A., Karr, A.F.: Global measures of data utility for microdata masked for disclosure limitation. J. Priv. Confid. **1**, 111–124 (2009)

Xu, J., Zhang, Z., Xiao, X., Yang, Y., Yu, G., Winslett, M.: Differentially private histogram publication. VLDB J. **22**(6), 797–822 (2013)

Zhang, J., Cormode, G., Procopiuc, C.M., Srivastava, D., Xiao, X.: PrivBayes: private data release via bayesian networks. ACM Trans. Database Syst. (TODS) **42**(4), 25 (2017)

Zhang, J., Zhang, Z., Xiao, X., Yang, Y., Winslett, M.: Functional mechanism: regression analysis under differential privacy. Proc. VLDB Endow. **5**(11), 1364–1375 (2012)

The Application of Genetic Algorithms to Data Synthesis: A Comparison of Three Crossover Methods

Yingrui Chen, Mark Elliot[✉], and Duncan Smith

University of Manchester, Manchester M13 9PL, UK
Yingrui.chen@postgrad.manchester.ac.uk,
{Mark.Elliot,Duncan.G.Smith}@manchester.ac.uk

Abstract. Data synthesis is a data confidentiality method which is applied to microdata to prevent leakage of sensitive information about respondents. Instead of publishing real data, data synthesis produces an artificial dataset that does not contain the real records of respondents. This, in particular, offers significant protection against reidentification attacks. However, effective data synthesis requires retention of the key statistical properties of (and respecting the multiple utilities of) the original data. In previous work, we demonstrated the value of matrix genetic algorithms in data synthesis [4]. The current paper compares three crossover methods within a matrix GA: parallelised (two-point) crossover, matrix crossover, and parametric uniform crossover. The crossover methods are applied to three different datasets and are compared on the basis of how well they reproduce the relationships between variables in the original datasets.

Keywords: Genetic algorithms · Data synthesis · Data privacy

1 Introduction

Published data are provided in many formats, although the underlying data are often microdata collected from some population [5]. Confidentiality protection techniques for microdata attempt to camouflage sensitive information in the original data while retaining its statistical properties for analysts. Data synthesis is a protection technique that produces a synthetic dataset that is designed to preserve the same statistical properties as the original data and provide sufficient variables to allow proper multivariate analyses [1].

The quality of synthetic data is strongly dependent on the design of the synthetic data generator [7]. Properties that are not explicitly included in the generator will not be present in the synthetic dataset (unless they are structurally or statistically related to properties that are, and therefore emerge from the synthesis process). Unforeseen analysis on fully synthetic data may therefore lead to different results from the same analysis on the original data [8].

© Springer Nature Switzerland AG 2018
J. Domingo-Ferrer and F. Montes (Eds.): PSD 2018, LNCS 11126, pp. 160–171, 2018.
https://doi.org/10.1007/978-3-319-99771-1_11

In this paper we use Genetic Algorithms (GAs) to generate synthetic data. GAs are iterative optimisation algorithms that simulate the process of natural evolution. They comprise of three main operators: selection, crossover and mutation. A group of candidate solutions are specified (the initial population). The fitnesses of these candidates are calculated and a selection operator selects a subset of the fitter candidates which are used to generate a new population. In crossover some pairs of these selected candidates are combined (using a variety of methods) to produce new candidate solutions. Some candidates are then subjected to mutation – random changes that will produce changes in fitness. After crossover and mutation we have the new population/generation. The process is repeated a number of times in order to (hopefully) generate fitter solutions than those in the initial population. Crossover and mutation rates can be varied from one iteration to the next, and tuning of these parameters can greatly influence performance.

GAs have been proposed as a potential method to protect respondents' from disclosures from published data. For example, Navarro-Arribas and Torra [9] mentioned that data protection could be treated as an optimisation problem with conflicting objectives and cite GAs as one approach to delivering this. Reasons for using GAs to produce synthetic data are: (i) they are designed to solve problems that have no observable solution space. The *a priori* knowledge required for setting up the initial population is minimal. (ii) GAs are interruptible so do not require complete *a priori* knowledge to set up objectives and, most crucially, (iii) GAs work well at optimising across competing constraints and therefore could, if well designed, have advantages over orthodox statistical model based synthesizers in: ameliorating overfitting, generating emergent properties and accommodating unforeseen analyses.

Matrix GAs are believed to capable of representing and solving more complex problem structures than the more orthodox bitstring GAs [12–14]. Although GAs have been used in various optimisation problems, the exploration of applications for matrix GAs has been limited. However, given that microdata are essentially matrices the production of synthetic microdata seems an obvious application. In previous work we have evaluated the potential for matrix GAs with promising initial results [3,4]. The current paper explores the performance of three different crossover methods for matrix GAs in producing synthetic data. We consider three datasets, with different data structures, and sampled from different survey populations.

Note that in this initial phase of this research, we are concerned only with optimising the utility of the synthesised data and not with the residual disclosure risk. The rationale for this is twofold: (i) optimising the utility of a synthetic dataset represents a difficult problem by itself and adding in the contrary constraint of disclosure control will introduce further complexity, and (ii) of the two elements the utility problem is the more significant for synthetic data; if this cannot be solved the efficiency of the risk optimisation will be irrelevant. Understanding the properties of the utility optimisation problem before introducing the complexity disclosure control as an objective is therefore the appropriate research strategy.

1.1 Microdata and Contingency Tables

A microdata set for n cases and m variables is usually represented as an n by m matrix indexed $i \in \{1, \ldots, n\}$ and $j \in \{1, \ldots, m\}$. Here we use Y to denote and original dataset and its synthetic version is denoted as X. X shares the same structure as Y as illustrated in Fig. 1.

$$
\begin{pmatrix}
y_{11} & y_{12} & \cdots & y_{1m} \\
y_{21} & y_{22} & \cdots & y_{2m} \\
y_{31} & y_{32} & \cdots & y_{3m} \\
y_{41} & y_{42} & \cdots & y_{4m} \\
\vdots & \vdots & \vdots & \vdots \\
y_{n1} & y_{n2} & \cdots & y_{nm}
\end{pmatrix}
\begin{pmatrix}
x_{11} & x_{12} & \cdots & x_{1m} \\
x_{21} & x_{22} & \cdots & x_{2m} \\
x_{31} & x_{32} & \cdots & x_{3m} \\
x_{41} & x_{42} & \cdots & x_{4m} \\
\vdots & \vdots & \vdots & \vdots \\
x_{n1} & x_{n2} & \cdots & x_{nm}
\end{pmatrix}
$$

Fig. 1. Microdata Y and its synthetic version X

For categorical variables the same information can be encoded in a contingency table, which captures the between-variate structure of the candidate. Assume our variables take values in finite sets I_j so that $I = \underset{j \in [1..m]}{\times} I_j$ denotes the possible configurations of the variables. Then a contingency table is an m-dimensional table containing a count for each member of I. For example, if we denote the jth column of a microdata set Y as $Y_{:,j}$, then the 2-dimensional contingency table constructed from distinct columns $Y_{:,j}$ and $Y_{:,k}$ is $CT(Y_{:,j}, Y_{:,k})$ with entries $n_{r,c}$ is,

$$
n_{r,c} = \sum_{i=1}^{n} [Y_{i,j} = (I_j)_r \wedge Y_{i,k} = (I_k)_c] \tag{1}
$$

where the square brackets are Iverson brackets and the levels of I_j and I_k are indexed $r \in [1..|I_j|]$ and $c \in [1..|I_k|]$ respectively.

1.2 Objectives

Respecting variable associations in the original data is an important aspect of producing high quality synthetic data. Thus, objective functions are designed based on the differences between synthetic (contingency) tables and original tables in low dimensions. A measure of the difference between a pair of contingency tables is the Jensen-Shannon distance D_{JS} between their normalised (to sum to 1) counterparts[1]. Suppose P and Q are two discrete probability distributions, then $D_{JS}(P||Q)$ is given by:

$$D_{JS}(P||Q) = (\frac{1}{2}D_{KL}(P||M) + \frac{1}{2}D_{KL}(Q||M))^{\frac{1}{2}} \tag{2}$$

where $M = \frac{1}{2}(P + Q)$ and D_{KL} is the well-known Kullback-Leibler divergence.

So our distance measure for a pair of 2-dimensional contingency tables is defined as:

$$\Delta(X, Y, \{j, k\}) = D_{JS}(\frac{1}{n}CT(X_{:,j}, X_{:,k})||\frac{1}{n}CT(Y_{:,j}, Y_{:,k})) \tag{3}$$

Our first objective function is the mean of these distances over all pairs of variables:

$$F_1(X, Y) = \binom{m}{2}^{-1} \sum_{j=1}^{m-1} \sum_{k=j+1}^{m} \Delta(X, Y, \{j, k\}) \tag{4}$$

Analogous measures are also considered for all 3-dimensional and all 4-dimensional contingency tables. So our other two objectives are defined as:[2]

$$F_2(X, Y) = \binom{m}{3}^{-1} \sum_{S \in P_3([1..m])} \Delta(X, Y, S) \tag{5}$$

$$F_3(X, Y) = \binom{m}{4}^{-1} \sum_{S \in P_4([1..m])} \Delta(X, Y, S) \tag{6}$$

where $P_k(Z)$ denotes the members of the powerset of Z of size K.

The fitness of each candidate is calculated by the Euclidean distance from the synthetic to the original data in the space delineated by the three objective functions. The fitness value is normalized to the range $[0, 1]$ by dividing by $\sqrt{3}$.[3] So our overall objective function is:

$$F = \sqrt{3}^{-1} \sqrt{(F_1(X, Y)^2 + F_2(X, Y)^2 + F_3(X, Y)^2} \tag{7}$$

---

[1] Regarding the choice of divergence measure. The Kullback-Leibler divergence cannot be used directly because of the requirement for absolute continuity. Aside from that constraint there was no prior compelling reason for picking any specific measure, and there is no specific empirical work to guide us. The Jensen-Shannon distance was chosen mainly on the basis that it is a true metric, unlike e.g. the Jensen-Shannon divergence. The impact of using alternative measures is another area which future research could explore.

[2] Clearly this is not a complete set of possible objectives but these are probably necessary to produce reasonable synthetic categorical data and provide sufficient complexity for our crossover experiments.

[3] So on this scale 0 is the best fitness possible and 1 is the worst.

2 Crossover Methods

A crossover operator produces variation in a GA population. The operators considered here will change a pair of individuals by swapping randomly selected sub-matrices. In the case of uniform crossover these sub-matrices will necessarily have dimension 1×1 and we will essentially be swapping individual elements of the matrices.

The three crossover methods presented here have been used previously in various application areas, but not usually compared and certainly not in the context of synthetic data generation. Two of them use the mechanism of two-point crossover where not all sub-matrices (or elements) have an equal probability of being swapped (positional bias). The third operator, uniform crossover, does not suffer from positional bias and is included in order to examine the impact of positional bias on the effectiveness of matrix GA data synthesizers.

Parallelised Crossover. The design of parallelised crossover is based on a two-point crossover method from linear GAs, which swaps the elements between two randomly selected crossover points a and b between a pair of bitstrings. Since solutions that GAs generate are operationally independent (in that a change in one individual has no direct effect on another), crossover and mutation can be parallelised [2]. Parallelised crossover occurs on a single variable and therefore it is possible to have m sub-processors working separately on different variables in the generator. The generator works by randomly choosing a sub-matrix from within a single data column and swapping with the corresponding sub-matrix in the paired candidate. Figure 2 illustrates parallelised crossover between a pair of candidates X^1 and X^2:

$$\begin{pmatrix} x^1_{11} & x^1_{12} & \dots & x^1_{1m} \\ x^1_{21} & x^1_{22} & \dots & x^1_{2m} \\ x^1_{31} & x^1_{32} & \dots & x^1_{3m} \\ x^1_{41} & x^1_{42} & \dots & x^1_{4m} \\ \vdots & \vdots & \vdots & \vdots \\ x^1_{n1} & x^1_{n2} & \dots & x^1_{nm} \end{pmatrix} \begin{pmatrix} x^2_{11} & x^2_{12} & \dots & x^2_{1m} \\ x^2_{21} & x^2_{22} & \dots & x^2_{2m} \\ x^2_{31} & x^2_{32} & \dots & x^2_{3m} \\ x^2_{41} & x^2_{42} & \dots & x^2_{4m} \\ \vdots & \vdots & \vdots & \vdots \\ x^2_{n1} & x^2_{n2} & \dots & x^2_{nm} \end{pmatrix}$$

$$\downarrow$$

$$\begin{pmatrix} x^1_{11} & x^1_{12} & \dots & x^1_{1m} \\ x^2_{21} & x^1_{22} & \dots & x^1_{2m} \\ x^1_{31} & x^2_{32} & \dots & x^2_{3m} \\ x^1_{41} & x^2_{42} & \dots & x^2_{4m} \\ \vdots & \vdots & \vdots & \vdots \\ x^1_{n1} & x^1_{n2} & \dots & x^2_{nm} \end{pmatrix} \begin{pmatrix} x^2_{11} & x^2_{12} & \dots & x^2_{1m} \\ x^1_{21} & x^2_{22} & \dots & x^2_{2m} \\ x^2_{31} & x^1_{32} & \dots & x^1_{3m} \\ x^2_{41} & x^1_{42} & \dots & x^1_{4m} \\ \vdots & \vdots & \vdots & \vdots \\ x^2_{n1} & x^2_{n2} & \dots & x^1_{nm} \end{pmatrix}$$

Fig. 2. X^1 and X^2 in parallelized crossover

$$\begin{pmatrix} \boxed{\begin{matrix} x_{11}^1 & x_{12}^1 & \dots & x_{1m}^1 \\ x_{21}^1 & x_{22}^1 & \dots & x_{2m}^1 \\ x_{31}^1 & x_{32}^1 & \dots & x_{3m}^1 \\ x_{41}^1 & x_{42}^1 \end{matrix}} \dots x_{4m}^1 \\ \vdots \quad \vdots \quad \vdots \quad \vdots \\ x_{n1}^1 & x_{n2}^1 & \dots & x_{nm}^1 \end{pmatrix} \begin{pmatrix} \boxed{\begin{matrix} x_{11}^2 & x_{12}^2 & \dots & x_{1m}^2 \\ x_{21}^2 & x_{22}^2 & \dots & x_{2m}^2 \\ x_{31}^2 & x_{32}^2 & \dots & x_{3m}^2 \\ x_{41}^2 & x_{42}^2 \end{matrix}} \dots x_{4m}^2 \\ \vdots \quad \vdots \quad \vdots \quad \vdots \\ x_{n1}^2 & x_{n2}^2 & \dots & x_{nm}^2 \end{pmatrix}$$

$$\downarrow$$

$$\begin{pmatrix} x_{11}^2 & x_{12}^2 & \dots & x_{1m}^1 \\ x_{21}^2 & x_{22}^2 & \dots & x_{2m}^1 \\ x_{31}^2 & x_{32}^2 & \dots & x_{3m}^1 \\ x_{41}^2 & x_{42}^2 & \dots & x_{4m}^1 \\ \vdots & \vdots & \vdots & \vdots \\ x_{n1}^1 & x_{n2}^1 & \dots & x_{nk}^1 \end{pmatrix} \begin{pmatrix} x_{11}^1 & x_{12}^1 & \dots & x_{1m}^2 \\ x_{21}^1 & x_{22}^1 & \dots & x_{2m}^2 \\ x_{31}^1 & x_{32}^1 & \dots & x_{3m}^2 \\ x_{41}^1 & x_{42}^1 & \dots & x_{4m}^2 \\ \vdots & \vdots & \vdots & \vdots \\ x_{n1}^2 & x_{n2}^2 & \dots & x_{nm}^2 \end{pmatrix}$$

Fig. 3. X^1 and X^2 in matrix crossover

Matrix Crossover. Matrix crossover was first proposed by Wallet et al. [14]. Unlike parallelised crossover, matrix crossover generates crossover points for the rows as well as columns. Thus it swaps elements from a randomly generated sub-matrix (as opposed to the column vectors swapped in parallelised crossover). Figure 3 illustrates matrix crossover.

Parametric Uniform Crossover (PUC). In PUC, the probability of crossover being applied to each element (1×1 sub-matrix) of the given candidate is determined by a user-specified parameter P_0. Figure 4 illustrates PUC.

2.1 Positional Bias

Both parallelised crossover and matrix crossover are based on the idea of two-point crossover. In parallelised crossover, the element with row index i will be swapped if, and only if, one of the selection points has index not greater than i while the other has index greater than i. Thus the swap probability is the hypergeometric probability:

$$P(\min(a,b) \le i < \max(a,b)) = i(n-i+1)\binom{n+1}{2}^{-1} \tag{8}$$

where a and b are the indices of a pair of (distinct) randomly chosen crossover points.

It is trivial to show that this is a montone increasing function of i where $i < \frac{n}{2}$ and a monotone decreasing function of i where $i > \frac{n}{2}$.

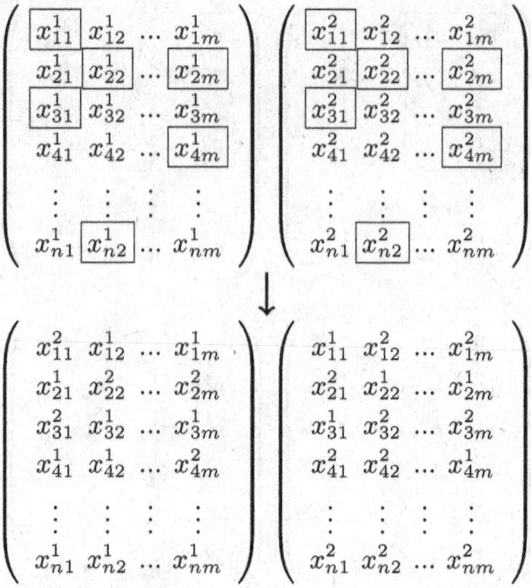

$$\begin{pmatrix} x^1_{11} & x^1_{12} & \cdots & x^1_{1m} \\ x^1_{21} & x^1_{22} & \cdots & x^1_{2m} \\ x^1_{31} & x^1_{32} & \cdots & x^1_{3m} \\ x^1_{41} & x^1_{42} & \cdots & x^1_{4m} \\ \vdots & \vdots & \vdots & \vdots \\ x^1_{n1} & x^1_{n2} & \cdots & x^1_{nm} \end{pmatrix} \begin{pmatrix} x^2_{11} & x^2_{12} & \cdots & x^2_{1m} \\ x^2_{21} & x^2_{22} & \cdots & x^2_{2m} \\ x^2_{31} & x^2_{32} & \cdots & x^2_{3m} \\ x^2_{41} & x^2_{42} & \cdots & x^2_{4m} \\ \vdots & \vdots & \vdots & \vdots \\ x^2_{n1} & x^2_{n2} & \cdots & x^2_{nm} \end{pmatrix}$$

$$\downarrow$$

$$\begin{pmatrix} x^2_{11} & x^1_{12} & \cdots & x^1_{1m} \\ x^1_{21} & x^2_{22} & \cdots & x^2_{2m} \\ x^2_{31} & x^1_{32} & \cdots & x^1_{3m} \\ x^1_{41} & x^1_{42} & \cdots & x^2_{4m} \\ \vdots & \vdots & \vdots & \vdots \\ x^1_{n1} & x^1_{n2} & \cdots & x^1_{nm} \end{pmatrix} \begin{pmatrix} x^1_{11} & x^2_{12} & \cdots & x^2_{1m} \\ x^2_{21} & x^1_{22} & \cdots & x^1_{2m} \\ x^1_{31} & x^2_{32} & \cdots & x^2_{3m} \\ x^2_{41} & x^2_{42} & \cdots & x^1_{4m} \\ \vdots & \vdots & \vdots & \vdots \\ x^2_{n1} & x^2_{n2} & \cdots & x^2_{nm} \end{pmatrix}$$

Fig. 4. X^1 and X^2 in PUC

For matrix crossover we also select a pair of crossover points for the columns and the probability of an element with index (i, j) being swapped is a product of hypergeometric probabilities. PUC, on the other hand contains no positional bias and it is therefore useful to provide us with an implicit evaluation of the effect of positional bias on the optimising ability of the matrix GA data synthesizer.

3   Empirical Study

3.1   Design

The three crossover methods were compared using three datasets that were each sampled from a different social survey. All three datasets contain 10 variables and 1000 cases. Dataset 1 was sampled from the Crime Survey for England and Wales, 2015–2016 [10] and has 10 binary variables. Dataset 2 was sampled from European Union Statistics on Income and Living Conditions, 2009 [11]. It has 6 binary variables, 1 three-category variable and 3 four-category variables. Dataset 3 was sampled from the Citizenship Survey, 2010–2011 [6]. It contains 1000 cases and 10 variables including 4 binary variables, 2 four-category variables, 2 six-category variables, 1 nine-category variable and 1 eleven-category variable.

For each dataset there was a fixed initial population of 100 candidates that was generated by independently sampling (with replacement) from the univariate distributions of the original data (Table 1). Deterministic tournament selection[4] was used to select candidates with tournament size $t = 2$.

---

[4] In generalised tournament selection, candidates are randomly selected into tournaments of size t (with or without replacement). The probability that a candidate wins the tournament and enters crossover is given by $p(1 - p)^r$ where p is a parameter (such that $1/t < p \leq 1$) and r is the rank of the candidate's fitness within the tournament. In deterministic tournament selection p is set to 1.

Table 1. Fitness values of the initial population of each of the three test dataset

Fitness of fixed population (size = 100) for each data			
	Best fitness	Mean	s.d
Data 1	0.0734	0.0800	0.0034
Data 2	0.2176	0.2259	0.0031
Data 3	0.2544	0.2610	0.0027

Synthetic data were generated using GAs with two distinct crossover rates. Matrix crossover used rates of 1.0 and 0.7. The corresponding crossover rates for parallelised crossover and PUC were chosen so that the probability of swapping individual elements was similar.

The synthetic data generator used a low mutation rate ($p_m = 0.01$) to reduce the noise in the final results.[5] Candidates chosen for mutation had a randomly selected sub-matrix swapped with data independently sampled from the original univariate distributions.

For each set of parameters we generated 10 synthetic populations. Each such trial was run for 100 generations.

3.2 Experimental Results

Table 2 shows the means and standard deviations of the fittest solutions in the final (100th) populations. The rightmost column shows the fitness of the best individual that was generated over the 10 trials.

The generator used the objective function in Eq. 7. The number of individual contingency tables compared depends on the number of variables and would increase substantially if we extended the measure to, say, 5-dimensional tables. Table 2 shows that the fitness of candidates for Dataset 1 is always closer to the original data compared with the other two no matter which crossover operator is used, followed by Dataset 2 and Dataset 3. This is monotonic with the complexity of the data structures of the three datasets. This issue will need further exploration to establish how the degree of complexity affects the viability of GA generated synthesis.

The experimental results also indicate that positional bias does impact the effectiveness of matrix GA generator. All the best means and individuals after 100 generations for the three datasets are generated by the synthesizer with the PUC operator that has $p_0 = 0.3818$. The second best mean and individuals are generated by the same synthesizer with $p_0 = 0.1911$.

---

[5] Mutation is another important operator in GA that helps find more promising candidates from the solution space and reduces the risk of becoming caught in local optima. However it can also reduce the fitness of a candidate. Here our focus is on comparing crossover operators so we selected a low mutation rate to reduce the noise in the final results. In future work will examine the relationship between the two operators.

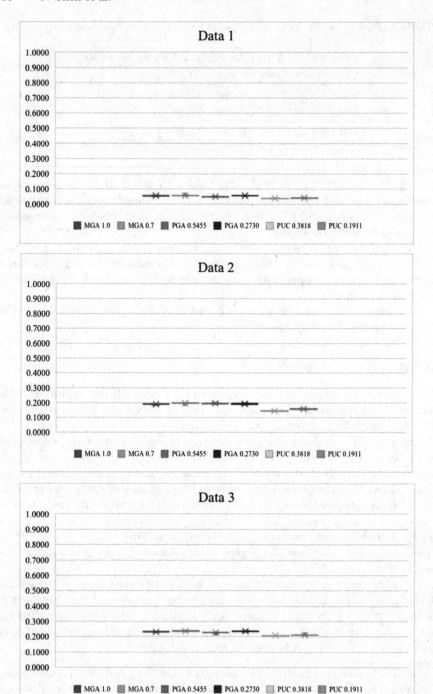

Fig. 5. Box plots of final fitness values of the best individual for Dataset 1, 2 and 3 from ten trials of matrix GA synthetic data generator using different crossover operators

Table 2. Summary statistics from ten trials of matrix GA data synthesizers equipped with three different operators: matrix crossover (MGA), parallelized crossover (PGA) and PUC.

Crossover type	Crossover rate	Data	Best fitness value		Best individual
			Mean	s.d	
MGA	1	Data 1	0.0564	0.0026	0.0517
		Data 2	0.1887	0.0030	0.1818
		Data 3	0.2319	0.0020	0.2281
	0.7	Data 1	0.0579	0.0022	0.0541
		Data 2	0.1958	0.0034	0.1889
		Data 3	0.2375	0.0032	0.2340
PGA	0.5455	Data 1	0.0497	0.0022	0.0472
		Data 2	0.1936	0.0038	0.1885
		Data 3	0.2259	0.0019	0.2213
	0.273	Data 1	0.0561	0.0015	0.0533
		Data 2	0.1929	0.0041	0.1867
		Data 3	0.2363	0.0025	0.2315
PUC	0.3818	Data 1	0.0393	0.0021	0.0362
		Data 2	0.1450	0.0025	0.1408
		Data 3	0.2060	0.0009	0.2048
	0.1911	Data 1	0.0429	0.0022	0.0397
		Data 2	0.1576	0.0038	0.1521
		Data 3	0.2112	0.0020	0.2092

Moreover, there was a significant improvement on the initial population no matter which crossover method was used. The increase of fitness (decrease in distance from the original data) indicates that matrix GA is efficient in generating synthetic data with real-coded data or even more complex data structures. Table 3 shows the mean improvement of the fitness of the population from the beginning to the 100th generation over all trials.

Table 3. Mean fitness improvement over ten trials on the best fitness value of initial population

	On mean of the best fitness values over all trials
Data 1	0.0296
Data 2	0.0470
Data 3	0.0362

4   Conclusions

Our experimental results indicate that PUC performs better than matrix and parallelised crossover in producing synthetic data for all three datasets. This is likely to be due to the lack of positional bias. Results also indicate that the performance of matrix GA on synthetic data generation strongly depends on the structure of data and the number of cases. For example, the optimisation of Dataset 1 is the most effective because it has the simplest data structure (containing only binary variables) compared to Dataset 2 and Dataset 3 (Fig. 5).

Beyond the issue of positional bias, the overall performance for all three crossover operators in producing synthetic data is reasonable. All approaches significantly improved the fitness of the 100 candidates from the initial population over 100 generations.

Our future research will focus on testing the effectiveness and practicality of the matrix GA generator by introducing adaptive crossover rates, more objectives and larger datasets. A key element missing from these initial experiments has been the assessment of disclosure risk. As outlined in the introduction, this was a rational approach to isolate the difficult problem of optimising utility. However, a full GA data synthesiser should incorporate risk. Therefore, in future work we will bring measures of disclosure risk into the GA framework. In many ways this is when the GA approach will come into its own. The risk utility trade-off is usually dealt with as a two step-process and optimising both within a single framework is likely to be more efficient.

Overall, these initial experiments using matrix GA generators to generate synthetic data show that matrix GA is of interest for the problem of data synthesis and for solving problems with higher-dimensional and complex structures in general.

References

1. Abowd, J.M., Lane, J.: New approaches to confidentiality protection: synthetic data, remote access and research data centers. In: Domingo-Ferrer, J., Torra, V. (eds.) PSD 2004. LNCS, vol. 3050, pp. 282–289. Springer, Heidelberg (2004). https://doi.org/10.1007/978-3-540-25955-8_22
2. Cantu-Paz, E., Goldberg, D.: Efficient parallel genetic algorithms: theory and practice. Comput. Methods Appl. Mech. Eng. **186**, 221–238 (2000)
3. Chen, Y., Elliot, M., Sakshaug, J.: A genetic algorithm approach to synthetic data production. In: Proceedings of the 1st International Workshop on AI for Privacy and Security (2016). Article no. 13
4. Chen, Y., Elliot, M., Sakshaug, J.: Genetic algorithms in matrix representation and its application in synthetic data, UNECE work session on statistical data confidentiality (2017). https://www.unece.org/fileadmin/DAM/stats/documents/ece/ces/ge.46/2017/2_Genetic_algorithms.pdf. Accessed 20 Dec 2017
5. Ciriani, V., di Vimercati, S.D.C., Foresti, S., Samarati, P.: Microdata protection. In: Yu, T., Jajodia, S. (eds.) Secure Data Management in Decentralized Systems, vol. 33, pp. 291–321. Springer, New York (2007). https://doi.org/10.1007/978-0-387-27696-0_9

6. Department for Communities and Local Government, Ipsos MORI: Citizenship Survey, 2010–2011. [data collection]. UK Data Service. SN: 7111 (2012). https://doi.org/10.5255/UKDA-SN-7111-1. Accessed 20 Dec 2017
7. Drechsler, J.: Synthetic data, where do we come from? Where do we want to go? In: Synthetic Data Workshop; Office of National (2014)
8. Maimon, O., Rokach, L.: Data Mining and Knowledge Discovery Handbook, p. 704. Springer, Heidelberg (2010)
9. Navarro-Arribas, G., Torra, V.: Advanced research on data privacy in the ARES project. In: Navarro-Arribas, G., Torra, V. (eds.) Advanced Research in Data Privacy. SCI, vol. 567, pp. 3–14. Springer, Cham (2015). https://doi.org/10.1007/978-3-319-09885-2_1
10. Office for National Statistics. Crime Survey for England and Wales, 2015–2016. [data collection]. UK Data Service. SN: 8140 (2017). https://doi.org/10.5255/UKDA-SN-8140-1. Accessed 11 Jan 2018
11. Office for National Statistics. Social Survey Division, Northern Ireland Statistics and Research Agency, Eurostat (2011). European Union Statistics on Income and Living Conditions, 2009. [data collection]. UK Data Service. SN: 6767 (2009). https://doi.org/10.5255/UKDA-SN-6767-1. Accessed 11 Jan 2018
12. Pongcharoen, P., Khadwilard, A., Klakankhai, A.: Multi-matrix real-coded genetic algorithm for minimising total costs, in logistics chain network. In: World Academy of Science, Engineering and Technology, vol. 1, no. 11, pp. 574–597 (2007). (Int. J. Econ. Manag. Eng.)
13. Sun, L., Zhang, Y., Jiang, C.: A matrix real-coded genetic algorithm to the unit commitment problem. Electr. Pow. Syst. Res. **76**, 716–728 (2006)
14. Wallet, B.C., Marchette, D.J., Solka, J.L.: A matrix representation for genetic algorithms. In: Proceedings of Automatic Object Recognition IV of SPIE Aerosense. Naval Surface Warfare Center Dahlgren, Virginia (1996)

Microdata and Big Data Masking

Multiparty Computation with Statistical Input Confidentiality via Randomized Response

Josep Domingo-Ferrer, Rafael Mulero-Vellido, and Jordi Soria-Comas[✉]

Department of Computer Science and Mathematics, UNESCO Chair in Data Privacy,
CYBERCAT-Center for Cybersecurity Research of Catalonia,
Universitat Rovira i Virgili, Av. Països Catalans 26,
43007 Tarragona, Catalonia, Spain
{josep.domingo,rafael.mulero,jordi.soria}@urv.cat

Abstract. We explore a setting in which a number of subjects want to compute on their pooled data while keeping the statistical confidentiality of their input. Statistical confidentiality is different from the cryptographic confidentiality guaranteed by cryptographic multiparty secure computation: whereas in the latter nothing is disclosed about the input, in statistical input confidentiality a noise-added version of the input is disclosed, which allows more flexible computations. We propose a protocol based on local anonymization via randomized response, whereby the empirical distribution of the data of the subjects is approximated. From that distribution, most statistical calculations can be approximated as well. Regarding the accuracy of the approximation, *ceteris paribus* it improves with the number of subjects. Large dimensionality (that is, a large number of attributes) decreases accuracy and we propose a strategy to mitigate the dimensionality problem. We show how to characterize the privacy guarantee for each subject in terms of differential privacy. Experimental work is reported on the attained accuracy as a function of the number of respondents, number of attributes and randomized response parameters.

Keywords: Multiparty anonymous computation
Randomized response · Local anonymization · Big data · Privacy

1 Introduction

There are several situations in which a number of distrusting parties wish to collaborate at evaluating functions that take as inputs private data from each party, in such a way that the *privacy* of those inputs is preserved. Two different notions of input privacy are conceivable:

- *Cryptographic input confidentiality.* The input of each party should not be disclosed to the other parties.

© Springer Nature Switzerland AG 2018
J. Domingo-Ferrer and F. Montes (Eds.): PSD 2018, LNCS 11126, pp. 175–186, 2018.
https://doi.org/10.1007/978-3-319-99771-1_12

– *Statistical input confidentiality.* A noise-added version of the input of each party is disclosed.

Multiparty computation with cryptographic input confidentiality can be easily motivated with the following example. Each of a set of companies has collected experimental data at some cost and possibly under privacy pledge to its respondents/customers/patients. Thus, no company wishes to share its data set with any other company (as data have costed money and are industrial property). At the same time, feeling that better conclusions could be drawn from their pooled data sets than from a single data set, companies would like to engage in joint computation on their pooled data. If no third party trusted by all companies is available (that can receive all data sets in confidence, perform the computations on the pooled data and return the results to all companies), this scenario is handled with secure multiparty computation [1,3,12].

The problem of multiparty computation in the above cryptographic sense is that a different protocol is needed for each type of required computation. This hampers exploratory analysis, which is more and more important in our big data world. Multiparty computation with statistical input confidentiality can be far more flexible at the cost of providing somewhat weaker input confidentiality. Virtually any statistical computation can be performed without requiring specific protocols. Furthermore, the set of collaborating parties can be much larger than in cryptographic multiparty computation: there can be as many parties as respondents in a data set, with each party holding just her own record.

Contribution and Plan of This Paper. In this paper, we propose an approach for multiparty computation with statistical input confidentiality based on randomized response [2,6,11]. Specifically, local anonymization via randomized response is used by the collaborating subjects to approximate the empirical distribution of their pooled data. From that distribution, most statistical calculations can be approximated as well. Furthermore, we show how to characterize the privacy guarantee for each subject in terms of differential privacy.

Section 2 gives background on randomized response. In Sect. 3, we describe the proposed approach for multiparty computation with statistical input confidentiality based on randomized response. In Sect. 4 we propose solutions to mitigate the curse of dimensionality, that is, the decreasing accuracy of the approximated empirical distribution as the number of attributes increases. Section 5 establishes the privacy guarantees for subjects. Experimental work is reported in Sect. 6. Finally, conclusions and future research directions are gathered in Sect. 7.

2   Background on Randomized Response

Randomized response [6,11] is a mechanism that respondents to a survey can use to protect their privacy when asked about the value of sensitive attribute (*e.g.* did you take drugs last month?). The interesting point is that the data collector can still estimate from the randomized responses the proportion of each of the possible *true* answers of the respondents. Randomized response is closely related to post-randomization (PRAM). They differ on who performs the randomization [8]: whereas in randomized response it is the respondent before delivering her response, in PRAM it is the data controller after collecting all responses (hence the name post-randomization).

Let us denote by X the attribute containing the answer to the sensitive question. If X can take r possible values, then the randomized response Y reported by the respondent instead of X follows a $r \times r$ matrix of probabilities

$$\mathbf{P} = \begin{pmatrix} p_{11} & \cdots & p_{1r} \\ \vdots & \vdots & \vdots \\ p_{r1} & \cdots & p_{rr} \end{pmatrix} \tag{1}$$

where $p_{uv} = \Pr(Y = v | X = u)$, for $u, v \in \{1, \ldots, r\}$ denotes the probability that the randomized response is v when the respondent's true attribute value is u.

Let $\pi_1, \ldots, \pi_r$ be the proportions of respondents whose true values fall in each of the r categories of X and let $\lambda_v = \sum_{u=1}^{r} p_{uv} \pi_u$ for $v = 1, \ldots, r$, be the probability of the reported value Y being v. If we define $\lambda = (\lambda_1, \ldots, \lambda_r)^T$ and $\pi = (\pi_1, \ldots, \pi_r)^T$, it holds that $\lambda = \mathbf{P}^T \pi$. Furthermore, if $\hat{\lambda}$ is the vector of sample proportions corresponding to λ and $\mathbf{P}$ is nonsingular, in Chap. 3.3 of [2] it is proven that an unbiased estimator π can be computed as

$$\hat{\pi} = (\mathbf{P}^T)^{-1} \hat{\lambda} \tag{2}$$

and they also provide an unbiased estimator of the dispersion matrix. In particular, the larger the off-diagonal probability mass in $\mathbf{P}$, the more dispersion (and the more respondent protection).

3 Randomized Response to Achieve Multiparty Computation With Statistical Input Confidentiality

Assume n subjects $i = 1, \ldots n$ each holding one record $\mathbf{x}_i = (x_{i1}, \ldots, x_{im})$ containing the values for m attributes. These subjects want to engage in secure multiparty computation with statistical input confidentiality on the data set $\mathbf{X} = \{\mathbf{x}_1, \ldots, \mathbf{x}_n\}$ that would be formed by their respective records. Statistical input confidentiality means that no subject wants to disclose her true record to the other subjects, even if she is ready to disclose a randomized version of it.

A possible way is for subjects to approximate the empirical distribution of $\mathbf{X}$ via randomized response. Once they have a reasonably good approximation of that distribution, most statistical calculations on $\mathbf{X}$ can be approximately computed based on the data set's approximate empirical distribution.

A first naive solution is for each subject i to separately deal with each attribute value x_{ij} for $j = 1, \ldots, m$ via randomized response. If the j-th attribute

A_j can take r_j different values, then an $r_j \times r_j$ probability matrix $\mathbf{P}_j$ (see Expression (1)) can be used for each subject to report a randomized value y_{ij} for A_j instead of her true value x_{ij}.

As mentioned in Sect. 2, this would allow all subjects to approximate the *marginal* empirical distribution $\pi^j = (\pi_1^j, \ldots, \pi_{r_j}^j)$ of each attribute A_j as

$$\hat{\pi}^j = ((\mathbf{P}^j)^T)^{-1} \hat{\lambda}^j$$

where $\lambda^j = (\mathbf{P}^j)^T \pi^j$.

The problem is that approximating the marginal empirical distributions of attributes does not yield an approximation of the joint empirical distribution of $\mathbf{X}$.

To approximate the joint distribution of $\mathbf{X}$ via randomized response, subjects must report their randomized response for the value of $A_1 \times A_2 \times \ldots \times A_m$ and proceed as above.

Once the empirical distribution of $\mathbf{X}$ has been approximated, the approximation can be made public and any subjects can perform statistical computations on it. At the same time, all subjects have preserved the confidentiality of their inputs.

However, this only works well if the number of subjects n is much larger than the number of possible values of the above Cartesian product, that is

$$n \gg |A_1| \times |A_2| \times \ldots \times |A_m|. \tag{3}$$

Otherwise, many elements of the Cartesian product are likely to have zero frequency in the empirical distribution of the reported values. Such a reported sparse distribution is unlikely to constitute a good approximation of the true empirical distribution of $\mathbf{X}$.

4   Mitigating the Curse of Dimensionality

If Constraint (3) does not hold, there are two alternatives:

1. Attempt to partition the set of attributes into clusters $C_1, \ldots, C_l$, for some l, such that

$$\bigcup_{i=1}^{l} C_i = \{A_1, \ldots, A_m\}$$

$C_i \cap C_j = \emptyset$ for $i \neq j$, and the attributes within each cluster are highly dependent/correlated and the attributes belonging to different clusters are weakly dependent or even independent. In this way, only the joint distribution of attributes within each cluster needs to be approximated. Therefore, Constraint (3) is relaxed to

$$n \gg \max_i \prod_{A_j \in C_i} |A_j|, \tag{4}$$

which is easier to satisfy. Furthermore, if within a cluster two or more attributes are very highly correlated, only one of them needs to be taken into account in the joint distribution approximation, with the rest being re-computed based on the approximated representative attribute. This may reduce the size of attribute clusters even more, and hence the bound on the right-hand side of Inequality (4).

2. If the above attribute clustering is not feasible (because all pairs of attributes are significantly dependent/correlated), an alternative solution is to coarsen the values of attributes $A_1, \ldots, A_m$, in such a way to reduce the right-hand side of Inequality (3). This obviously will reduce the accuracy of the approximation to the empirical distribution. Hence, it should be only used as a fallback solution.

We now describe in more detail attribute clustering:

1. Compute an approximation of all bivariate empirical distributions, by using randomized response on the Cartesian product $A_i \times A_j$, for all pairs (A_i, A_j) of attributes.
2. Construct a complete graph such that:
   (a) Nodes are attributes $A_1, \ldots, A_m$.
   (b) The edge between each pair of attributes A_i and A_j is labeled with a measure of independence between A_i and A_j.
3. Cluster attributes according to their distances in the graph. A possibility is to use the power iteration clustering (PIC) algorithm [7].

The specific measure of independence to be used must take into account the type of the attributes, as follows. If A_i and A_j are numerical and/or ordinal, we can take as a measure of independence

$$1/|r_{ij}|, \tag{5}$$

where r_{ij} is Pearson's correlation coefficient between A_i and A_j.

If one of A_i and A_j is nominal (without an order relationship between its possible values) and the other is nominal or ordinal, we can take as a measure of independence

$$1/V_{ij}, \tag{6}$$

where V_{ij} is Cramér's V statistic [4], that gives a value between 0 and 1, with 0 meaning complete independence between A_i and A_j and 1 meaning complete dependence. Cramér's V_{ij} is computed as

$$V_{ij} = \sqrt{\frac{\chi_{ij}^2/n}{\min(c_i - 1, c_j - 1)}},$$

where c_i is the number of categories of A_i, c_j is the number of categories of A_j, n is the total number of subjects/records and χ_{ij}^2 is the chi-squared independence statistic defined as

$$\chi_{ij}^2 = \sum_{a=1}^{c_i} \sum_{b=1}^{c_j} \frac{(o_{ab}^{ij} - e_{ab}^{ij})^2}{f_{ab}^j}, \tag{7}$$

with o_{ab}^{ij} the observed frequency of the combination $(A_i = a, A_j = b)$ and e_{ab}^{ij} the expected frequency of that combination under the independence assumption for A_i and A_j. This expected frequency is computed as

$$e_{ab}^{ij} = \frac{n_a^i n_b^j}{n},$$

where n_a^i and n_b^j are, respectively, the number of subjects who have reported $A_i = a$ and $A_j = b$.

Finally, if one of A_i, A_j is nominal and the other is numerical, the latter must be discretized, for example by rounding or by replacing values by intervals. After that, the contingency table between A_i and A_j can be constructed, and the measure of independence given by Expression (6) can be computed.

Since the denominators in Expressions (5) and (6) are bounded in $[0, 1]$, the outputs of both expressions are comparable when trying to cluster the nodes in the graph.

5 Privacy Guarantees

The confidentiality guarantee given by randomized response results from the fact that each individual may misrepresent her data by randomly drawing from a previously fixed distribution. Thus, given the individual's randomized response, we are uncertain about what her true response would have been.

In spite of the previous intrinsic guarantee of randomized response, given the popularity of differential privacy, it may be interesting to analyze the privacy guarantees of randomized response in terms of differential privacy.

A randomized query function κ gives ϵ-differential privacy [5] if, for all data sets D_1, D_2 such that one can be obtained from the other by modifying a single record, and all $S \subset Range(\kappa)$, it holds

$$\Pr(\kappa(D_1) \in S) \leq \exp(\epsilon) \times \Pr(\kappa(D_2) \in S). \tag{8}$$

In plain words, the presence or absence of any single record is not noticeable (up to $\exp(\epsilon)$) when seeing the outcome of the query. Hence, this outcome can be disclosed without impairing the privacy of any of the potential respondents whose records might be in the data set. A usual mechanism to satisfy Inequality (8) is to add noise to the true outcome of the query, in order to obtain an outcome of κ that is a noise-added version of the true outcome. The smaller ϵ, the more noise is needed to make queries on D_1 and D_2 indistinguishable up to $\exp(\epsilon)$.

In [9,10], a connection between randomized response and differential privacy is established: randomized response is ϵ-differentially private if

$$e^\epsilon \geq \max_{v=1,\ldots,r} \frac{\max_{u=1,\ldots,r} p_{uv}}{\min_{u=1,\ldots,r} p_{uv}}. \tag{9}$$

The rationale is that the values in each column v ($v \in \{1, \ldots, r\}$) of matrix **P** correspond to the probabilities of the reported value being $Y = v$, given that the true value is $X = u$ for $u \in \{1, \ldots, r\}$. Differential privacy requires that the maximum ratio between the probabilities in a column be bounded by e^ϵ, so that the influence of the true value X on the reported value Y is limited. Thus, the reported value can be released with limited disclosure of the true value.

6   Empirical Results

Randomized response is usually performed independently for each attribute. As a result, marginal distributions are well preserved but multivariate distributions are not (see Sect. 3). If the aim is to preserve the joint distribution of all attributes $A_1, \ldots, A_m$, we should perform a single randomized response over $A_1 \times \ldots \times A_m$. However, this is usually unfeasible due to the curse of dimensionality explained at the end of Sect. 3.

In this section we empirically evaluate the technique proposed to solve the previous difficulties, which is based on clustering attributes in several groups (so that attributes in different groups have low correlation) and running randomized response independently for each of the groups.

6.1   Dataset

Experiments are based on the Adult dataset. This is a data set with over 32,500 records and a combination of numerical and categorical attributes. For the experiments, we only take categorical attributes into account. These attributes are: Work-class (with 9 categories), Education (with 16 categories), Marital-status (with 7 categories), Occupation (with 15 categories), Relationship (with 6 categories), Race (with 5 categories), Sex (with 2 categories), Native-country (with 42 categories) and Income (with 2 categories).

6.2   Methodology

In the test dataset there are 76,204,800 possible combinations of attribute values. This makes the clustering approach to randomized response (see Sect. 4) indispensable to get useful results.

We clustered attributes using the PIC algorithm with $k = 3$. The graph was constructed following the procedure described in Sect. 4. That is, for each pair of attributes:

- we obtained randomized responses on the Cartesian product of the two attributes, and
- the distance between the nodes corresponding to these attributes was computed as $1/V$, where V is Cramer's V statistic computed over the randomized data.

To avoid having to manually build the randomized response matrix for a Cartesian product (which would be burdensome because there are potentially many categories), we automatically build the matrix as follows:

- The probability of the cells in the main diagonal is set to a fixed value $p \in [0, 1]$.
- The probability of the off-diagonal cells is set to be inversely proportional to the number of attribute changes that the cell accounts for:

$$p_{uv} = (1 - p) \frac{d_{uv}}{\sum_k d_{uk}},$$

where d_{uv} is the inverse of the number of attributes whose values differ between u and v.

It would seem that, to compute $\sum_k d_{uk}$, we need to loop through each of the possible combinations of attribute values and do the sum. This is not necessary, because we can compute that sum by just considering the number of attributes whose categories change between u and k:

$$\sum_k d_{uk} = \sum_{a_1 \in \{1,\dots,r\}} (|A_{a_1}| - 1)$$

$$+ \frac{1}{2} \sum_{1 \leq a_1 < a_2 \leq m} (|A_{a_1}| - 1)(|A_{a_2}| - 1)$$

$$\dots$$

$$+ \frac{1}{m} \sum_{1 \leq a_1 < \dots < a_m \leq m} (|A_{a_1}| - 1) \dots (|A_{a_m}| - 1), \qquad (10)$$

where the first sum on the right-hand side of Expression (10) corresponds to changes of a single attribute (for each category a_1, we count the number $|A_{a_1}| - 1$ of alternative categories of the attribute A_{a_1} to which a_1 belongs); the second sum corresponds to the changes of two attributes and uses the same notation; and so on. We can rewrite Expression (10) in a more compact way as:

$$\sum_k d_{uk} = \sum_{w=1}^{r} \frac{1}{r} \sum_{1 \leq a_1 < \dots < a_w \leq m} (|A_{a_1}| - 1) \dots (|A_{a_w}| - 1). \qquad (11)$$

Notice that Eq. (11) does not depend on u. This means that, the value of $\sum_k d_{uk}$ is constant across all u; thus, we only need to do this computation once.

6.3   Risk Evaluation

Table 1 shows the levels ϵ_1, ϵ_2 and ϵ_3 of differential privacy attained in each of the three clusters for $p = 0.9$, $p = 0.8$, and $p = 0.7$, respectively. The overall level of differential privacy is the sum $\epsilon = \epsilon_1 + \epsilon_2 + \epsilon_3$, that is, the sum of levels across the three clusters. Note that the composition of clusters changed when p changed.

Table 1. Cluster evaluation results

p = 0.9			p = 0.8			p = 0.7		
Cluster	Attributes	Epsilon	Cluster	Attributes	Epsilon	Cluster	Attributes	Epsilon
C1	2	7.309	C1	4	11.844	C1	3	7.337
C2	2	7.735	C2	2	5.171	C2	3	8.569
C3	5	11.975	C3	3	6.234	C3	3	5.695

The overall ϵ decreases when p decreases, which means that privacy increases as p decreases. This was to be expected: the less probability mass in the diagonal of the randomized response matrix, the more privacy. It must be pointed out here that the randomized response matrix was not designed with differential privacy in mind; that is, we did not seek to minimize Eq. (9). Seeking such minimization would yield yet smaller ϵ, but would probably impinge on utility.

6.4 Utility Evaluation

We measured the utility of the generated randomized dataset by measuring the difference in the number of records for combinations of attribute values between the original dataset (X) and the randomized dataset (Y). In particular,

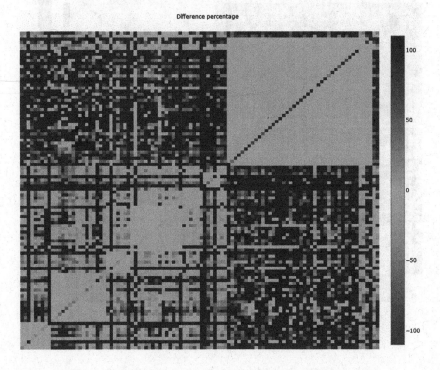

Fig. 1. Relative errors in randomized response for $p = 0.9$

we computed the difference for all combinations of values of two attributes. If $a_i \in A_i$ and $a_j \in A_j$ are attribute values, we computed the relative error as:

$$e_{ij} = \frac{X_{a_i a_j} - Y_{a_i a_j}}{X_{a_i a_j}} \times 100,$$

where $X_{a_i a_j}$ and $Y_{a_i a_j}$ are the number of records with attribute values a_i and a_j in the original dataset and in the randomized dataset.

Figures 1, 2 and 3 show the values of $e_{a_i a_j}$ when randomized response was run with parameter p equal to 0.9, 0.8 and 0.7, respectively. Both, in the x-axis and in the y-axis, we represent all possible attribute values; that is, in each axis we represent the set $\{a : a \in A_i, 1 \leq i \leq N\}$. In the intersection between column a_i and row a_j, we represent e_{ij}. From the histograms, we observe that for higher values of p the difference between the original and the randomized dataset is smaller. The light gray squares on the bottom-left top-right diagonal of the histogram represent the cases in which both attribute values are categories of the same attribute (which are impossible combinations, as we are interested in combinations of values of two attributes).

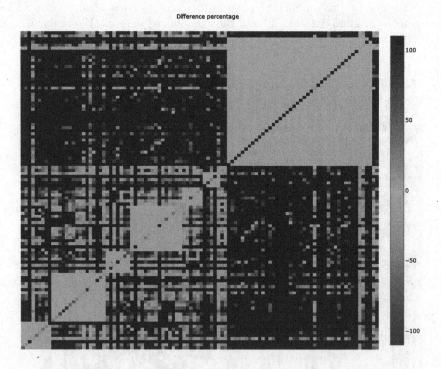

Fig. 2. Relative errors in randomized response for $p = 0.8$

Difference percentage

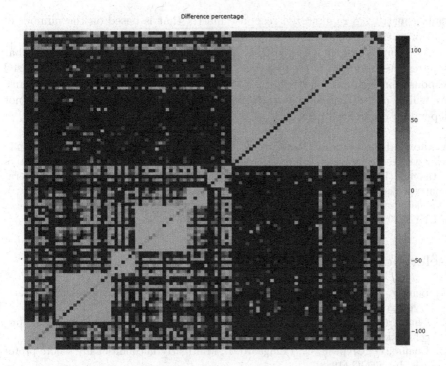

Fig. 3. Relative errors in randomized response for $p = 0.7$

7   Conclusions and Future Research

We have proposed a methodology to perform computations on a dataset that offers statistical input confidentiality. Each respondent can keep her input (true answer) confidential by giving to the data collector a reported answer via randomized reponse. Doing so still allows the data collector to approximate the empirical distribution of the pooled true answers of the set of respondents. After that, statistical computations can be performed on the approximated distribution.

Randomized response is only feasible when the number of possible categories is small compared to the number of records. For this reason, this technique is usually applied on an attribute-by-attribute basis. However, separately dealing with each attribute does not allow approximating the joint empirical distribution of the data. In this work, we have proposed a way to overcome this issue. We cluster attributes so that attributes in different clusters are independent (or nearly so) from each other. In this way, we can perform randomized response independently for each cluster without severely impairing the approximation of the joint empirical distribution.

We have experimentally validated the proposed methodology on a standard data set. In the experimental section, we have also described how to automatically construct a randomized response matrix that is based on the number of categories that are altered in the randomization process.

As future research, we plan to develop a new clustering procedure that requires less information (the current procedure needs users to run randomized response for each pair of attributes to measure the dependency between them). We will also investigate a quantification of the privacy guarantees that does not depend on differential privacy.

Acknowledgments and Disclaimer. The following funding sources are gratefully acknowledged: European Commission (H2020-700540 "CANVAS"), Government of Catalonia (ICREA Acadèmia Prize to J. Domingo-Ferrer) and Spanish Government (projects TIN2014-57364-C2-1-R "SmartGlacis" and TIN2015-70054-REDC). The views in this paper are the authors' own and do not necessarily reflect the views of UNESCO or any of the funders.

References

1. Ben-Or, M., Goldwasser, S., Wigderson, A.: Completeness theorems for non-cryptographic fault-tolerant distributed computation. In: STOC (1988)
2. Chaudhuri, A., Mukerjee, R.: Randomized Response: Theory and Techniques. Marcel Dekker, New York (1988)
3. Chaum, D., Crépeau, C., Damgaard, I.: Multiparty unconditionally secure protocols. In: STOC (1988)
4. Cramér, H.: Mathematical Methods of Statistics. Princeton University Press, Princeton (1946)
5. Dwork, C.: Differential privacy. In: Bugliesi, M., Preneel, B., Sassone, V., Wegener, I. (eds.) ICALP 2006. LNCS, vol. 4052, pp. 1–12. Springer, Heidelberg (2006). https://doi.org/10.1007/11787006_1
6. Greenberg, B.G., Abul-Ela, A.-L.A., Simmons, W.R., Horvitz, D.G.: The unrelated question randomized response model: theoretical framework. J. Am. Stat. Assoc. **64**(326), 520–539 (1969)
7. Lin, F., Cohen, W.W.: Power iteration clustering. In: Proceedings of the 27th International Conference on Machine Learning-ICML 2010 (2010)
8. Van den Hout, A.: Analyzing misclassified data: randomized response and post randomization. Ph.D. thesis, University of Utrecht (2004)
9. Wang, Y., Wu, X., Hu, D.: Using randomized response for differential privacy preserving data collection. Technical report DPL-2014-003. University of Arkansas (2014)
10. Wang, Y., Wu, X., Hu, D.: Using randomized response for differential privacy preserving data collection. In: EDBT/ICDT 2016 Joint Conference, Bordeaux, France (2016)
11. Warner, S.L.: Randomised response: a survey technique for eliminating evasive answer bias. J. Am. Stat. Assoc. **60**(309), 63–69 (1965)
12. Yao, A.: Protocols for secure computations. In: FOCS (1982)

Grouping of Variables to Facilitate SDL Methods in Multivariate Data Sets

Anna Oganian[1(✉)], Ionut Iacob[2], and Goran Lesaja[2]

[1] National Center for Health Statistics, 3311 Toledo Rd, Hyattsville, MD 20782, USA
aoganyan@cdc.gov
[2] Department of Mathematical Sciences, Georgia Southern University,
P.O. Box 8093, Statesboro, GA 30460-8093, USA
{ieiacob,goran}@georgiasouthern.edu

Abstract. Data sets that are subject to Statistical Disclosure Limitation (SDL) often have many variables of different types that need to be altered for disclosure limitation. To produce a good quality public data set, the data protector needs to account for the relationships between the variables. Hence, ideally SDL methods should not be univariate, that is, treating each variable independently of others, but multivariate, handling many variables at the same time. However, if a data set has many variables, as most government survey data do, the task of developing and implementing a multivariate approach for SDL becomes difficult. In this paper we propose a pre-masking data processing procedure which consists of clustering the variables of high dimensional data sets, so that different groups of variables can be masked independently, thus reducing the complexity of SDL. We consider different hierarchical clustering methods, including our version of hierarchical clustering algorithm, that we call *K-Link*, and outline how the data protector can define an appropriate number of clusters for these methods. We implemented and applied these methods to two genuine multivariate data sets. The results of the experiments show that *K-Link* has a potential to solve this problem efficiently. The success of the method, however, depends on the correlation structure of the data. For the data sets where most of the variables are correlated, clustering of variables and subsequent independent application of SDL methods to different clusters may lead to attenuated correlation in the masked data, even for efficient clustering methods. Thereby, the proposed approach is a trade-off between the computational complexity of multivariate SDL methods and data utility loss due to independent treatment of different clusters by SDL methods.

Keywords: Statistical Disclosure Limitation (SDL)
Hierarchical clustering · Dimensionality reduction

J. Domingo-Ferrer and F. Montes (Eds.): PSD 2018, LNCS 11126, pp. 187–199, 2018.
https://doi.org/10.1007/978-3-319-99771-1_13

1   Introduction

Data sets that are released to the public by the data collecting organizations often contain many variables of different types. For example, U.S. government surveys such as the National Health Interview Survey, the Behavioral Risk Factor Surveillance System, the Current Population Survey and American Community Survey are high dimensional. Data collecting organizations have an obligation by law to protect the privacy and confidentiality of responses provided by individuals or enterprises. This is usually accomplished by altering—we use the term *masking*—the original data before release, for example, by aggregating categorical values, swapping data values for selected records, or adding noise to numerical values. See [10,11] for more details. Data can also be synthesized, however, to do so one needs to come up with a good data generation model which is a complex task. As the dimensionality of the data increases, model estimation becomes more and more difficult. In case of the big governmental surveys mentioned above, model estimation can become extremely difficult and time consuming. Finding the best strategy for joint masking of many variables at a time is not a straightforward task either. Whatever approach for SDL is chosen, the organizations that disseminate the data strive to release data products with high utility - a goal competing with confidentiality protection, because any data alteration done to thwart identification will negatively impact at least some statistical properties of the data.

In this paper we propose a pre-masking procedure of clustering the variables into groups with the objective of increasing the separation between the groups as much as possible. Separation is viewed in terms of how related the variables in different groups are and we want to make the variables in different groups as unrelated as possible, so that SDL can be applied independently to different clusters with minimal loss of data utility.

1.1   Contribution and Plan of the Paper

The main contribution of the paper is a pre-masking procedure of clustering variables in the data set that can help government agencies reduce the complexity of SDL methods. In Sect. 2 we describe our clustering approach. We propose a variant of hierarchical clustering method, that we call *K-Link*, which can serve this purpose. In Sect. 3 we present numerical experiments with genuine data sets. Our results show that *K-Link* compares favorably to other hierarchical clustering methods. Concluding remarks are given in Sect. 4.

2   Clustering of Variables for Disclosure Limitation

In order to design any clustering procedure, first it is necessary to define measures of proximity of the objects being clustered. In case of clustering variables these are the measures of similarity/dissimilarity between the variables which are often based on some form of correlation [2,5,18]. Different types of variables require different metrics. For quantitative variables some function of the correlation coefficients may be used while for categorical variables many association measures exists, such as χ^2, Jaccard, Rand and others. When there are both types of variables in the data, a metric that can be computed for both types of variables is necessary. Similar to [2] we will use squared canonical correlation as such a metric. It can be computed as a first eigenvalue of the product $X'YY'X$ for two data matrices $X_{n \times d_1}$ and $Y_{n \times d_2}$ for which $min(n, d_1, d_2) = d_1$. As shown in [2] this metric is equivalent to a squared Pearson correlation for two quantitative variables. In the case of one quantitative X and one categorical variable Y, it is a correlation ratio. For the case of two categorical variables squared canonical correlation does not correspond to any well known association measure, but nevertheless, it can be interpreted geometrically according to [2]: the closer to one it is, the closer are the two linear subspaces spanned by the matrices representing these categorical variables, which means that the two qualitative variables bring similar information.

Hence, the dissimilarity matrix of variables is created as a lower triangular matrix DM with elements $DM[i, j] = 1 - r[i, j]$ where $r[i, j]$ is a squared canonical correlation between variables x_i and x_j.

Once the dissimilarity matrix is established a clustering method that fits our goals can be chosen. In [2] a hierarchical clustering method suited for clustering variables based on homogeneity criteria is proposed. In the sequel we will refer to this method as *Homclust*. Homogeneity of a cluster is calculated for this method as the sum of squared canonical correlations between each variable of the cluster and the central synthetic variable of that cluster. Such a synthetic variable of the cluster is defined as a quantitative variable "most linked" to other variables in the cluster and computed as the first principal component of the variables in the cluster. The goal of this method is to produce the most homogeneous clusters, so that the variables within the cluster are strongly related to each other. However, in our case when the purpose of clustering the variables is to find groups of variables to which SDL can be applied independently with minimal loss of correlation in the masked data, the objective will be different: the variables in different groups should be as uncorrelated as possible, so that independent masking of different groups of variables would not lead to significant correlation loss comparative to joint masking of all the variables at the same time. On the other hand, from a utility prospective it is not problematic if some, but not all, variables in the same group have little association. Indeed, if the multivariate SDL method preserves correlation structure, application of such method to the cluster of variables in the original data will produce masked cluster with similar associations, strong or week. Thereby, our goal is not necessarily to produce homogenous clusters, that is, clusters with highly correlated variables, but to maximize the separation between the clusters. Because the method described in [2] is focused on the homogeneity and not the separation, the clusters in the resulting partition may not be very far apart, so the variables assigned to different clusters may still be highly correlated. This will be demonstrated in Sect. 3. Hence, the methods based on the dissimilarities between the variables in different clusters may better suite our goals then the methods based on the proximities within the clusters such as *Homclust* or k-means. We will call such methods tentatively "separation clustering methods".

One type of such methods is divisive or "top down" hierarchical clustering. This approach starts with all the objects in one cluster and at each subsequent step, the largest available cluster is split into two clusters until finally all clusters comprise of single objects. One of the well known divisive methods is *Diana*(DIvisive ANAlysis) [12] implemented in R [3] as well as in other packages. *Diana* starts from finding a data point that has the highest average dissimilarity to all other objects. This object initiates a new cluster, that is called a splinter group. Then remaining objects are assigned either to the splinter group or to the complementary group based on average distance to the objects in these groups. Splitting clusters continues until all the objects end up in different clusters.

Contrary to divisive clustering, the agglomerative hierarchical approach starts with every object being a separate cluster and at each iteration the closest clusters are merged together building clustering hierarchy until all the objects end up in the same cluster. Some simple examples of such methods are *Single-Link*, *Complete-Link*, *Average* (see [6] and references therein). These algorithms differ in the way how they define distance between clusters. For example, for *Single-Link* it's a distance between the two closest objects in the respective clusters. Distances between the clusters increase from iteration to iteration, so for *Single-Link* the separation is guaranteed to increase as one goes up the dendrogram tree. For *Complete-Link* method the distance between clusters is defined as the distance between the two farthest points in the respective clusters. In some sense *Complete-Link* method is complementary to the *Single-Link* method. For *Average* method it is measured as an average of pairwise distances of all points between two clusters.

The difference in definition of distances between the clusters may have significant effect on the form, size and, especially on the separation between the clusters, as it can be seen in Fig. 1 of the Appendix. In particular, *Single-Link* may be the best choice if the goal is to achieve good separation between the clusters. Indeed, in the case of *Single-Link* two clusters S and L for which the gap between the closest points $i \in L$ and $j \in S$ is the smallest are joined at each iteration. Thus, unlike *Complete-Link* or *Average*, *Single-Link* will not create a partition where the gap between the borders of the clusters is smaller than the gap between the points of the same cluster (see Fig. 1 in Appendix).

However, the shortest distance between points $i \in L$ and $j \in S$ may not always be a good measure of a gap between two clusters. Points i and j can be relatively close to each other but far away from the rest of the points in their respective clusters, so with the exception of these two points the gap between L and S may be larger than it is assessed by *Single-Link*. To mitigate this issue we propose a simple modification of *Single-Link*: we measure the distance between clusters L and S as an average of k shortest distances between points in L and S. We call such a distance k-distance. This approach is in some way 'midway' between *Single-Link* and *Average* clustering method. The *Average* takes the average of pairwise distances of all the points between two clusters which may be big simply because two clusters are spread out, while the actual gap between the clusters may be small. Therefore, *Single-Link* may be 'too little' while the *Average* may be 'too much'. Since we want to focus on the gap between two clusters, it makes sense to concentrate on the few points that are close to the boundary. We call this intermediate approach *K-Link*.

Once the hierarchy of clusters is built we need to cut the dendrogram at some height to obtain an actual partition. In our clustering application, cutting height and the number of clusters may be determined based on the data protector preferences for the maximal utility loss due to independent masking of clusters. Indeed, the vertical axis of the dendrogram is a measure of closeness between the clusters. In other words, cutting the dendrogram at a particular height h sets up a lower bound on the distance between pairs of clusters in the partition which, in turn, corresponds to the upper bound on the allowed correlation between different clusters. Correlation between the variables in different clusters may be attenuated or lost after the SDL method is applied independently to different clusters. Thus, the data protector can set up an upper bound on maximal loss of correlation by choosing the acceptable value of h. The exact interpretation of h may differ for different clustering methods as it is based on the definition of a distance between the clusters. For example, for the *Single-Link* method this is a maximal correlation between two variables in different clusters. For *K-Link* method, h is an average of a few largest correlations that can be observed between the variables in different clusters, and so on. However, for all of the methods it is essentially a summary of the observed correlation between the variables in different clusters.

It should be noted that for *K-Link*, cutting the dendrogram at a particular height may in some rare cases lead to several solutions, that is, several clustering partitions with different number of clusters. This might happen because the sequence of k-distances is not strictly monotone as in *Single-Link*, although, there is a clear overall increasing trend. In particular, the k-distance may slightly decrease from one iteration to another which results in merging of next closest clusters slightly lower in the dendrogram tree.

We would also like to note, that one of the reasons why in this paper we haven't considered such algorithms as k-means, k-medoids or some model-based clustering algorithms, such as [7], is due to the fact that all these methods require the number of clusters as an input parameter. To successfully apply these algorithms, one often needs to compare many clustering partitions corresponding to different numbers of clusters. Another reason, is that many of these algorithms, by design specifically target the homogeneity of clusters. For example, k-means minimizes the sum of squares within the cluster on each iteration, and, thus, may create a partition with poor separation between the clusters. We believe, that the approach outlined above for the hierarchical agglomerative clustering methods allows for a more straightforward way of determining the number of clusters. It is important to mention that this approach produces clusters of variables that are a suitable input for the subsequent use of SDL methods.

3   Numerical Experiments

We applied our approach for clustering the variables to two real multivariate data sets. One of them is the National Health Interview Survey 2015 fourth quarter sample adult component public file [16]. In the sequel we will refer to it as NHIS. This is a public use file that has already undergone disclosure limitation. It has 6213 records. For our experiments we selected 86 variables of different types, continuous and categorical. The summary description is given in the Appendix. When the correlations were computed for NHIS data, sampling weights and design structure were taken into account.

Our second data set was downloaded from the UCI Machine Learning Repository [4]. This is a sample drawn from the Public Use Microdata Samples (PUMS) person 1990 US Census file. We will refer to this file as Census in the paper. It has 68 categorical variables and about 2.5 million records. Full description of the variables can be found in [1].

We applied *Diana, Single-Link, Average, Complete-Link, Homclust* as well as our *K-Link* method to these data sets.

In order to assess and compare the quality of partitions obtained by different methods we need to choose appropriate clustering criterion. Because clustering of variables is the first step and application of SDL to clustered data is the second step, ideally the clustering criterion should be in concordance with the SDL procedure in order to produce masked data with good utility. In regard to data utility, we want to note that the clustering procedure and subsequent independent application of SDL methods to different clusters will not have any effect on univariate statistics of the masked data. These statistics will depend only on the properties of SDL method applied to the variables. Furthermore, clustering does not affect any relationships between the variables that belong to the same cluster. The only influence clustering may have is on the relationships between the variables that belong to different clusters. The worst case scenario or the worst output corresponds to the case when all the correlations between the variables that belong to different clusters are lost in the masked data because of the independent application of SDL methods to these clusters. That is why we base the assessment criterion on the separation between the clusters, which measures correlation between the variables in different clusters - the correlation which can be lost in the worst case scenario. The smaller the correlation between the variables in different clusters - the better the output from the utility prospective.

Many clustering criteria were proposed in the literature, some examples are [8,13–15]. Many of them, however, are focused on the compactness of the clusters. However, as we mentioned above, compactness of the clusters is not an important quality for masking or synthesis of the variables in the clusters, however, separation between the clusters is. Several separation indexes were proposed in the literature. For example, [9] mentions an index that is computed as the ratio of the shortest distance between two clusters S and L (computed as the shortest distance between two points $i \in S$ and $j \in L$) and the maximal cluster diameter in the partition. A similar metric was proposed in [17], in particular, the gap between two clusters is divided by the total spread of both clusters. However, the spread of a cluster or it's diameter is a measure of cluster compactness. It was incorporated in the aforementioned separation indexes in order to give preference to the partitions with compact and well separated clusters, which makes sense if the ultimate goal of clustering is to detect the "true" cluster structure of the data. However, when diameters of clusters increase, such separation indexes become smaller indicating that the quality of the partition becomes worse. But for the purpose of masking clustered data such a partition is not worse than a partition with compact clusters if the gap between the clusters is the same. Therefore, we will include only the separation component into our clustering criterion, but not the spread. Thus, the separation criterion that we will use to compare different methods is as follows: first, separation between any two clusters S, L is computed as an average of the smallest s distances/dissimilarities $d(i, j)$ between the elements $i \in L$ and $j \in S$. Next, minimum of separations between all pairs of clusters in a partition is found.

The elements, located in different clusters at shortest distances from each other essentially represent the boarders of these two clusters and s can be thought as a parameter of "thickness" of the boarder. In general, s is data dependent and can be set to different values, for example, s can be equal approximately to the 5th percentile of the number of distances between the elements in different clusters $i \in S$ and $j \in L$. One of the topics of our future research is to investigate further the best ways of defining s. For simplicity, in our experiments we set s to be equal to 5 for all the pairs of clusters with 5 or more pairwise distances between the elements in different clusters. For the most populated pairs of clusters, 5 is approximately a 5th percentile of the number of distances between the pairs of clusters for our data sets. For the pairs of small clusters, for which there are less than 5 distances between the variables in different clusters, we consider all the distances. Further in the text when we refer to the distance between two clusters, we mean average of the s shortest distances between the variables in different clusters.

Table 1 shows the results of the comparison of *Diana, Single-Link, Average, Complete-Link, Homclust* and *K-Link* methods which were applied to NHIS data. For *K-Link* method we experimented with different values of the parameter k. The best results in terms of separation were obtained for $k = 3$ and we present these results in Table 1.

The minimal distance between any two clusters in the partition for each of the methods does not give a full picture of the partition. We get a more realistic impression of the composition of the partition by considering several closest distances between pairs of clusters, not just the shortest one. In Table 1 we listed distances between 10 closest pairs of clusters for *3-Link, Single-Link, Complete-Link, Average, Homclust* and *K-Link*. Column *3-Link-10* denotes *3-Link* method where the size of the cluster was enforced not to exceed a predefined limit, in this case, 10 variables per cluster. This method will be discussed later in the paper.

For all the methods we partitioned the data into 25 clusters which corresponds to cutting the dendrograms approximately at height 0.8, so that the maximum correlation loss between two clusters does not exceed 0.2.

Results for the Census data are shown in Table 2. Parameters were set the same as for the NHIS data. Both tables show similar patterns in terms of relative closeness between the clusters for different methods.

Table 1. NHIS data set: minimal separation between the clusters for *3-Link*, *3-Link-10*, *Single-Link*, *Complete-Link*, *Average*, *Homclust* and *Diana*. $Min_1, Min_2 \cdots Min_{10}$ are the distances between the ten closest pairs of clusters.

	3-Link	3-link-10	Single-Link	Complete-Link	Average	Homclust	Diana
Min_1	0.8190	0.7501	0.7901	0.5734	0.7501	0.5717	0.6923
Min_2	0.8245	0.7583	0.7955	0.6321	0.7583	0.6128	0.7411
Min_3	0.8282	0.7701	0.8196	0.7501	0.7701	0.6764	0.7600
Min_4	0.8302	0.7802	0.8243	0.7599	0.7792	0.7432	0.7681
Min_5	0.8351	0.7900	0.8261	0.7701	0.7891	0.7523	0.7809
Min_6	0.8431	0.7925	0.8267	0.7735	0.7925	0.7606	0.7834
Min_7	0.8525	0.8001	0.8299	0.7810	0.8001	0.7783	0.7845
Min_8	0.8590	0.8019	0.8307	0.7880	0.8019	0.7801	0.7881
Min_9	0.8635	0.8038	0.8341	0.7903	0.8025	0.7834	0.7916
Min_{10}	0.8691	0.8200	0.8408	0.7999	0.8154	0.7999	0.8032

As it can be seen from Tables 1 and 2, *3-Link* has the largest separation for all ten closest pairs of clusters. It is followed by *Single-Link* and then by *Average*. Consistently worse are *Diana* and *Complete-Link*. The worst separation among the methods that have no limits on the cluster size is observed for *Homclust* which agrees with our assumption that the algorithms based on clusters' homogeneity, which groups the most correlated variables in the same clusters, may not be appropriate for those cases when the objective is to create maximal separation, or minimal correlation, between the variables in different clusters.

Moreover, it is worth noting that we were not able to apply *Homclust* to the Census data set, which has 2.5 million records. The implementation of *Homclust* provided in package "ClustOfVar" [2] by its authors was not able to handle data set of this size. Thus, poor scalability is an additional issue of the *Homclust* method.

The entries in Tables 1 and 2 are the averages over the shortest s distances between two clusters. We also compared the actual correlations between the variables in different clusters. We observed that some of the variables which were placed in different clusters by *Complete-Link*, *Homclust* and *Diana* were very close in terms of correlation. For example, for the NHIS data set, the shortest distance between the variables in the two closest clusters is 0.81 for *K-Link*, 0.78 for *Single-Link*, 0.70 for *Average*, however, they are about 0.1 for *Complete-Link* and *Homclust* and $2.33E - 16$ for *Diana*. The same was true for the next closest

Table 2. Census data set: minimal separation between the clusters for *3-Link*, *3-Link-15*, *3-Link-25*, *3-Link-28*, *Single-Link*, *Complete-Link*, *Average*, *Homclust* and *Diana*. $Min_1, Min_2 \cdots Min_{10}$ are the distances between the ten closest pairs of clusters.

	3-Link	3-Link15	3-Link-25	3-Link-28	Single-Link	Complete-Link	Average	Diana
m1	0.5587	0.0228	0.1463	0.1883	0.2443	0.1967	0.1883	0.0000
m2	0.5914	0.0716	0.1883	0.2503	0.2503	0.2503	0.2503	0.1883
m3	0.5982	0.1883	0.4110	0.2790	0.2790	0.2790	0.2790	0.2790
m4	0.6250	0.3892	0.5914	0.4110	0.3314	0.4110	0.4110	0.3825
m5	0.6331	0.4110	0.6250	0.5210	0.4400	0.5148	0.5210	0.4731
m6	0.6351	0.5210	0.6392	0.5971	0.5587	0.5725	0.5971	0.5085
m7	0.6443	0.6405	0.6443	0.6392	0.5657	0.5971	0.6392	0.5631
m8	0.6627	0.7034	0.7034	0.6405	0.5971	0.6104	0.6405	0.5710
m9	0.6853	0.7050	0.7067	0.6740	0.6019	0.6378	0.6740	0.5971
m10	0.7034	0.7067	0.7080	0.7067	0.6351	0.6578	0.7067	0.6579

pair of clusters as well, that is, the actual distances between the variables were considerably smaller for *Diana*, *Complete-Link* and *Homclust* comparative to *K-Link* and *Single-Link*. A similar pattern was observed for the Census data.

Regarding partitions obtained by the applications of the clustering methods mentioned above, we observed that *Single-Link* and *K-Link* may lead to a partition where one of the clusters contains many (or a majority) of the variables and a number of small clusters with one or two variables, while for methods that lead to more compact and homogenous clusters such as *Complete-Link*, *Average*, *Diana* and *Homclust*, the largest cluster has less variables and overall partition is slightly more balanced. For example, when we partitioned the NHIS data into 25 clusters, the largest cluster contains 9 variables for *Diana*, 11 for *Average* and 11 variables for *Complete-Link*. However, for *Single-Link* and *3-Link* the largest clusters contain 50 and 58 variables respectively. In the case of Census data the largest cluster contains 14 variables for *Diana*, 28 variables for *Complete-Link*, *Average*, but 38 variables for *K-Link* and 35 for *Single-Link*.

Since the main reason of clustering the variables here is to reduce the complexity of joint masking or joint synthesis, clustering partitions with one or few very big clusters may not serve the main purpose very well. That is why we implemented a modification of *K-Link* method that incorporates an upper bound on the cluster size: as soon as the cluster size reaches n variables, the cluster cannot "accept" any new members. We will refer to this modification as *k-Link-n* further in the text. Another possibility to solve a "big cluster" problem is to split the biggest cluster in two or three smaller ones. There is, however, no guarantee that the obtained partition would not result in one big and another small cluster again. Moreover, incorporation of the restriction n during the merging process, as opposed to cutting the biggest cluster after clustering hierarchy is complete, may lead to better results because the variables that cannot join the cluster any longer that reached the maximum number of variables, can still join any other cluster which is closest to it. This will occur in the earlier stages of clusters formation, that is, as soon as the limit n was reached for the big cluster. Partitioning the biggest cluster in two or three after the dendrogram was finished, would limit the possibilities of grouping the variables only with those that are part of this cluster while all other clusters remain unchanged, which may not be the best solution.

For NHIS data K-$Link$ with the limit of $n = 10$ variables per cluster still compares reasonably well with other methods that do not have restrictions. It is the second method after $Single$-$Link$. Recall, however, that $Single$-$Link$ creates a cluster of 50 variables, while 3-$Link$-10 has only 10. Performance of 3-$Link$-10 is very similar now to $Average$, for which the largest cluster in the partition has 10 variables as well.

For the Census data, enforcing the limit on cluster size had a larger effect on the separation between the clusters than for NHIS data. In fact, in the Census data there is a big group of correlated variables. Thus, $Single$-$Link$, $Average$, $Complete$-$Link$ form a big cluster in their corresponding partitions. The size of the largest cluster varies from 28 to 38 variables among these methods. Thus, by enforcing the limits on the cluster size for K-$Link$ we inevitably reduced the separation between the clusters in the obtained partition. Columns 3-$Link$-15, 3-$Link$-25 and 3-$Link$-28 of Table 2 show the separation for 3-$Link$ with limits $n = 15, 25$ and 28. It can be seen that separation is not very good especially for the lower values of n.

It is important to observe that, it is not possible to enforce n till the top of the dendrogram. Our implementation of k-$Link$-n reports the minimal number of clusters when n can still be enforced. After that, to complete the hierarchy, clustering process continues as in the original version without restriction until the dendrogram is complete.

We conclude that clustering of variables can help reduce the complexity of SDL methods. However, there is a trade-off between complexity reduction and data utility, which depends on the correlation structure of the original data.

4   Concluding Remarks and Future Work

In this paper we propose a pre-masking clustering procedure that can be used by data publishing organizations that release data sets with many attributes of different types, such as big government surveys. Joint masking of data sets with many variables may be complicated and computationally involved. To reduce the complexity of the problem we outline a procedure of grouping variables into clusters in such a way that data utility loss due to independent application of SDL methods to these groups is limited. An upper bound on utility loss can be set up by the data protector. The value of this bound determines the parameters of the clustering procedure. Furthermore, we present a hierarchical clustering method, that we call K-$Link$, that can be suitable for the purpose of subsequent independent application of SDL to these clusters of variables. In our experiments K-$Link$ compares favorably with a number of existing hierarchical agglomerative and divisive clustering methods. In our future research we plan to consider a wider range of clustering methods that may be used for this purpose.

It is worth mentioning that we focus on the correlation-based utility loss due to clustering. In the future research we plan to expand the study of utility loss by considering other types of associations between the variables in the masked data.

In this paper we do not specify, neither do we focus on any particular SDL method as we believe that in general our clustering approach for variables should help to reduce complexity of any multivariate SDL method which preservers correlation structure of the data. As we mentioned earlier in the paper, it may be particularly beneficial for synthetic methods. Clustering of variables can also be helpful for developing multivariate analogs of some commonly used univariate SDL procedures, for example top-coding. Extreme values of some continuous variables are often top coded. For example, weight, height or income can be top-coded. However, if the upper bound of top-coded variable is determined independently from other variables, protection may be inadequate for different groups of individuals. For example, assume that the data protector sets the upper bound for weight to be equal to 300 pounds for all the respondents. However, a female respondent with such a top-coded weight whose race/ethnicity is Asian is more extreme as opposed to respondent with the same weight who is male Caucasian. So, ideally top-coding should be multivariate. While grouping race/ethnicity, gender, weight, height together may seem intuitive, there may be other, much less obvious combinations of variables especially in big survey data sets with hundreds of variables. Clustering of variables can clearly be helpful for finding such groups. Development of multivariate top-coding preceded by clustering of variables is an interesting topic for future research.

Acknowledgments. The authors would like to thank Van Parsons from the National Center for Health Statistics (US) for providing cleansed version of NHIS public use sample file for our experiments and for his valuable suggestions. Also the authors would like to express their appreciation to Donald Malec also from the National Center for Health Statistics for his careful reading of the paper and many useful suggestions. The findings and conclusions in this paper are those of the authors and do not necessarily represent the official position of the Centers for Disease Control and Prevention.

Appendix

Part A: Different Partitions Obtained by *Single-Link*, *Average*, *Complete-Link* and *Diana*

Figure 1 illustrates how differences in definition of distance between the clusters for *Single-Link*, *Average*, *Complete-Link* and *Diana* may influence the form and separation between the clusters. For this data set *Single-Link* was able to capture the structure of the data and created the most separated clusters. Separation between the clusters for partitions obtained by *Complete-Link*, *Average* and *Diana* is poor. These methods cut the vertical cluster in two or three parts very close to each other. On the other hand, a distant group of four point to the right of vertical cluster is merged with it.

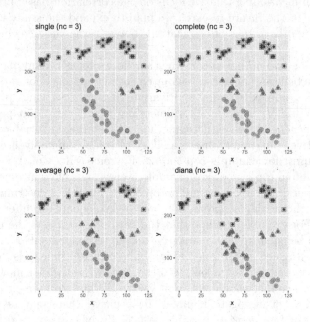

Fig. 1. Partitions in three clusters ($nc = 3$) for *Single-Link, Complete-Link,Average* and *Diana* for an artificial data set of points with coordinates (x, y).

Part B: Summary Description of NHIS Variables

The NHIS data set contains 86 variables. The variables are the respondents' answers to the questions in the following categories: health conditions, mental and emotional health, health behavior, affordability and accessibility of health care services, health insurance options, food availability and accessibility, employment status, income and education. The group of health related variables include presence or absence of asthma, diabetes, bronchitis, other pulmonary diseases and high blood pressure. Health behavior group of variables are the answers to the questions about cigarette smoking, alcohol use, leisure-time physical activity and exercising. Mental and emotional health variables are the answers about feeling hopeless, nervous, restless and fidgety, and also feeling worthless and sad. NHIS file also includes height, weight, body mass index of the respondents as well as demographic variables, such as race, age, marital status and region.

Most of the categorical variables are binary (Yes or No answers). There are also few continuous variables in the file, for example, age, weight, height and BMI.

References

1. Census: US census (1990) data set. UCI Machine Learning Repository (2017). https://archive.ics.uci.edu/ml/datasets/US+Census+Data+%281990%29
2. Chavent, M., Kuentz-Simonet, V., Liquet, B., Saracco, J.: ClustOfVar: an R package for the clustering of variables. J. Stat. Softw. **50**(i13), 1–16 (2012)
3. Cluster R package. https://cran.r-project.org/web/packages/cluster/cluster.pdf
4. Dheeru, D., Karra Taniskidou, E.: UCI machine learning repository. University of California, Irvine, School of Information and Computer Sciences (2017). http://archive.ics.uci.edu/ml
5. Dhillon, I., Marcotte, E., Roshan, U.: Diametrical clustering for identifying anti-correlated gene clusters. Bioinformatics **19**(13), 1612–1619 (2003)
6. Everitt, B., Landau, S., Leese, M., Stahl, D.: Cluster Analysis. Series in Probability and Statistics, 5th edn. Wiley, Hoboken (2011)
7. Fraley, C., Raftery, A.: MCLUST version 3 for R: Normal mixture modeling and model-based clustering. Technical report, Department of Statistics, University of Washington (2006). http://cran.r-project.org/web/packages/mclust/index.html
8. Halkidi, M., Batistakis, Y., Vazirgiannis, M.: On clustering validation techniques. J. Intell. Inf. Syst. **17**, 107–145 (2001)
9. Höppner, K., Klawonn, F., Runkler, T.: Fuzzy Cluster Analysis: Methods for Classification, Data Analysis and Image Recognition. Wiley, New York (1999)
10. Hundepool, A., et al.: Handbook on Statistical Disclosure Control (version 1.2). ESSNET, SDC project (2010). http://neon.vb.cbs.nl/casc
11. Hundepool, A., et al.: Statistical Disclosure Control. Wiley, Hoboken (2012)
12. Kaufman, L., Roussew, P.: Finding Groups in Data - An Introduction to Cluster Analysis. Wiley, Hoboken (1990)
13. Kim, J.J.: A method for limiting disclosure in microdata based on random noise and transformation. In: Proceedings of the ASA Section on Survey Research Methodology, pp. 303–308 (2002)
14. Lin, C., Chen, M.: A robust and efficient clustering algorithm based on cohesion selfmerging. In: Proceedings of Eighth ACM SIGKDD International Conference on Knowledge Discovery and Data Mining, Edmonton, Alberta, Canada, pp. 582–587 (2002)
15. Milligan, G.: A Monte Carlo study of thirty internal criterion measures for cluster analysis. Psychometrika **46**(2), 187–199 (1981)
16. NHIS: National Health Interview Survey. National Center for Health Statistics (2015). https://www.cdc.gov/nchs/nhis/nhis_2015_data_release.htm
17. Qiu, W.: Separation index, variables selection and sequential algorithm for cluster analysis. Ph.D. thesis, The University of British Columbia (2004)
18. Vigneau, E., Qannari, E.: Clustering of variables around latent components. Commun. Stat. Simul. Comput. **32**(4), 1131–1150 (2003)

Comparative Study of the Effectiveness of Perturbative Methods for Creating Official Microdata in Japan

Shinsuke Ito[1]([✉]), Toru Yoshitake[2], Ryo Kikuchi[3], and Fumika Akutsu[4]

[1] Faculty of Economics, Chuo University,
742-1 Higashinakano, Hachioji-shi, Tokyo 192-0393, Japan
ssitoh@tamacc.chuo-u.ac.jp
[2] National Statistics Center, 19-1 Wakamatsu-cho, Shinjuku-ku,
Tokyo 162-8668, Japan
tyoshitake@nstac.go.jp
[3] NTT Secure Platform Laboratories,
3-9-11 Midorimachi, Musashino, Tokyo 180-8585, Japan
kikuchi.ryo@lab.ntt.co.jp
[4] Statistics Bureau of Japan, 19-1 Wakamatsu-cho, Shinjuku-ku,
Tokyo 162-8668, Japan
fakutsu@soumu.go.jp

Abstract. The Statistics Bureau of Japan currently creates and provides anonymized microdata for six surveys. Anonymized microdata from the Population Census of Japan are created using non-perturbative methods such as recoding, top coding, and record deletion as well as the perturbative method of swapping.

This paper analyzes several types of anonymized microdata created based on individual data from the Population Census, and explores the potential of using perturbative methods such as swapping and PRAM to create anonymized microdata from Japanese Census Data. Results suggest that perturbative methods can increase data quality, but should be selected according to the properties of the microdata that are to be anonymized. Perturbative methods have the potential to help further enhance statistical methodologies in Japan.

Keywords: Census microdata · Targeted data swapping
Random data swapping · PRAM · Distance-based record linkage
True match rate

S. Ito—was a research fellow at the National Statistics Center until the end of March 2018 and conducted research on disclosure limitation methods for microdata in co-ordination with officials at the National Statistics Center and the Statistics Bureau of Japan.
R. Kikuchi—is a research fellow at the National Statistics Center.

J. Domingo-Ferrer and F. Montes (Eds.): PSD 2018, LNCS 11126, pp. 200–214, 2018.
https://doi.org/10.1007/978-3-319-99771-1_14

1 Introduction

In recent years, the "statistical revolution" has attracted attention in Japan, as combining official statistical data with private-sector big data and government records can further the use of evidence-based policy making (EBPM). The "Basic Plan Concerning the Development of Official Statistics" was enacted in March 2018, and details future measures for the provision and use of survey data and other data sources. The "statistical revolution" has increased the focus on creating and providing official statistical data in Japan, and started a debate on the possibility of providing multiple types of official statistical microdata, including public use files.

Currently, the Statistics Bureau of the Ministry of Internal Affairs and Communication creates and provides anonymized microdata for six surveys: the Housing and Land Survey, the National Survey of Family Income and Expenditure, the Employment Status Survey, the Survey on Time Use and Leisure Activities (Surveys A and B), the Labor Force Survey, and the Population Census. Anonymized microdata from the Japanese Population Census are (1) available for the years 2000–2005; (2) classified by prefecture, with cities having populations over 500,000 broken out; (3) data covers 1% of the total population, and is extracted in sets of households; (4) only one type of anonymized microdata is created and provided; and (5) data is anonymized using non-perturbative methods such as recoding, top coding, and record deletion as well as the perturbative method of swapping.

The practical application of anonymized microdata from the Population Census requires methods of anonymization that are effective also in cases of special uniques (e.g., when a cell of the results at the nationwide level has value one), which reflects the growing need for anonymized microdata for small regions. This situation merits further investigation into the use of perturbative methods for anonymizing microdata.

Existing research on creating anonymized microdata (i.e., microdata to which anonymization methods have been applied) has evaluated data swapping and the merging of categories in recoding and top coding for individual data from the Population Census (see [11–13]). Further improvements in anonymized microdata can be achieved by further exploring perturbation while taking into account confidentiality, and investigating the usability of anonymized microdata to which perturbation has been applied.

This paper aims to examine the potential of perturbative methods to create anonymized microdata based on individual data records from the Population Census, and to compare the effectiveness of perturbative methods including data swapping and PRAM.

2 The Methodology of Data Swapping

Studies on the potential of data swapping as a method to limit disclosure of microdata via anonymization include those by [3, 8, 16, 17, 18]. Takemura [19] and [9, 10, 14] have conducted empirical research on the effectiveness of data swapping specifically for Japanese microdata. The U.S. Census Bureau has developed n-cycle swapping and conducted empirical research on this method [4].

As part of this research, an empirical study of data swapping – including record swapping between different geographical areas – was conducted using data from the 2010 Population Census of Japan. Three sets of census data, each containing a different number of records, were used as test data. The data sets were created from data on more than 50,000 residents of a specific Japanese prefecture. Data were grouped by distinct geographic areas, with about 10,000 records from one area (Data A), about 20,000 from a different area (Data B), and about 20,000 from an area different from either of these (Data C).

Data swapping was conducted as follows. First, population uniques were identified for every combination of patterns for the following key variables:

Type and tenure of dwelling (8 categories)
Sex (2 categories)
Marital status (4 categories)
Nationality (2 categories)
Type of (work) activity (8 categories)
Employment status (8 categories)
Age (114 categories)
Industry (21 categories)
Occupation (12 categories).

Second, records that can be uniquely represented by any combination of the nine key variables were selected as target records for data swapping. To determine the degree of priority for data swapping, cross tabulation was conducted for all combinations of key variables. The number of times a specific record corresponded to a unique cell was counted for each combination of cross tabulations, and this score was added to each record in the test data. Records for which the score was high were classified as "risky" records and assigned a higher priority for data swapping [7].

Third, targeted data swapping was performed for records with a score of 1 or higher. Targeted data swapping was performed for records that corresponded to the top p% (with the value of p set between 0 and 25 in this study) of the group and in order of descending score. Random data swapping was performed for records that corresponded to the p% of the group and with data selected randomly.

Fourth, the distances between each target record and all donor file records were calculated using the above nine key variables, and the nearest donor file record was swapped for each target record. In case of multiple records with identical distances, the donor file record was selected randomly from among equidistant records.

3   The Methodology of Post Randomization

The post randomization method (PRAM) was first proposed by [15]. In official statistics, PRAM has been used as a method for controlling statistical disclosure. A detailed description and empirical study of PRAM appears in [5, 6]. Similar approaches have been studied in the context of privacy-preserving data mining by [1, 2].

PRAM consists of two procedures: perturbation and reconstruction. Perturbation changes values according to a transition probability matrix, while reconstruction estimates the distribution of the original data from the perturbed one.

3.1 Perturbation

Let the original data be microdata in which each row and column represent single individual and attribute, respectively, and $\mathbb{V}$ be the set of all possible attribute values. Perturbation changes values according to a transition probability matrix $\mathbf{A}$. The transition probability matrix contains the probabilities at which each value in a private microdata will be changed into another (not necessarily distinct) value. $A_{u,v}$ denotes the probability of $u \in \mathbb{V}$ being changed into $v \in \mathbb{V}$. For example, $A_{male,female}$ means the probability of "male" becoming "female".

Retention–replacement perturbation [2] is a specific instance of the perturbation. This is also known as "fully filled matrices with equal off-diagonal element" [6]. In the retention-replacement perturbation, a value is retained with *retention probability* ρ. If the value is unretained, it is replaced with a value chosen uniformly at random from the attribute domain (including the original value). Therefore, transition probability matrix $\mathbf{A}$ is defined as $A_{v,v'} = \rho + \frac{1-\rho}{|\mathbb{V}|}$ if $v = v'$ and $A_{v,v'} = \frac{1-\rho}{|\mathbb{V}|}$ otherwise for $v, v' \in \mathbb{V}$, where $|\mathbb{V}|$ denotes the cardinality of $\mathbb{V}$.

3.2 Reconstruction

The statistics of the original microdata can be estimated from a perturbed microdata and the retention probability matrix. Kooiman et al. [15] and [1] proposed reconstruction methods for estimating the cross tabulation of the original microdata from a released (i.e., perturbed) one. Retention–replacement perturbation results in cross tabulation values being close to those for a uniform distribution, so the original cross tabulation values are estimated in those methods by increasing gaps among the values of the cross tabulation.

Although reconstruction can improve the accuracy of cross tabulation, we did not use reconstruction in this research due to the difficulty of managing decimals. If we had employed reconstruction, we would have perturbed and reconstructed a microdata with every attribute iteratively. In each iteration, we would have had to obtain an intermediate microdata from (reconstructed) cross tabulation, where some attributes were perturbed and reconstructed but the others were not. Since the reconstructed cross tabulation could have contained decimals, it would have been difficult to convert the cross tabulation into the corresponding microdata. A naïve solution to manage the decimal would have been to round up/down each value, but this would have caused a change in the number of records. Even when rounding up/down while maintaining the number of records, it would have caused the change of the distribution of non-perturbed attributes. Therefore, reconstruction was not applied as part of this research.

4   Comparative Analysis of Perturbative Methods Applied to Japanese Microdata

This research focused on perturbed records created by random data swapping or application of PRAM to the original records. Specifically, we measured the distance among each of the perturbed records using the following variables and searched for the records with the nearest distance.

Sex (2 categories)
Age in bins of 5 years (16 categories)
Marital status (2 categories)
Nationality (2 categories)
Type of (work) activity (2 categories)
Employment status (2 categories)
Industry major category (3 categories)
Occupation major category (3 categories)
Type of dwelling (2 categories)
Dwelling building type (2 categories)
Number of floors in building (2 categories)
Floor on which household lives (2 categories)
Total floor area (2 categories)
Means of transport used (Variable with 9 patterns and 2 categories for each variable)
Region (3 categories) (in cases where region is also included among key variables).

If a record can be linked to the original record with a 1-to-1 relation, then this relation is called a "true match". When a 1-to-many relation is possible, the relation is not treated as a "true match". In this research, the proportion of "true matches" that occurred among all records was called the "true match rate".

First, we examined the relation between perturbative methods and the true match rate. Figure 1 shows the relation between the swapping rate and the true match rate for random data swapping. Figure 2 shows the relation between transition probability and the true match ratio for PRAM. PRAM was applied to three variables: age in bins of 5 years, industry, and occupation. The results show that for random data swapping, the true match rate decreases as the swapping rate increases. Figure 1 shows that the true match rate decreases almost linearly as the swapping rate increases to 25%. For PRAM, the true match rate similarly decreases as the transition probability increases. The true match rate decreases almost linearly as the transition probability increases to 25%.

Next, we compared information loss between random data swapping and PRAM. The differences in information loss were measured by focusing on the same level of true match rate (the true match rate was set to three patterns of approximately 60%, 55%, and 50%). For this, swapping rates were set to approximately 2%, 6%, and 9% for the true match rates of 60%, 55%, and 50%, respectively. Retention probabilities for PRAM, applied based on age, industry, and occupation, were set at 95%, 85%, and 75%, respectively, for the above three true match rates.

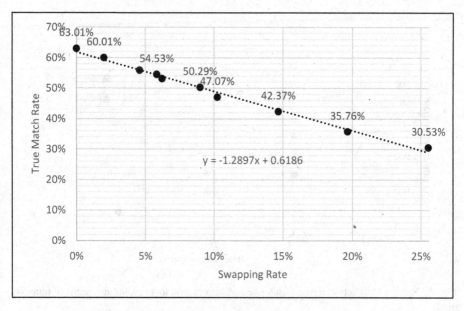

Fig. 1. Relation between swapping rate and true match rate

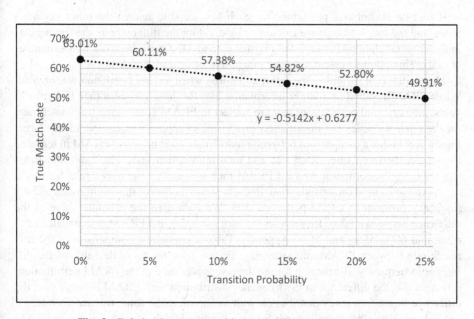

Fig. 2. Relation between transition probability and true match rate

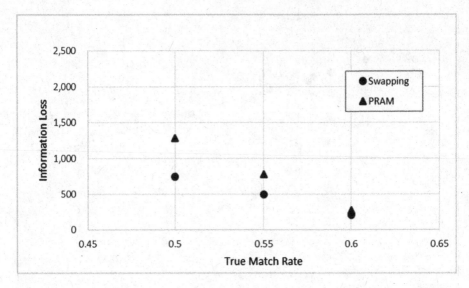

Fig. 3. Relationship between true match rate and information loss (excluding means of transport used)

Cross tabulation was performed for each key variable against the other key variables, and the total values were calculated by dividing the difference in the frequencies in the cross tabulations between the original data and the perturbed data by the number of cells. This allowed calculating the magnitude of information loss.

Appendix Table 1 shows the information loss when a cross tabulation was created. While PRAM resulted in a larger information loss for three variables (age in bins of 5 years, industry, and occupation) that are subject to PRAM at the same true match rate, for the other variables the information loss in for swapping was comparatively large. Figure 3 contains a comparison between random data swapping and PRAM in terms of the relation between true match rate and total information loss. The figure illustrates that the overall information loss for PRAM increases as the true match rate decreases.

Next, we focused on the univariate frequency distribution of the original data and the frequency distribution of the perturbed data. We compared the absolute value of the difference between relative frequencies for both swapping and PRAM at true match rates of around 60%, 55%, and 50%. Appendix Table 2 contains a comparison of the true match rate between PRAM and swapping. Focusing on the absolute value of the difference in frequency distribution for the three variables used in the PRAM perturbation, it is clear that the difference in the frequency distribution with PRAM is larger than the difference in the frequency distribution with swapping at the same true match rate.

This indicates that there is room for investigating which variables to use in PRAM and what proportion of PRAM and swapping should be applied in order for key variables to remain confidential and reduce the number of special uniques.

Next, we compared information loss for targeted data swapping and random data swapping for the same true match rate and considering two cases: applying PRAM to three variables (age in bins of 5 years, industry, and occupation) and applying PRAM

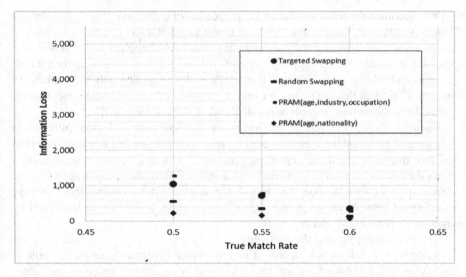

Fig. 4. Relationship between true match rate and information loss

to only two variables (age in bins of 5 years and nationality). Figure 4 shows the relationship between true match rate and information loss for perturbed data using random data swapping, targeted data swapping and PRAM. As swapping was limited to unique records, the information loss for swapping decreased slightly.

This analysis shows that while information loss for PRAM with three variables was comparatively large (as described earlier), information loss was comparatively small for PRAM with only two variables. This result indicates that anonymized microdata with high usability can be created using PRAM when the variables to which PRAM is applied are carefully selected.

The above discussion is based on the assumption that "true matches" do not occur in records replaced by swapping, since the swapped records are replaced with records from another region. In the following discussion, the assumed strategy of intruders (i.e., those seeking to identify individual records) is changed, and it is assumed that there is a possibility of identification through external information possessed by the intruder even if a record is replaced with a record from another region through swapping. In this case, region information is among the key variables that an intruder can use to identify an individual, and we compared the effectiveness of swapping and PRAM depending on the level of importance the intruder gives to region information as a key variable to identify a specific person.

Specifically, we examined the effectiveness of swapping and PRAM by setting the following two intruder strategies:

(1) The intruder has partial information about a particular individual who resides in region A, and searches for that person among the perturbed data for region A.
(2) The intruder has partial information about a particular individual who resides in region A, and searches for that person among not only region A but also region B that was selected for data swapping.

We examined the relation between the proportion of individuals that an intruder can identify and the information loss for three cases: targeted data swapping (T), random data swapping (R), and targeted data swapping with PRAM applied to age in bins of 5 years and nationality (P). Swapping was applied to records that are unique for the nine variables and arbitrary combinations thereof. PRAM was applied only to "particularly dangerous" records that are unique for only 1 variable or 2 variables (the proportion of "particularly dangerous" records was calculated as 4.0% in region A).

Figure 5 shows the results for swapping applied with a region information weighting of 1000 and for intruder type (1). The results indicate that it is not possible to find any true matches if the region information is changed by swapping. Figure 6 shows the results for swapping applied with a region information weighting of 0.5 and intruder type (2). Results show the cases where an intruder who assumes that swapping is performed between region A and region B can attempt to identify an individual by targeting both region A and region B.

The two figures suggest the following:

Since type (1) intruders look only at perturbed data in region A, they are unable to identify records that have been replaced with those from another region by swapping. As a result, the decrease in true match rate as the swapping rate increases is comparatively large, which indicates that swapping is more effective.

Since type (2) intruders look at the perturbed data of both region A and the swapping region for region A (i.e. region B), they are able to identify records that have been replaced by swapping, and swapping is less effective.

Therefore, PRAM is much stronger against type (2) intruders than against type (1) intruders.

The true match rate can be decreased by using PRAM against type (2) intruders. This effect is the same as the decrease in true match rate obtained by performing 2–3% targeted data swapping at the point where 2–3% PRAM is used.

Although the effectiveness of random data swapping and targeted data swapping are virtually the same against type (1) intruders, targeted data swapping is weak against type (2) intruders. This is because it is easy to create true matches based on uniqueness even if a relatively unique record is moved to another region.

Next, we looked at the case where the number of swapping regions is increased from one to two. Specifically, replacement was performed with the most similar record among the combination regions B and C. This was expected to have the following opposing effects:

(A) Since the intruder can treat the target record as being "somewhere among the 3 regions", then since the weighting by differences between region information is reduced and the effectiveness of changing region information by swapping is reduced, attacks on swapped records will succeed more easily.

(B) Since the intruder needs to find the target record from among a larger number of records, attacks will succeed less easily.

A comparison of Figs. 6 and 7 shows the following: The true match rate increases when swapping is applied to two regions. This suggests a higher possibility that hypothesis (A) applies rather than hypothesis (B), when the intruder recognizes that

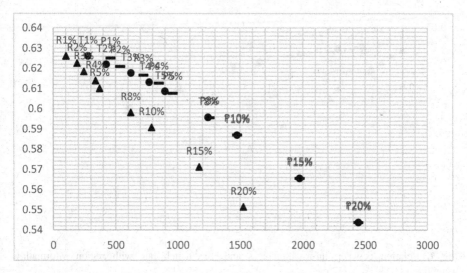

Fig. 5. Relation between information loss and true match rate with region information, weighting of 1000

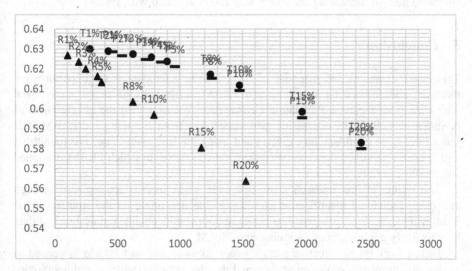

Fig. 6. Relation between information loss and true match rate with region information, weighting of 1/2

swapping is conducted with two or more regions then other key variables are more effective than region information for identifying an individual.

Although in the case of swapping with two regions the number of records to search increases, avoiding true matches was difficult for relatively unique records.

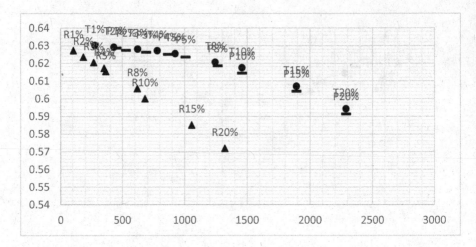

Fig. 7. Relation between information loss and true match rate with region information, weighting of 1/3

Therefore, even when swapping was performed with two regions, a slightly stronger effect of decreasing the true match rate was found with PRAM.

5   Conclusion

This paper examines the potential of anonymizing microdata through perturbative methods, and focuses on swapping and PRAM as methods to create anonymized microdata from Japanese Population Census data. We examined the effectiveness of perturbative methods by comparing information loss from swapping and PRAM after setting the same confidentiality criteria and using the true match rate. We found that highly useful perturbed data can be created if the variables to which PRAM is applied are chosen carefully.

The results of this research show that even when records are replaced through swapping, individual identification is still possible depending on the strategy of the intruder. In such cases, it is possible to ensure both confidentiality and usability of data by effectively applying both swapping and PRAM.

Future work on perturbative methods such as swapping and PRAM is needed to explore the properties of the microdata, and thereby enhance statistical methodologies in Japan.

Note
The opinions expressed in this paper do not necessarily reflect those of organizations to which the authors belong or those of the Statistics Bureau of Japan or the National Statistics Center.

Appendix

(See Appendix Tables 1 and 2)

Table 1. Information loss with each variable as the pivot

True match rate	60%	55%	50%	60%	55%	50%	
Method	Swapping			PRAM			Total
Sex	70.81	129.03	235.43	30.49	89.82	150.18	705.75
Age in bins of 5 years	10.03	23.65	32.86	44.90	110.60	223.79	445.83
Marital status	39.24	110.52	178.74	16.34	46.85	75.19	466.89
Nationality	60.06	167.53	239.53	29.09	85.43	143.33	724.97
Employment status (8 categories)	16.89	40.02	73.43	7.51	21.77	36.47	196.09
Position at job (8 categories)	27.35	60.54	81.29	9.31	25.16	40.56	244.22
Industry major category	11.75	30.10	47.93	162.48	458.02	795.13	1505.41
Occupation major category	21.32	52.99	56.02	206.30	621.49	949.03	1907.16
Dwelling type	41.32	99.33	151.02	9.01	24.27	38.20	363.16
Dwelling building type	71.96	161.07	217.17	17.18	46.97	74.80	589.16
Number of floors in building	6.47	17.44	26.38	1.20	3.18	5.07	59.74
Floor on which household lives	4.64	12.82	15.46	1.16	3.14	5.00	42.22
Total floor area (14 categories)	36.46	92.78	126.39	6.01	15.05	23.17	299.87
Means of transport used 1	212.35	457.23	746.60	30.28	88.02	143.49	1677.97
Means of transport used 2	89.20	158.21	271.68	29.72	86.16	144.80	779.78
Means of transport used 3	60.80	137.25	202.49	29.22	85.37	143.45	658.57
Means of transport used 4	73.80	134.14	204.56	29.12	85.97	143.88	671.47
Means of transport used 5	222.31	464.65	750.53	30.26	88.73	143.18	1699.67
Means of transport used 6	57.01	126.80	193.26	28.93	85.19	142.81	634.00
Means of transport used 7	66.11	135.38	232.29	29.97	85.44	144.37	693.56
Means of transport used 8	76.17	130.93	207.28	29.63	85.94	144.08	674.04
Means of transport used 9	61.30	129.14	196.33	29.59	85.68	143.10	645.14
Total number	668.68	1435.79	2243.34	403.85	1164.13	1926.55	
Total number (excluding means of transport used)	209.16	498.92	740.83	270.49	775.88	1279.96	

Table 2. Comparison of original data and perturbed data with PRAM and swapping

Age

PRAM	True match rate			Swapping	True match rate		
Age in bins of 5 years	Around 60%	Around 55%	Around 50%		Around 60%	Around 55%	Around 50%
	Difference in relative frequency (absolute value)				Difference in relative frequency (absolute value)		
Unknown	0.00%	0.00%	0.00%		0.00%	0.00%	0.00%
Under 15	0.00%	0.00%	0.00%		0.00%	0.00%	0.00%
15 to 19	0.00%	0.00%	0.00%		0.01%	0.00%	0.00%
20 to 24	0.06%	0.26%	0.51%		0.01%	0.01%	0.00%
25 to 29	0.06%	0.26%	0.51%		0.00%	0.01%	0.01%
30 to 34	0.11%	0.13%	0.31%		0.00%	0.00%	0.02%
35 to 39	0.11%	0.13%	0.31%		0.00%	0.00%	0.01%
40 to 44	0.05%	0.07%	0.21%		0.00%	0.00%	0.01%
45 to 49	0.05%	0.07%	0.21%		0.00%	0.00%	0.01%
50 to 45	0.00%	0.04%	0.16%		0.00%	0.01%	0.01%
55 to 59	0.00%	0.04%	0.16%		0.00%	0.01%	0.01%
60 to 64	0.09%	0.35%	0.74%		0.00%	0.01%	0.00%
65 to 69	0.09%	0.35%	0.74%		0.00%	0.01%	0.00%
70 to 74	0.05%	0.09%	0.23%		0.00%	0.00%	0.00%
75 to 79	0.05%	0.09%	0.23%		0.00%	0.00%	0.00%
80 to 84	0.03%	0.01%	0.04%		0.00%	0.00%	0.00%
85 or older	0.03%	0.01%	0.04%		0.00%	0.00%	0.00%

Industry major category

PRAM	True match rate			Swapping	True match rate		
Industry major category	Around 60%	Around 55%	Around 50%		Around 60%	Around 55%	Around 50%
	Difference in relative frequency (absolute value)				Difference in relative frequency (absolute value)		
Unknown	0.00%	0.00%	0.00%		0.00%	0.01%	0.00%
A Agriculture	0.06%	0.18%	0.30%		0.00%	0.02%	0.01%
U Forestry	0.04%	0.08%	0.16%		0.00%	0.01%	0.00%
B Fisheries	0.03%	0.10%	0.14%		0.00%	0.00%	0.01%
C Mining, quarrying, gravel extraction	0.76%	2.30%	3.84%		0.00%	0.00%	0.07%
D Construcion	0.51%	1.28%	2.06%		0.01%	0.01%	0.28%
E Manufacturing	1.27%	3.58%	5.90%		0.03%	0.07%	0.02%
F Electricity, gas, heat, water	0.15%	0.40%	0.71%		0.00%	0.00%	0.05%
G Information and communication	0.15%	0.37%	0.67%		0.02%	0.03%	0.05%
H Transportation and post	0.08%	0.18%	0.63%		0.00%	0.02%	0.03%
I Wholesale and retail	0.13%	0.46%	0.95%		0.00%	0.03%	0.03%
J Finance and insurance	0.14%	0.35%	0.79%		0.02%	0.02%	0.04%
K Real estate and goods rental	0.13%	0.36%	0.80%		0.00%	0.02%	0.14%
L Academic research, expert and technical services	0.65%	1.84%	3.15%		0.01%	0.07%	0.01%
M Lodging, food and beverage services	0.03%	0.26%	0.32%		0.02%	0.02%	0.04%
N Lifestyle related services, leisure	0.06%	0.11%	0.36%		0.01%	0.01%	0.01%
O Education and study support	0.06%	0.19%	0.16%		0.02%	0.05%	0.04%
P Medical and welfare	0.06%	0.11%	0.03%		0.00%	0.03%	0.00%
Q Compound services	0.09%	0.46%	0.97%		0.02%	0.02%	0.01%
R Services (does not fall into other categories)	0.03%	0.05%	0.19%		0.00%	0.04%	0.00%
S Public service (does not fall into other categories)	0.02%	0.09%	0.11%		0.00%	0.03%	0.00%
T Industry that cannot be categorized	0.00%	0.00%	0.00%		0.00%	0.00%	0.00%

Occupation major category

PRAM	True match rate			Swapping	True match rate		
	Around 60%	Around 55%	Around 50%		Around 60%	Around 55%	Around 50%
Occupation major category	Difference in relative frequency (absolute value)				Difference in relative frequency (absolute value)		
Unknown	0.00%	0.00%	0.00%		0.00%	0.01%	0.00%
0 Administrative worker	0.58%	1.74%	2.40%		0.01%	0.03%	0.01%
1 Expert or technical worker	0.57%	1.62%	2.54%		0.05%	0.08%	0.12%
2 Office worker	0.02%	0.13%	0.14%		0.01%	0.03%	0.04%
3 Sales worker	0.21%	0.26%	0.45%		0.03%	0.06%	0.05%
4 Service professional worker	0.01%	0.33%	0.47%		0.00%	0.05%	0.01%
5 Protective services worker	0.22%	0.69%	1.09%		0.00%	0.03%	0.01%
6 Agriculture, forestry, fisheries worker	0.22%	0.74%	1.50%		0.00%	0.02%	0.01%
7 Manufacturing process worker	1.01%	3.23%	5.07%		0.07%	0.14%	0.13%
8 Transportation or machine operator worker	0.19%	0.52%	0.83%		0.03%	0.04%	0.02%
9 Construction or mining worker	0.09%	0.22%	0.35%		0.00%	0.00%	0.01%
V Transportation, cleaning, or packing worker	0.09%	0.47%	0.38%		0.00%	0.05%	0.02%
T Occupation that cannot be categorized	0.00%	0.00%	0.00%		0.00%	0.00%	0.00%

References

1. Agrawal, R., Srikant, R.: Privacy-preserving data mining. In: SIGMOD 2000 (2000)
2. Agrawal, R., Srikant, R., Thomas, D.: Privacy preserving OLAP. In: SIGMOD. ACM (2005)
3. Dalenius, T., Reiss, S.P.: Data-swapping: a technique for disclosure control (extended abstract). In: Proceedings of Section on Survey Research Methods, pp. 191–194. American Statistical Association, Washington, D.C. (1978)
4. DePersio, M., Lemons, M., Ramanayake, K.A., Tsay, J., Zayatz, L.: n-cycle swapping for the american community survey. In: Domingo-Ferrer, J., Tinnirello, I. (eds.) PSD 2012. LNCS, vol. 7556, pp. 143–164. Springer, Heidelberg (2012). https://doi.org/10.1007/978-3-642-33627-0_12
5. de Wolf, P.P., Gouweleeuw, J.M., Kooiman, P., Willenborg, L.C.R.J.: Reflections on PRAM, Statistical Data Protection (1998)
6. de Wolf, P.P., van Gelder, I.: An empirical evaluation of PRAM, Discussion paper 04012, Statistics Netherlands (2004)
7. Elliot, M., Manning, A.M., Ford, R.W.: A computational algorithm for handling the special uniques problem. Int. J. Uncertain. Fuzziness Knowl. Based Syst. 10(5), 493–509 (2002)
8. Gomatam, S., Karr, A.F.: Distortion measures for categorical data swapping. Technical report, No. 131. National Institute of Statistical Sciences (2003)
9. Ito, S., Hoshino, N.: The potential of data swapping as a disclosure limitation method for official microdata in Japan: an empirical study to assess data utility and disclosure risk for Census microdata. Paper presented at Privacy in Statistical Databases 2012, Palermo, Sicily, Italy, pp. 1–13 (2012)

10. Ito, S., Hoshino, N.: Assessing the effectiveness of disclosure limitation methods for Census microdata in Japan. Paper presented at Joint UNECE/Eurostat Work Session on Statistical Data Confidentiality, Ottawa, Canada, pp. 1–10 (2013)

11. Ito, S., Hoshino, N.: Data swapping as a more efficient tool to create anonymized Census microdata in Japan. Paper presented at Privacy in Statistical Databases 2014, Ibiza, Spain, pp. 1–14 (2014)

12. Ito, S., Hoshino, N., Akutsu, F.: A quantitative assessment of data confidentiality and data utility to create anonymized Census microdata in Japan. Paper presented at Joint UNECE/Eurostat Work Session on Statistical Data Confidentiality, Helsinki, Finland, pp. 1–14 (2015)

13. Ito, S., Hoshino, N., Akutsu, F.: Potential of disclosure limitation methods for Census microdata in Japan. Paper presented at Privacy in Statistical Databases 2016, Dubrovnik, Croatia, pp. 1–14 (2016)

14. Ito, S., Hoshino, N., Akutsu, F., Kikuchi, R.: Investigating new methods for creating anonymized microdata based on Japanese Census data. Paper presented at Joint UNECE/Eurostat Work Session on Statistical Data Confidentiality, Ministry of Foreign Affairs, Skopje, Macedonia, 2017, pp. 1–16 (2017)

15. Kooiman, P., Willenborg, L., and Gouweleeuw, J.: PRAM: a method for disclosure limitation of microdata, Research Paper, No. 9705, Statistics Netherlands, Voorburg (1998)

16. Moore, R.A.: Controlled data-swapping techniques for masked public use microdata sets. Statistical Research Division Report Series, RR 96-04, U.S. Bureau of the Census, Statistical Research Division, Washington, D.C. (1996)

17. Nin, J., Herranz, J., Torra, V.: Rethinking rank swapping to decrease disclosure risk. Data Knowl. Eng. **64**(1), 346–364 (2008)

18. Shlomo, N., Tudor, C., Groom, P.: Data swapping for protecting census tables. In: Domingo-Ferrer, J., Magkos, E. (eds.) PSD 2010. LNCS, vol. 6344, pp. 41–51. Springer, Heidelberg (2010). https://doi.org/10.1007/978-3-642-15838-4_4

19. Takemura, A.: Local recoding and record swapping by maximum weight matching for disclosure control of microdata set. J. Off. Stat. **18**(2), 275–289 (2002)

A General Framework and Metrics for Longitudinal Data Anonymization

Nicolas Ruiz[⊠]

Department of Computer Science and Mathematics,
CYBERCAT-Center for Cybersecurity Research of Catalonia UNESCO
Chair in Data Privacy, Universitat Rovira i Virgili, Tarragona, Spain
nicolas.ruiz@oecd.org

Abstract. The bulk of methods in statistical disclosure control primarily deal with individual data from a cross-sectional perspective, i.e. data where individuals are observed at one single point in time. However, nowadays longitudinal data, i.e. individuals observed over multiple periods, are increasingly collected. Such data enhance undoubtedly the possibility of statistical analysis compared to cross-sectional data, but also come with some additional layers of information that have to remain practically useful in a privacy-preserving way. Building on the recently proposed permutation paradigm as an overarching approach to data anonymization, this paper establishes a general framework for the formulation of longitudinal data anonymization and proposes some universal metrics for the assessment of disclosure risk and information loss. We illustrate the application of these new tools using an empirical example.

Keywords: Statistical disclosure control · Longitudinal data
Permutation paradigm

1 Introduction

Data on individual subjects are increasingly collected and exchanged. By their nature, they provide a rich amount of information that can inform statistical and policy analysis in a meaningful way. However, due to the legal obligations surrounding these data, this wealth of information is often not fully exploited in order to protect the confidentiality of respondents. In fact, such requirements shape the dissemination policy of micro data at national and international levels. The issue is how to ensure a sufficient level of data protection to meet releasers' concerns in terms of legal and ethical requirements, while offering to users a reasonable richness of information. Moreover, over the last decade the role of micro data has changed from being the preserve of National Statistical Offices and government departments to being a vital tool for a wide range of analysts trying to understand both social and economic phenomena. As a result, more parties, often very heterogeneous in their privacy and information requirements, are now involved in micro data transactions. This has opened a new range of questions and pressing needs about the privacy/information trade-off and the quest for best practices that can be both useful to users but also respectful of respondents' privacy.

© Springer Nature Switzerland AG 2018
J. Domingo-Ferrer and F. Montes (Eds.): PSD 2018, LNCS 11126, pp. 215–230, 2018.
https://doi.org/10.1007/978-3-319-99771-1_15

Statistical disclosure control (SDC) research has a rich history in addressing those issues, by providing the analytical apparatus through which the privacy/information trade-off can be assessed and implemented. SDC consists in the set of tools that can enhance the level of confidentiality of any data while preserving to a lesser or greater extent its level of information (see [5] for an authoritative survey). Over the years, it has burgeoned in many directions. In particular, techniques applicable to micro data, which are the focus of this paper, offer a wide variety of tools to protect the confidentiality of respondents while maximizing the information content of the data released, for the benefits of society at large.

There are several types of individual data that can be published in a privacy–preserving way for fulfilling analysis needs, e.g. relational data, transaction data, sequence data, trajectory data, and graph data... These data types differ in structure, properties and the information they contain about individuals. The dissemination of any specific type entails its own privacy risks and information preservation requirements, which should ideally be considered by the SDC approach selected to perform anonymization. Among these different types, longitudinal data are of particular interest in many areas, e.g. economics, medical research, sociology, finance, marketing... A dataset is longitudinal if it contains information on the same variables of interest about an individual at several points in time. For example, the information collected in clinical trials to evaluate the impact of treatments, or the dynamic of an individual's income, is longitudinal data. They are built from the pooling of observations on a cross-section of individuals over several time periods, achieved by surveying a number of individuals and following them over time.

However, despite the fact that the SDC literature offer a wide variety of tools suited to different contexts and data types [4], there have been very few attempts to deal with the challenges posed by longitudinal data. To the best of the author knowledge, only one approach, formulated in the context of medical data and based on global suppression and generalization, has been proposed so far [8]. Hence, the objective of this paper is to contribute in filling this gap by proposing a general framework and some associated metrics of disclosure risk and information loss tailored to the specific challenges posed by longitudinal data anonymization, notably by building on the recently proposed permutation paradigm as an overarching approach to data anonymization [2].

The rest of this paper is structured as follows. Section 2 gives some background concepts on longitudinal data and the permutation paradigm in data anonymization needed later on. Section 3 proposes a new framework for conceptualizing longitudinal data anonymization and then proposes some universal measures of disclosure risk and information loss applicable to such data. Section 4 presents some empirical results. Conclusions and future research directions are gathered in Sect. 5.

2 Background Concepts

2.1 Longitudinal Data

Longitudinal data are repeated observations of the same respondents that are published at different points in time and are ubiquitous in a wide range of fields: medicine, public health, education, business, economics, psychology, biology, and more. Economists generally refer to it as panel data. They vary from cross-sectional data, i.e. where individuals are observed at a single point in time, and from time-series data, i.e. where *one* single entity is observed along generally a long time-span, in the sense that the defining feature of longitudinal data is that the multiple observations within several individuals can be ordered across time. Longitudinal surveys generally use calendar time, months or years, as the dimension separating observations on the same subject. Although the notion of time in longitudinal data can be quite intricate [9], in this paper we will focus on repeatedly measured attributes that can be ordered along a line to describe the sequence of measurement.

Compared to cross-sectional data, longitudinal data provide some clear advantages as they are generally more informative. Cross-sectional distributions that look relatively stable can in fact hide a multitude of changes that can only be captured if the same set of individuals is followed over time. For example, spells of unemployment, job turnover, residential and income mobility are better studied with longitudinal data. Longitudinal data are also well suited to study states durations, e.g. disease, unemployment and poverty, and if the time dimension is long enough, they can shed light on the speed of adjustments to medical treatments or policy changes. For instance, in measuring unemployment, cross-sectional data can estimate what proportion of the population is unemployed at a point in time. Repeated cross-sections can show how this proportion changes over time. But only longitudinal data can estimate what proportion of those who are unemployed in one period can remain unemployed in another period.

However, longitudinal data can be potentially plagued by several problems, the main specific one being attrition. While nonresponse from individuals is a standard issue in cross-sectional data, it is a more serious problem in longitudinal data because different periods of the data can be subject to varying rates of nonresponse from individuals. This issue generally leads to what is called an unbalanced longitudinal data set, i.e. *not every* individual is observed *every* year, while in the case of a balanced data set all individuals are observed at all periods. While the former case may appear as more realistic, it remains however barely considered in practice and unbalanced data are generally made *de facto* balanced by not considering as relevant information individuals not observed across all periods. For example, econometric analysis techniques are much easier to implement and more developed on balanced than unbalanced data [10]. In this paper, we will assume that the longitudinal data set to be anonymized is balanced, albeit anonymization on unbalanced data remains a path for future research.

Now, it is clear that the anonymization of longitudinal data poses some specific challenges. While it is beyond the scope of the present paper to investigate exhaustively the possible forms of an attacker's background knowledge specific to longitudinal data, we can outline the main ones. Indeed, such knowledge may be thought of with its own characteristics compared to other types of data, and in particular cross-sectional data, and thus will carry specific privacy challenges. For example, an adversary may know that someone has transitioned from unemployment to employment between two time periods. Thus, while the employment status can be considered as a quasi-identifier in cross-sectional data, the change in employment status over time is also in itself a quasi-identifier in longitudinal data and can be used as additional background knowledge for the attacker.

Along the same line, changes in confidential attributes, such as salary, can also be viewed as a quasi-identifier: an attacker may for example not know the salary of an individual at two periods, but may know that it has increased significantly between the two and can use that information to conduct the attack. Thus, the individual may consider as a privacy risk the fact that someone can learn about his salary variation, even if his salaries at the two time periods are not disclosed, e.g. the two salary values have been masked enough to avoid attribute disclosure but the masked values can still go up over time, providing insights for the intruder. Thus, longitudinal data generally enlarge privacy threats.

Now, this widening is also a widening of information specific to longitudinal data. This is in fact what make them specifically valuable in the first place and must be preserved to a lesser or greater extent for the dissemination of longitudinal data to be useful. The trade-off between privacy and information is thus very direct in longitudinal data: the information on the dynamics of several variables at the individual level is valuable but is also problematic from a privacy perspective. The metrics developed later in this paper for the measures of disclosure risk and information loss in the context of longitudinal data will rely on such direct link.

2.2   The Permutation Paradigm in Data Anonymization

The permutation paradigm is a recent contribution in the literature that proposed a general functional equivalence based on permutations to describe any data masking method (see [2, 6] and its subsequent development in [7]). It starts from the observation that any anonymized data set can be viewed as a permutation of the original data plus a non-rank perturbative noise addition. Thus, it establishes that all masking methods can be thought of in terms of a single ingredient, i.e. permutation. This result clearly has far reaching conceptual and practical consequences, in the sense that it provides a single and easily understandable reading key, independent of the model parameters, the risk measures or the specific characteristics of the data, to interpret the utility/protection outcome of an anonymization procedure.

To illustrate this equivalence, we use a toy example which consists (without loss of generality) of five records and three attributes $X = (X_1, X_2, X_3)$ generated by sampling $N(10, 10^2)$, $N(100, 40^2)$ and $N(1000, 2000^2)$ distributions, respectively. Noise is then added to obtain $Y = (Y_1, Y_2, Y_3)$, the three anonymized version of the attributes, from $N(0, 5^2)$, $N(0, 20^2)$ and $N(0, 1000^2)$ distributions, respectively. One can see that the masking procedure generates a permutation of the records of the original data (Table 1).

Table 1. Illustration of the permutation paradigm

Original dataset X			Masked dataset Y		
X_1	X_2	X_3	Y_1	Y_2	Y_3
13	135	3707	8	160	3248
20	52	826	20	57	822
2	123	-1317	-1	122	248
15	165	2419	18	135	597
29	160	-1008	29	164	-1927
Rank of the original attribute			Rank of the masked attribute		
X_{1R}	X_{2R}	X_{3R}	Y_{1R}	Y_{2R}	Y_{3R}
4	3	1	4	2	1
2	5	3	2	5	2
5	4	5	5	4	4
3	1	2	3	3	3
1	2	4	1	1	5

Now, as long as the attributes' values of a data set can be ranked, which is obvious in the case of numerical and categorical ordinal attributes, but also feasible in the case of nominal ones [3], it is always possible to derive a data set Z that contains the attributes X_1, X_2 and X_3, but ordered according to the ranks of Y_1, Y_2 and Y_3, respectively, i.e. in Table 1 re-ordering (X_1, X_2, X_3) according to (Y_{1R}, Y_{2R}, Y_{3R}). This can be done following the post-masking reverse procedure outlined in [6]. Finally, the masked data Y can be fully reconstituted by adding small noises (E_1, E_2, E_3) (small in the sense that they cannot re-rank Z while they can still be large in absolute values) to each observation in each attribute (Table 2).

Table 2. Equivalence in anonymization: post-masking reverse mapping plus noise addition

Original dataset X			Reverse mapped dataset Z		
X_1	X_2	X_3	Z_1	Z_2	Z_3
13	135	3707	13	160	3707
20	52	826	20	52	2419
2	123	-1317	2	123	-1008
15	165	2419	15	135	826
29	160	-1008	29	165	-1317
Noise E			Masked dataset Y(=Z+E)		
E_1	E_2	E_3	Y_1	Y_2	Y_3
-5	0	-459	8	160	3248
0	5	-1597	20	57	822
-3	0	1256	-1	122	248
2	0	-229	18	135	597
0	-1	-610	29	164	-1927

By construction, Z has the same marginal distribution as X, which is an appealing property. Moreover, under a maximum-knowledge intruder model of disclosure risk evaluation, the small noise addition turns out to be irrelevant [2]: re-identification via record linkage can only come from permutation, as by construction noise addition cannot alter ranks. Reverse mapping thus establishes permutation as the overarching principle of data anonymization, allowing the functioning of any method to be viewed as the outcome of a permutation of the original data, independently of how the method operates. This functional equivalence leads to the following proposition (see [7] for the original and full proposal):

Proposition 1: For a dataset X with n records and p attributes $(X_1,..,X_p)$, its anonymized version Y can always be written, regardless of the anonymization methods used, as:

$$Y = \left(P_1 X_1, \ldots, P_p X_p\right) + E \tag{1}$$

where $P_1 = A_1^T D_1 A_1, .., P_p = A_p^T D_p A_p$ is a set of p permutation matrices and E is a matrix of small noises. $A_1,..,A_p$ is a set of p permutation matrices that sort the attributes in increasing order, $A_1^T,..,A_p^T$ a set of p permutation matrices that put back the attribute in the original order, and $D_1,..,D_p$ is a set of permutation matrices for anonymizing the data.

This proposition characterizes permutation matrices as an encompassing tool for data anonymization: the analytical framework of anonymization mechanisms can in fact be viewed as functionally equivalent to a set of permutation matrices. Proceeding attribute by attribute, each is first permuted to appear in increasing order, then the key is injected, and finally it is re-ordered back to its original form by applying the inverse of the first step (which in the case of a permutation matrix is simply its transpose). Clearly, this formalizes the common basis of comparison for different mechanisms that the permutation paradigm originally proposed. Whatever the differences in the natures of the methods to be compared and the distributional features of the original data, the methods can fundamentally always be viewed as the application of different permutation matrices to the original data, but independently of them.

3   A General, Permutation-Based Approach to Longitudinal Data Anonymization

3.1   Backward Mapping of Attributes in Longitudinal Data

We first start by an observation on the relationship between two attributes follow-up over time and over the same set of individuals, i.e. data are balanced, as assumed above. In fact, and while the context and the goal are different, one attribute observed during two periods t and t + 1 can also always be reverse mapped in a way to express the attribute in t + 1 as a function of itself in t. This approach, general in its scope, will lead to a simple characterization of the essential information and privacy risks specifically contained in longitudinal data. It can be noted that this is equivalent to considering time as an anonymization method, where the attribute in t + 1 is the anonymized version of the attribute in t.

By definition, to be followed-up over time, an attribute must keep the same form and definition, e.g. if it is categorical in t it must remain categorical in t + 1 and track the same categories; if it is numerical in t it must remain numerical in t + 1 and capture the same variable. Let denote by $X_{j,t} = (x_{1,j,t}, \ldots, x_{n,j,t})$ the values taken by attribute j in t and $X_{j,t+1} = (x_{1,j,t+1}, \ldots, x_{n,j,t+1})$ its values taken in t + 1. As said above, n is assumed to remain constant between t and t + 1. Note that no assumption is made on the nature of the attribute j apart that it can always be ranked: it can be numerical, categorical or nominal[1]. The knowledge of $X_{j,t}$ and $X_{j,t+1}$ allows expressing the later as a function of the former by disentangling the nature of information in longitudinal data, using the following algorithm:

> **Algorithm**: backward mapping of attributes in longitudinal data
>> **Require**: attribute in t $X_{j,t} = (x_{1,j,t}, \ldots, x_{n,j,t})$
>> **Require**: attribute in t+1 $X_{j,t+1} = (x_{1,j,t+1}, \ldots, x_{n,j,t+1})$
>> **For** i=1,...,n **do**
>>> Compute k=Rank($x_{i,j,t+1}$)
>>> Set z_i=Rank($x_{k,j,t}$) (where $x_{k,j,t}$ is the value of $X_{j,t}$ of rank k)
>> **End for**
>> **Return** $Z_{j,t} = (z_{1,j,t}, \ldots, z_{n,j,t})$

The resulting backward mapped attribute $Z_{j,t}$ expresses $X_{j,t+1}$ as a permutation of $X_{j,t}$. Because the values of the attribute may change over time, particularly in the case of numerical attribute, one must also add $E_{j,t,t+1}$, the difference between $X_{j,t+1}$ and $Z_{j,t}$, to get an exact recomposition of $X_{j,t+1}$ as a function of $X_{j,t}$. Then, and because $Z_{j,t}$ is a permutation of $X_{j,t}$, it always hold that (with Q_{Tj} denoting a permutation matrix):

$$X_{j,t+1} = Q_{Tj} X_{j,t} + E_{j,t,t+1} \text{ with } Q_{Tj} = C_{Tj}^T K_{Tj} C_{Tj} \tag{2}$$

It must be noted that the backward mapping procedure used here is analytically similar to the reverse mapping procedure developed in [6] (and outlined above), but serves a completely different purpose. In fact, it does not deal with anonymization but allows characterizing the two types of temporal information available in longitudinal data.

Indeed, Eq. (2) disentangles the effect of time over an attribute, leading to two entities. First, time modifies an attribute by changing the ranks of the individuals in a distribution. Because $Z_{j,t}$ is a permutation of $X_{j,t}$, the change of ranks through time can always be captured by the permutation matrix Q_{Tj} (or more precisely by the temporal key K_{Tj}, as C_{Tj} and its inverse just re-order the data as in Eq. (1)). In fact, and following *Proposition 1*, this means that the main feature of longitudinal data can

---

[1] For this last case, it can also be ranked using semantic distance metrics (see above).

always be represented by the same entities used to express any anonymization method. As will be apparent below, this will turn out to be convenient for thinking about longitudinal data anonymization in a very general way.

The second type of information produced by time is what can be qualified as residual trajectories, i.e. changes in the attribute's values within two ranks, and is captured by $E_{j,t,t+1}$. Such information is contextual in nature. For a categorical attribute, $E_{j,t,t+1}$ will be by definition null. In the case of a numerical attribute, it will capture the effect of time on an attribute not due to rank changes. For example, if the salary of an individual moves from rank 4 to rank 7 in the salary distribution, then his residual trajectory will be such that his salary will still be contained between the values (also altered) of ranks 6 and 8. By nature this information is less relevant than the permutation patterns contained in Q_{Tj}: the major effect of time is rank changes. However, it cannot be entirely discarded: if for instance the salaries in an economy grow at the same pace for everyone between two periods and no rank changes occur[2], this overall increase can only be seized by $E_{j,t,t+1}$. Thus, $E_{j,t,t+1}$ will notably capture how the entire distribution shifts through time, while Q_{Tj} will always capture how individuals move within the distribution over time.

3.2 The Effect of Anonymization on Temporal Information

Now, following *Proposition 1* and using Eq. (1), the anonymized versions of $X_{j,t}$ and $X_{j,t+1}$, denoted respectively by $X_{j,t}^A$ and $X_{j,t+1}^A$, can always be written, whatever the anonymization methods considered for the two periods, as:

$$X_{j,t}^A = P_{j,t}X_{j,t} + E_{j,t} \tag{3}$$

$$X_{j,t+1}^A = P_{j,t+1}X_{j,t+1} + E_{j,t+1} \tag{4}$$

where $P_{j,t}$ and $P_{j,t+1}$ are, following the permutation paradigm, the matrices used to describe the core functioning of the anonymization method used for the attribute observed in t and t + 1 respectively, and $E_{j,t}$ and $E_{j,t+1}$ are the eventual matrices of small noises.

From an information perspective, it is clear that Eq. (2) has to remain exactly conserved for the specific temporal information conveyed by the longitudinal data to stay untouched. Now, by substituting (2) in (4), using the expression of $X_{j,t}$ in (3) as a function of its anonymized version and keeping in mind that the inverse of a permutation matrix is its transpose, one gets after rearrangements:

$$X_{j,t+1}^A = P_{j,t+1}Q_{Tj}P_{j,t}'X_{j,t}^A + \left[P_{j,t+1}\left(E_{j,t,t+1} - Q_{Tj}P_{j,t}'E_{j,t}\right) + E_{j,t+1}\right] \tag{5}$$

As a result, if the two anonymization methods used in t and t + 1 don't alter temporal information, it must hold, by comparison of (2) and (5), that:

---

[2] In that case P_{Tj} will be the identity matrix.

$$P_{j,t+1} Q_{T,j} P'_{j,t} = P_{T,j} \tag{6}$$

$$P_{j,t+1} \left(E_{j,t,t+1} - Q_{T,j} P'_{j,t} E_{j,t} \right) + E_{j,t+1} = E_{j,t,t+1} \tag{7}$$

Equations (6) and (7) describe how the two anonymization methods in t and t + 1 must be related to preserve the temporal information. First, the principal source temporal of information $Q_{T,j}$ appears to be encased by the two permutation matrices of each method. Thus, for $Q_{T,j}$ to remain unaltered in the anonymized version of the data set, we see by (6) that the product of the anonymizing permutation matrix used in t + 1, the permutation matrix capturing the effect of time, and the transpose of the anonymizing permutation matrix used in t, must be equal to the permutation matrix capturing the effect of time itself (note that because it is a product of matrices the terms cannot be rearranged conveniently).

Second, using the fact that small noises turn out to be irrelevant to describe the core functioning of an anonymization method [2], we can simplify Eq. (7), which becomes:

$$P_{j,t+1} E_{j,t,t+1} = E_{j,t,t+1} \tag{8}$$

Thus, for the residual trajectories to be preserved $P_{j,t+1}$ must be the identity matrix, i.e. no anonymization at all must take place on the attribute in period t + 1. Therefore, for Eq. (6) to be verified then $P_{j,t}$ must also be the identity matrix, i.e. no anonymization at all must also take place in period t. This rather pointless and unsafe setting can be ignored given the fact that residual trajectories do not constitute the bulk of the relevant longitudinal information. In the remainder of this paper, we will thus focus on Eq. (6) and its implication for longitudinal data anonymization.

3.3   Universal Measures of Disclosure Risk and Information Loss for Longitudinal Data Anonymization

What precedes outlined a general way to conceive longitudinal data anonymization. It can be applied to any kind of attributes and characterizes that, compared to cross-sectional data, longitudinal data offer an essential but specific feature, i.e. the trans-position matrix $Q_{T,j}$ describing the effect of time on one attribute. This matrix contains the main source of information that must be preserved somehow but which at the same time entails some privacy risks. Thus, and as stated above, the flip side of disclosure risk in longitudinal data is information. A data user will esteem the fact of knowing how the attributes' values of some individuals change over time, but a data releaser may worry that such information could contribute to the knowledge of an intruder and being operationalized for re-identification. As a result, any modification of $Q_{T,j}$ will decrease disclosure risk but will also induce some information losses. The information/privacy trade-off is thus of a very direct kind in longitudinal data.

For data anonymization to take place Eq. (6) can never hold in practice. The question is thus more about how $P_{j,t+1} Q_{T,j} P'_{j,t}$ will depart from $Q_{T,j}$. Bearing in mind that the result of the product of some permutation matrices is always a permutation matrix, this question can be assessed by the fact that the encasing of $Q_{T,j}$ by $P_{j,t+1}$ and $P'_{j,t}$ will lead to a different pattern of rank changes over time.

Now, using the procedure developed in [7] in the case of cross-sectional data, assume the two following permutation matrices:

$$P_1 = \begin{pmatrix} 0 & 0 & 0 & 0 & 1 \\ 0 & 1 & 0 & 0 & 0 \\ 0 & 0 & 1 & 0 & 0 \\ 1 & 0 & 0 & 0 & 0 \\ 0 & 0 & 0 & 1 & 0 \end{pmatrix} \quad P_2 = \begin{pmatrix} 1 & 0 & 0 & 0 & 0 \\ 0 & 0 & 0 & 1 & 0 \\ 0 & 0 & 0 & 0 & 1 \\ 0 & 1 & 0 & 0 & 0 \\ 0 & 0 & 1 & 0 & 0 \end{pmatrix}$$

It can be easily retrieved from these matrices some vectors of rank displacement r_1 and r_2, i.e. vectors measuring the amount of rank shifting that a permutation matrix contains[3]. To build r_1 and r_2 one can count, columns by columns of P_1 and P_2, how many times the 1 s have been moved, using the identity matrix as a starting point (which is a particular case of a permutation matrix with no permutation applied), then assigning a negative (resp. positive) sign if the 1 has been moved up (resp. down). As a result, one gets:

$$r_1 = \begin{pmatrix} 3 \\ \varepsilon \\ \varepsilon \\ 1 \\ -4 \end{pmatrix} \quad r_2 = \begin{pmatrix} \varepsilon \\ 2 \\ 2 \\ -2 \\ -2 \end{pmatrix}$$

In the context of data anonymization, these vectors can be used to evaluate how anonymization has permuted individuals, and in particular by how far in terms of rank [7]. In the context of the backward mapping procedure developed above, such vector derived from $P_{T,j}$ will characterize how and by how far time has permuted individuals. Consequently, measuring disclosure risk and information loss in longitudinal data anonymization can be conceived as measuring the differences between the rank shifting induced by $P_{j,t+1}Q_{T,j}P'_{j,t}$ and the rank shifting induced by $Q_{T,j}$. It remains now to determine the aggregative structure for measurement.

A natural choice for gauging r_1 and r_2 is for example to take their Euclidean norms and adopting the rule that the higher the norm, the lower the disclosure risk (as the larger will be the permutation distances contained in the vectors). But other cases are possible. In general, any L(p)-norm is acceptable: for example, the ∞-norm (or Chebyshev distance) is also a valid candidate. This variety of choice to evaluate vectors generally depends on the problem at hand, as one will select a L(p)-norm adapted to the meaning of the object that is meant to be quantified. However, we argue that in the context of longitudinal data, such choice can be given an intuitive interpretation in term of disclosure risk and information loss, but in a way that appears to be different than from cross-sectional data [7].

---

[3] To avoid some unnecessary technical difficulties, in what follows zero values in these vectors will be assigned, without loss of generality, a infinitesimally small value $\varepsilon > 0$.

As mentioned, disclosure risk and information loss are tightly linked in longitudinal data. For instance, assume that between t and t + 1 an individual moved 4 ranks up in the distribution, i.e. in the rank displacement vectors derived from $P_{T,j}$ this individual is assigned +4. Assume also that after anonymization of the attribute in t and t + 1 the same individual is characterized by having moved 5 ranks up, i.e. in the rank displacement vectors derived from $P_{j,t+1}Q_{T,j}P'_{j,t}$, this individual is assigned +5. Anonymization has altered information but in a small way, as the individual is now characterized by a move between t and t + 1 close to his ex-ante anonymization move. However, it implies that this individual is not equipped with sufficient protection against disclosure risk, because his move in the anonymized data is very close to his move in the original data, and such closeness can still lead to a privacy threat by enlarging, albeit now imperfectly, the background knowledge of an intruder.

Now, assume that the same individual is, after anonymization, characterized by having moved 100 ranks up. Here anonymization has altered information in a major way as the individual is now characterized by a move between t and t + 1 quite dissimilar to his real, ex-ante anonymization move. But it implies also that this individual is now equipped with sufficient protection against disclosure risk, as his move in the anonymized data is far from his move in the original data. Such dissimilarity can now only poorly enlarged the background knowledge of an intruder, if not fool him.

As a result, small differences between the rank shifting vectors derived from $P_{j,t+1}Q_{T,j}P'_{j,t}$ and $P_{T,j}$ mean high disclosure risk and low information loss for the anonymization of longitudinal data, while large differences mean low disclosure risk and high information loss. Thus, the values in the vector of differences between the rank shifting vectors retrieved from $P_{j,t+1}Q_{T,j}P'_{j,t}$ and $P_{T,j}$ will account both for disclosure risk and information loss. How to evaluate this vector of differences leads to the following proposition:

Proposition 2: Denote by $r_{T,j}$ and $r_{A,j}$ the rank shifting vectors retrieved from $Q_{T,j}$ and $P_{j,t+1}Q_{T,j}P'_{j,t}$ respectively, and by $r_{T,A,j,i} = r_{T,j} - r_{A,j} = (r_{T,A,j,1}, \ldots, r_{T,A,j,n})$ the vector of differences between $r_{T,j}$ and $r_{A,j}$ over the n individuals for which the attribute j is available in t and t + 1. The following aggregative structure:

$$J(\alpha) = \begin{cases} \left(\frac{1}{n} \sum_{i=1}^{n} (abs(r_{T,A,j,i}))^{\alpha} \right)^{\frac{1}{\alpha}} & for\ \alpha \neq 0 \\ \prod_{i=1}^{n} (abs(r_{T,A,j,i}))^{\frac{1}{n}} & for\ \alpha = 0 \end{cases}$$

forms a class of both disclosure risk and information loss measures for the evaluation of longitudinal data anonymization.

Proposition 2 makes use of a power mean for aggregating of the components of $r_{T,A,j,i}$, with the parameter α acting as a zooming lens. The arithmetic mean becomes a special case ($\alpha = 1$) of $J(\alpha)$, which forms a dividing line by computing the average level of the rank changes differences between the original data and the anonymized data. From this benchmark, the more α *decreases*, the more emphasis is given to the *smallest* rank changes. In fact, $J(0)$ is the geometric mean and $J(-1)$ the harmonic mean and the more α approaches $-\infty$, the more $J(\alpha)$ converges towards the smallest rank change in $r_{T,A,j,i}$. As a result, for a given $r_{T,A,j,i}$ and $\alpha' < \alpha$, we have $J(\alpha') \leq J(\alpha)$: the lower is α, the stronger is the emphasis on the smallest rank changes between the original data and the anonymized data.

Conversely, from the arithmetic mean $J(1)$, the more α *increases*, the more emphasis is given to the *largest* rank changes. And through the same reasoning as above, the more α approaches $+\infty$, the more $J(\alpha)$ converges towards the largest rank change in $r_{T,A,j,i}$. Note also that $J(2)$ and $J(+\infty)$ are respectively the Euclidean norm and the Chebyshev distance up to the factor $\sqrt[\alpha]{n}$.

$J(\alpha)$ aims at measuring the extent of dissimilarity that anonymization introduced on temporal information, with α capturing the different emphasis on the rank changes. It is a measure very general in its scope as it encompasses different measures already quite popular in the anonymization literature [5]. It makes also use of the fact that data anonymization methods all boil down to applying permutation patterns, which greatly simplifies evaluation [7]. When using current methods, protection against disclosure risk and information loss occur at the intersection of two features: the appropriateness and the parametrization of the method selected, and the distributional properties of the data to be anonymized. But, when data anonymization is viewed as permutation, then only the alteration of ranks matters. This is why *Proposition 2* inherits the universal properties of similar measures of disclosure risk and information loss developed in the context of cross-sectional data [7], by making abstraction of the interplay between the distributional features of the data and the analytics of the methods. As a result, it can in fact by applied to any kind of longitudinal data and for the ex-post evaluation of any anonymization methods applied on any attribute followed over time.

4 Experimental Investigation

The objective of this section is to illustrate the use and effectiveness of the universal measures of disclosure risk and information loss developed above. The experimental data set used is one attribute of the Census data set, observed over 1080 individuals. This data set has been used many times in the literature to evaluate the properties of anonymization techniques in terms of disclosure risk and information loss [1].

The experiment is the following, assuming that the attribute from the original data is considered observed in period t:

  i. Time scenario 1: Given that in period t the attribute is closely distributed as a normal law, we randomly generated the attribute for t + 1 from a normal law with the same standard error than in t but with a mean of 2% more, assuming that overall the attribute's value has increased.
  ii. Time scenario 2: We randomly generated some growth rates for each individual, constrained between −20% and 20%.
  iii. Anonymization methods: for each time scenario, the attribute in t has been anonymized using additive noise with a standard deviation equal to 50% of the standard error of the original values in t. For the attribute in t + 1, we considered two versions: noise addition with half of the standard error in t + 1 or the same standard error than in t + 1.

iv. We then computed $r_{T,A,j,i}$, the values in the vector of differences between the rank shifting vectors derived from $P_{j,t+1}Q_{Tj}P'_{j,t}$ and Q_{Tj}, for each time scenario and anonymization procedures.

v. Finally, from these values we computed $J(\alpha)$ for a quasi-continuum of α parameters, that is by increments of 0.01. The results are displayed directly under the form of curves with the α parameters on the x-axis and the value of $J(\alpha)$ on the y-axis.

In this experiment, the purpose of having two time scenario aims at setting different longitudinal data configurations. In the first, the movements of individuals between t and t + 1 are of larger magnitudes in terms of rank changes, while it is the reverse in the second. This can be seen by the in Fig. 1, which are the curves derived from applying power means under the same range of α to $abs(r_{Tj})$, i.e. the absolute values of the rank shifting vector derived from Q_{Tj}. These curves show how time has moved individuals between t and t + 1 and are a display of the essential time information contained in the longitudinal data, following the backward mapping procedure. In fact, for both curves a large chunk of individuals kept the same ranks between t and t + 1, as both curves flat out at zero for α around -0.5. However, in the first time scenario the average level of rank changes (i.e. for $\alpha = 1$) is higher than for the second time scenario. When the focus is made on large rank changes (i.e. for $\alpha > 1$), scenario 1 also shows far greater magnitudes of rank changes.

The effect of anonymization on longitudinal information can be seen in Figs. 2 and 3. The curves displayed are the outcomes of anonymization on *both* disclosure risk and information[4]. Indeed, individual trajectories through the attribute space (that can

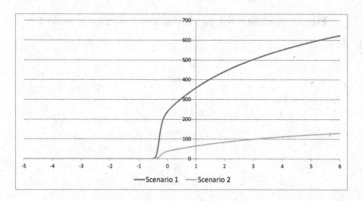

Fig. 1. Temporal information: time rank changes

---

[4] To avoid confusion, it must be noted that despite similar profiles Fig. 1 and Figs. 2 and 3 cannot directly be compared. In particular, the individuals with no time rank change in Fig. 1 are not necessarily the same than in Figs. 2 and 3: in the latter, individuals contributing to the flat portion of the curves at zero may have moved through time, but anonymization in t and t + 1 in fact didn't alter their moves.

overall be appraised by a permutation matrix), represent the essential source of information brought by longitudinal data, but are also a specific source of disclosure risk. Thus, a curve close to the x-axis means that anonymization didn't alter time rank changes: disclosure risk is high but information loss is low. Conversely, a curve far above the x-axis means that time rank changes have been substantially distorted: disclosure risk is low but information loss is high.

One alternative way to consider this is viewing Figs. 2 and 3 as two panels taking $\alpha = 1$ as a dividing line: on the left one is looking at disclosure risk first (by focusing on measures according relatively more weight to less altered time rank changes but thus with less information loss), while on the right one is looking at information first (by focusing on measures according relatively more weight to more altered time rank changes but thus with less risk of disclosure).

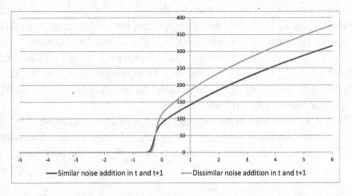

Fig. 2. Disclosure risk and information loss: time scenario 1

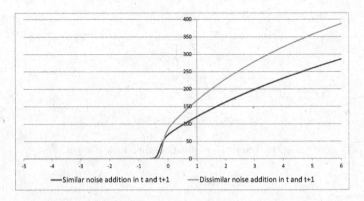

Fig. 3. Disclosure risk and information loss: time scenario 2

It appears that anonymization, when performed in a similar way between t and t + 1, appears to lead to less information loss and low protection against disclosure risk. This a rather intuitive finding. When the attribute is anonymized with noise addition set as half of the standard error of the original data in t and t + 1, the resulting curves are consistently lower than when the attribute in t + 1 has been anonymized with the same standard error (Figs. 2 and 3). Thus, it is clear that the dissimilarity in anonymization methods or parametrization through time will lead to better protection (but more information loss) of longitudinal data. However, and whatever the dissimilarity in methods, a large chunk of individuals is left with their time rank changes unmodified: across time scenario and anonymization methods, all curves are beating flat when crossing the geometric mean (i.e. for $\alpha = 0$) and below.

Finally, the dissimilarity in anonymization methods delivers the same outcomes whatever the time scenario considered. In Fig. 2, time rank changes are altered in close ways whether half or the same standard error of the original data is used to generate noise in t + 1. This is also the case in Fig. 3, albeit the differences are larger for the second time scenario when one is putting relatively more weight to largest disruption in time rank changes. We believe this issue deserves further scrutiny.

5 Conclusions and Future Work

The objective of this paper was to investigate longitudinal data anonymization. We first presented a backward mapping procedure that allows expressing any kind of attribute observed in t + 1 as a function of its values in t, i.e. by considering time as an anonymization procedure. Obviously, this procedure has nothing to do with anonymization *per se* but allows viewing the supplementary information contained in longitudinal data, in particular compared to cross-sectional data, mainly as a permutation matrix. Thus, given the recently established insights of the permutation-based paradigm in data anonymization, which describes the outcomes of any anonymization methods performed on any type of data as permutation, the backward mapping procedure appears to analytically align the specificities of longitudinal data with the overarching tool of data anonymization.

From this general view on longitudinal data, we then characterized the effect of anonymization on temporal information: anonymization of an attribute over two periods always appears to encase temporal information, leading to a specific alteration of time rank changes. This alteration can be then evaluated using a class of universal disclosure risk and information loss, two outcomes that are tightly linked in longitudinal data. This paper established such measures using a power-mean based aggregative structure, as recently proposed in the case of cross-sectional data, and provided some illustrations.

Meant to be very general in its scope, we hope that this framework for longitudinal data anonymization will allow to foster a research question that has so far been overlooked in the statistical disclosure control literature. As additional future work, we plan to: (i) extend this framework to more than two time periods through a generalization of the backward mapping procedure and of the measures of disclosure risk and information loss proposed; (ii) enlarge the scope of the experimental work, notably by using longitudinal data and testing a variety of anonymization methods at different time periods beyond the noise-based ones considered in this paper; (iii) provide a deeper assessment of the notion of disclosure risk in longitudinal data anonymization, and in particular how disclosure risk from time-variant attributes relates and combines with disclosure risk steaming from time-invariant attributes, as generally longitudinal data sets contain these both kind of attributes.

References

1. Brand, R., Domingo-Ferrer, J., Mateo-Sanz, J.M.: Reference data sets to test and compare SDC methods for the protection of numerical microdata. Deliverable of the EU IST-2000-25069 "CASC" Project (2003)
2. Domingo-Ferrer, J., Muralidhar, K.: New directions in anonymization: permutation paradigm, verifiability by subjects and intruders, transparency to users. Inf. Sci. **337**, 11–24 (2016)
3. Domingo-Ferrer, J., Sánchez, D., Rufian-Torrell, G.: Anonymization of nominal data based on semantic marginality. Inf. Sci. **242**, 35–48 (2013)
4. Fung, B.C.M., Wang, K., Chen, R., Yu, P.S.: Privacy-preserving data publishing: a survey of recent developments. ACM Comput. Surv. (CSUR) **42**, 1–53 (2010)
5. Hundepool, A., et al.: Statistical Disclosure Control. Wiley, Hoboken (2012)
6. Muralidhar, K., Sarathy, R., Domingo-Ferrer, J.: Reverse mapping to preserve the marginal distributions of attributes in masked microdata. In: Domingo-Ferrer, J. (ed.) PSD 2014. LNCS, vol. 8744, pp. 105–116. Springer, Cham (2014). https://doi.org/10.1007/978-3-319-11257-2_9
7. Ruiz, N.: A general cipher for individual data anonymization, under review for Information Sciences. https://arxiv.org/abs/1712.02557 (2018)
8. Sehatkar, M., Matwin, S.: HALT: hybrid anonymization of longitudinal transactions. In: Eleventh Conference on Privacy, Security, Trust (PST), pp. 127–134 (2013)
9. Weiss, R.E.: Modelling Longitudinal Data. Springer, New York (2005). https://doi.org/10.1007/0-387-28314-5
10. Wooldridge, J.M.: Econometric Analysis of Cross Section and Panel Data, 2nd edn. The MIT Press, Cambridge (2010)

Reviewing the Methods of Estimating the Density Function Based on Masked Data

Yan-Xia Lin[(✉)] and Pavel N. Krivitsky

School of Mathematics and Applied Statistics,
National Institute for Applied Statistics Research Australia,
University of Wollongong, Wollongong, Australia
yanxia@uow.edu.au

Abstract. Data privacy is an issue of increasing importance for big data mining, especially for micro-level data. A popular approach to protecting the such is perturbation. Therefore, techniques used to recover the statistical information of the original data from the perturbed data become indispensable in data mining.

This paper reviews and exams the existing techniques for estimating (alternatively, reconstructing) the density function of the original data based on the data perturbed using the additive/multiplicative noise method. Our studies show that the techniques developed for noise-added data cannot replace the techniques for noise-multiplied data, though the two types of masked data could be mutually converted through data transformation. This conclusion might attract data providers' attention.

Keywords: Confidential data · Masked data
Multiplicative noise method · Additive noise method

1 Introduction

Data privacy seeks to simultaneously achieve two goals: useful statistical inference for micro-level data about the individuals in the population of interest, while protecting the micro-data themselves from disclosure. Stochastically perturbing the microdata achieves the latter, but requires techniques that can take this perturbation into account.

This study considers two types of data masking schemes: additive and multiplicative. We define them here. Let X be a univariate continuous sensitive random variable and the original data $\{x_i\}_{i=1}^{n}$ are independent realisations of X. The noise random variable C is independent of X, used to protect the values of X through data masking schemes. The random variable C can work as an additive noise or a multiplicative noise, depending on the data provider. We assume that the density function of C and of the masked data are publicly available.

© Springer Nature Switzerland AG 2018
J. Domingo-Ferrer and F. Montes (Eds.): PSD 2018, LNCS 11126, pp. 231–246, 2018.
https://doi.org/10.1007/978-3-319-99771-1_16

Noise-Added (Data) Masking Scheme. Simulate a sample $\{c_i\}_{i=1}^n$ from C and add c_i to x_i, $i = 1, \ldots, n$, respectively. The dataset $\{x_i^* = x_i + c_i\}_{i=1}^n$ is called the noise-added dataset of $\{x_i\}_{i=1}^n$. For x_i^* to be an unbiased estimator of x_i, $i = 1, \cdots, n$, this scheme requires that $E(C) = 0$.

Noise-Multiplied (Data) Masking Scheme. Simulate a sample $\{c_i\}_{i=1}^n$ from C and multiply x_i by c_i, $i = 1, \ldots, n$, respectively. The dataset $\{x_i^* = x_i c_i\}_{i=1}^n$ is called the noise-multiplied dataset of $\{x_i\}_{i=1}^n$. For x_i^* to be an unbiased estimator of x_i, $i = 1, \ldots, n$, this scheme requires that $E(C) = 1$. It usually also requires that C is a positive random variable.

The probability density function of a continuous random variable X can uniquely describe the statistical information of the random variable. Developing methods for estimating[1] the density function of X based on masked data is one of the approaches for recovering the statistical information of the original data.

To our best knowledge, the four papers, [1–4], are the first to independently introduce the four fundamental methods for estimating the density function of the original data based on masked data, respectively. [5] introduced a computational algorithm and built an R package `MaskDensity14` for implementing the method of [4]. Later, [6] proposed another computational algorithm for improving `MaskDensity14`. [7] followed the technique of [1], and used an iterative approach for estimating the joint density function of the original data based on noise-added data. [8] developed a computational method for estimating the joint density function of the original data based on noise-multiplied data by combining the method of [4] and the Nataf [8, for example] method. Currently, the existing techniques for estimating the joint density function are developed on the top of the techniques of [1–4], respectively.

Protecting the original data by additive noise was introduced in the literature slightly earlier than by multiplicative noise [9–12]. The multiplicative noise method could be an appropriate method for perturbing observations in the business data as it offers uniform protection, in terms of the coefficient of variation of the noise, to all values in the data set [10, 13, 14].

In this paper, we review the techniques of [1–4]. The techniques of [1, 2] are widely cited in the literature, developed for noise-added data. Bringing the data providers' attention in using these two methods, we point out some weakness of the techniques in this paper and also point out that sometimes the method of [4] can perform better.

The rest of the paper proceeds as follows. In Sect. 2, we briefly introduce the existing techniques for estimating the density function of the original data based on masked data. We use simulation studies to show the limitations of each method. In Sect. 3, we apply the techniques to simulation data and compare their performance. The conclusion is in Sect. 4.

---

[1] Some literature uses the term reconstructing. We will use them interchangeably in this paper.

2   Reviewing Existing Techniques of Estimating Density Function Based on Masked Data

2.1   Techniques Associated with Noise-Added Data

The AS2000 Approach.   [1] proposed an iterative algorithm in the process of reconstructing the density function of X based on noise-added data $\{x_i^*\}$.

Based on the Bayes' theorem, for each x in the range of X,

$$\mathrm{E}_{X^*}\left[f_{X|X^*}(x|X^*)\right] = \mathrm{E}_{X^*}\left[\frac{f_C(X^* - x)f_X(x)}{f_{X^*}(X^*)}\right] = f_X(x), \tag{1}$$

where $f_{X|X^*}$ is the conditional density function of X given X^* and the notation "E_{X^*}" denotes the expectation on the probability space of X^*.

The sample mean of a random variable is an unbiased estimator of the expectation of the random variable. From (1), for each x in the range of X,

$$\hat{f}_X(x) = \frac{1}{n}\sum_{i=1}^{n}\frac{f_C(x_i^* - x)f_X(x)}{f_{X^*}(x_i^*)} = \frac{1}{n}\sum_{i=1}^{n}\frac{f_C(x_i^* - x)f_X(x)}{\int_{-\infty}^{\infty}f_C(x_i^* - z)f_X(z)dz} \tag{2}$$

is an unbiased estimator of $f_X(x)$. Given this fact, [1] conducted an approach for reconstructing the density function of X based on noise-added data.

Noting that the unknown density function f_X also appears on the right-hand-side of (2), [1] suggested the following algorithm for estimating f_X:

1. Initialise $f_X^0 := \mathrm{Unif}(a, b)$, for a and b plausible range of X.
2. Initialise $j := 0$.
3. $f_X^{j+1}(x) := \frac{1}{n}\sum_{i=1}^{n}\frac{f_C(x_i^* - x)f_X^j(x)}{\int_{-\infty}^{\infty}f_C(x_i^* - z)f_X^j(z)dz}$.
4. $j := j + 1$.
5. Repeat from Step 3 to convergence.

[1] presented examples to demonstrate the algorithm proposed. In the examples of [1], the sequences of the functions $\{f_X^j\}$ were convergent, and the limits were close to the density functions of the associated underlying original data. But they did not formally prove whether the iterative process of the algorithm proposed is always convergent. Define a mapping

$$H(g)(x) = \int_{-\infty}^{\infty}\left[\frac{f_C(x^* - x)g(x)}{\int_{-\infty}^{\infty}f_C(x^* - z)g(z)dz}\right]f_{X^*}(x^*)dx^*, \quad x \in R,$$

which maps a density function g from the density functions space $\mathcal{G}$ to a density function $H(g)$ in the same space. Equation (1) indicates that the density function f_X is a fixed-point in the density function space $\mathcal{G}$ under the mapping H. Density function f_X is not necessarily a unique fixed-point under the mapping H in general. The convergence of the iterative process is an open question for the approach. This issue was not investigated in [1]. [2] commented the AS2000 Approach that the approach might not converge. Even if it converges, the sequence of the reconstructed density functions does not necessarily converge to the density function of the original data.

Table 1. The probability distributions of the original data and the additive noise

Scenario	PDF of X	PDF of C	SNR
I	$N(3, 5)$	$\text{Unif}(-10, 10)$	0.75
II	$\text{Gamma}(\text{shape} = 1.5, \text{rate} = 0.5)$	$\text{Unif}(-10, 10)$	0.18
III	$\text{Gamma}(\text{shape} = 1.5, \text{rate} = 0.5)$	$N(0, 10)$	0.06

Example 1. We implemented the AS2000 Approach in R and applied it to eight sets of simulated data. Some of them have symmetrical distribution, and some of them do not. The additive noise used to mask the data is uniformly distributed or normally distributed, or mixture distributed. In the interest of space, we only report three simulation studies. We list the information of the distributions of the original data and the additive noises in the studies in Table 1 (SNR is a measurement for disclosure risk, discussed in the KEtal2003 Approach later).

The size of each dataset is 1000. Denote by $\{a_i\}_{i=1}^{50}$ the positions which divide the interval $[\min\{x_j\} - 2 \text{ s.d.}(X), \max\{x_j\} + 2 \text{ s.d.}(X)]$ into 49 equal length subintervals. In the iterative process of the AS2000 Approach, the difference between the currently reconstructed density function with the previous one are evaluated through the differences of the two functions at positions $\{a_i\}_{i=1}^{50}$. In the R program we developed, we set the maximum number of iterations to 200. The iteration process will be stopped if the number of iterations exceeds 200 or if $\sum_{i=0}^{50}[f_X^j(a_i) - f_X^{j+1}(a_i)]^2/50 < 0.000001$. The numbers of iterations of Scenario I –III discussed in Table 1 are 7, 11 and 9, respectively. The plots of the sequences of reconstructed density functions for each dataset are presented in Figs. 1.

The original data of Scenario I are symmetrically distributed. The data in Scenarios II and III are the **same**, which are asymmetrically distributed. We use two types of distribution of additive noise to mask the data of Scenarios II and III, respectively. Our experience on the AS2000 Approach shows that if the original data are symmetrically distributed, the output of the reconstructed density function is more likely acceptable (see the output for Scenario I). It might need caution for the reconstructed density function if the original data are asymmetrically distributed (see the outputs for Scenarios II and III). Sometimes, the reconstructed density function does not capture the character of the density function of the original data. Our simulation studies also show that the variance of the additive noise might have less impact on the estimated density function. However, the distribution of the additive noise has a significant impact on the estimated density function (see Scenarios II and III). In the algorithm of the AS2000 Approach, using a uniform density function as the initial density function in the iterative process might be inappropriate for some data.

The AA2001 Approach. [2] introduced an EM reconstruction algorithm (the AA2001 Approach for short), for reconstructing the density function of the original data based on noise-added data.

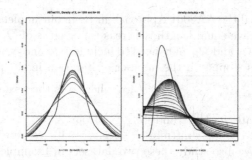

Fig. 1. Solid line: the density function of X; dash line: the reconstructed density function given by the AS2000 Approach. The result of Scenario I is on the left panel. On the right one, the reconstructed density function with a lower peak is the result of Scenario III. The other is the result of Scenario II.

Assume that the density function f_X is an analytic function. Then, f_X can be well approximated by a sequence of step functions $\{f_{X;\theta;K}\}_{K=1,2,\cdots}$, where

$$f_{X;\theta;K}(x) = \sum_{i=1}^{K} \theta_i I_{\Omega_i}(x), \qquad x \in R, \qquad \theta = (\theta_1,\cdots,\theta_K) \in \Theta. \tag{3}$$

and $\Omega_1,\cdots,\Omega_K$ are mutually exclusive intervals, partitioning the range Ω_X of X into K intervals. Θ is a K-dimensional parameter space; $I_{\Omega_i}(x) = 1$ if $x \in \Omega_i$ and 0 otherwise; there is a restriction on $\theta = (\theta_1,\cdots,\theta_K)$ where $\sum_{i=1}^{K} \theta_i m(\Omega_i) = 1$ and $m(\Omega_i)$ is the length of the interval Ω_i. In theory, the larger the K is and the finer the partition is, the closer the function $f_{X;\theta;K}$ will be to the density function f_X. Once the value of K is decided, the issue of estimating f_X becomes the issue of estimating $(\theta_1,\cdots,\theta_K)$. [2] introduced the EM method to estimate $(\theta_1,\cdots,\theta_K)$ based on the noise-added data $\{x_j^*\}_{j=1}^N$. The EM reconstruction algorithm of the AA2001 Approach consists of the following steps:

1. Initialise $\theta_i^{(0)} = 1/K$, $i = 1, 2, \cdots, K$; $k = 0$.
2. Update θ as follows: for each $i = 1, 2, \cdots, K$, $\theta_i^{(k+1)} = \psi_i(\{x_j^*\}_{j=1}^N; \theta^{(k)})/(m_i N)$, and

$$\psi_i(\{x_j^*\}_{j=1}^N; \theta^{(k)}) = \theta_i^{(k-1)} \sum_{j=1}^{N} \frac{\int_{(C \in x_j^* - \Omega_i)} f_C(c)dc}{f_{X^*;\theta^{(k-1)}}(x_j^*)}$$

where $m_i = m(\Omega_i)$; $c \in x_j^* - \Omega_i$ if $z_i - c \in \Omega_i$, and $f_{X^*;\theta^{(k-1)}}(x_j^*) = \sum_{l=1}^{K} \theta_l^{(k-1)} P(C \in x_j^* - \Omega_l)$.
3. $k = k + 1$.
4. If termination criterion not met, then return to Step 2.

The termination criterion for the algorithm is based on how much $\theta^{(k)}$ had changed since the last iteration. The sequence of estimates $\{\theta^{(k)} = (\theta_1^{(k)},\cdots,\theta_K^{(k)})\}$ is called the sequence of EM estimates.

We implemented the AA2001 Approach in R. In our implementation, Ω_is for $i = 2, \ldots, K - 1$ have equal length, whereas the lengths of Ω_1 and Ω_K may be different, because Ω_1 and Ω_K may need to include $-\infty$ and ∞, respectively. The iterative process is stopped if the number of iteration in the process is greater than 1000 or $\sqrt{\sum_{i=1}^{K}(\theta_i^{(k-1)} - \theta_i^{(k)})^2} < \alpha$, where α is the termination criterion.

Example 2. Simulate $\{x_i\}_{i=1}^{1000}$ and $\{c_i\}_{i=1}^{1000}$ from $X = 2 + \text{Gamma}(\text{shape} = 2, \text{scale} = 0.2)$ and $C = \text{Unif}(\text{min} = 3, \text{max} = 6)$, respectively, and obtain two sets of data. We also study these two datasets in Example 5, in Sect. 3.

In this example, we take data $\{\log x_i + \text{mean}(\log(C))\}$ to be the original data, a sample from $X' = \log(X) + \text{mean}(\log(C))$. We use the sample $\{\log(c_i) - \text{mean}(\log(C))\}$ of the additive noise $C' = \log(C) - \text{mean}(\log(C))$ to mask the original data. Then, we apply the AA2001 Approach to the masked data $\{\log x_i + \log c_i\} = \{[\log x_i + \text{mean}(\log(C))] + [\log c_i - \text{mean}(\log(C))]\}$. In the AA2001 Approach, the values of K and α need to be decided beforehand. We independently apply the approach to the **same set** of masked data by using different combinations pair values of K and α, where $K = 17, 18, 19,$ and 20, and $\alpha = 0.001, 0.01$ and 0.1. We only report the plot of the density function of the original data and the plots of the reconstructed density function given by the AA2001 Approach with $K = 20$ and $\alpha = 0.001$ and 0.1 (Fig. 2). The iterations in the estimating process are 1000 and 16 for $\alpha = 0.001$ and 0.1, respectively.

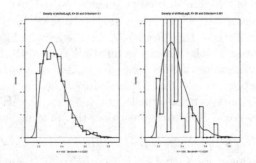

Fig. 2. The density function of the original data is in solid line. Reconstructed density function is in step-line. On the left panel, $K = 20$ and $\alpha = 0.1$. On the right panel, $K = 20$ and $\alpha = 0.001$.

Example 3. We apply the AA2001 Approach to the set of masked data of Scenario III discussed in Example 1. We set $\Omega_1 = (-\infty, \min\{x_i\} - 2 \text{ s.d.}(X))$ and $\Omega_K = (\max\{x_i\} + 2 \text{ s.d.}(X), \infty)$ in the AA2001 Approach. We consider different combinations of K and α, where $K = 10, 20, 30, 40, 50, 60$ and $\alpha = 0.0001, 0.001, 0.01, 0.1$ in applying the AA2001 Approach. Here, we only report the plots of the reconstructed density functions given by the AA2001 Approach when $K = 60$, $\alpha = 0.1$ and 0.0001 (Fig. 3).

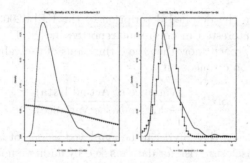

Fig. 3. The density function of the original data is in solid line. Reconstructed density function is in step-line. On the left panel, $K = 60$ and $\alpha = 0.1$. On the right panel, $K = 60$ and $\alpha = 0.0001$.

In theory, once K and the intervals $\{\Omega_i\}_{i=1}^K$ are determined, the ML estimate $\hat{\theta}_{ML}$ of the $\theta = (\theta_1, \cdots, \theta_K)$ in (3) can be uniquely determined. [2] proved that the sequence of EM estimates would converge to $\hat{\theta}_{ML}$. [2] also claimed that, by choosing K large enough, $f_{X;\theta;K}$ can approximate f_X with arbitrary precision. Our experience shows that the declaration is not correct. In fact, the larger the K is, the more parameters need to be estimated and the more sampling variation affects the reconstructed density function. The reconstructed density function is unlikely close to the density function of the original data if the termination criterion α used in the approach is big. However, our simulation studies show that sometimes using a smaller value of the termination criterion α in the approach does not necessarily lead to an accurately estimated density function. We demonstrate these evidence in Examples 2 and 3. In Example 2, when α is large, the approach gives a more accurate density estimation. It becomes different in Example 3. Our experience shows that, with an appropriate pairing of K and α, the AA2001 Approach works well for symmetrically distributed data. However, the approach might not work well for skewed data sometimes (See Examples 2 and 3).

The KEtal2003 Approach. [3] introduced the random matrix-based spectral filtering technique for estimating the values of the original data based on noise-added data (the KEtal2003 Approach for short). If the percentage of the values of the original data which can be correctly estimated from the noise-added data is high, the estimated density function of the original data can be reconstructed by using these accurately predicted values of the original data.

One can conclude that, if the density function of a set of original data can be successfully estimated by using the KEtal2003 Approach, it will mean that the noise-added data do not protect the values of the original data well. From the aspect of protecting data, the data provider should prevent the KEtal2003 Approach being successfully used in estimating the density function of the original data.

We do not give the detailed description of the KEtal2003 Approach here. However, we are interested in one result reported by [3]. Their experiences on the KEtal2003 Approach show that the estimations of the values of the original data will become too erroneous if

$$\text{SNR} = \frac{(\text{Variance of Actual Data})}{(\text{Additive Noise Variance})}$$

falls below 1. It means that SNR can be employed as one of the measurements of the predictive disclosure risk for data masked by additive noise. Before letting the noise-added data be available to the public, the data provider must make sure the value of SNR $\ll 1$.

Experience shows that, if the density function of the original data is symmetric, both the AS2000 Approach and the AA2001 Approach could be methods for estimating the density function of the original data based on noise-added data. One caution in using the AS2000 Approach is that the sequence of the reconstructed density functions might not converge, and even if the sequence is convergent, it might not converge to the actual density function of the original data. Using the uniform distribution density function as an initial density function in the AS2000 Approach might be not appropriate for some data. The difficulty in using the AA2001 Approach is about the choice of the values of K and α. It is a challenge to identify an appropriate pair (K, α) without the density function the original data as the reference.

2.2   Techniques Associated with Noise-Multiplied Data

The L2014 Approach. [15] showed that, if X is bounded by $[a, b]$, f_X can be well approximated by

$$f_{X,K}(x) = \frac{2}{b-a} \sum_{k=0}^{K} \lambda_k P_k \left(\frac{2x - (a+b)}{b-a} \right) = \sum_{k=0}^{K} a_k(x, a, b) \mu_X(k)$$

if the upper order of moments K is appropriate. Where $x \in$ the range of X, $\mu_X(k) = \mathrm{E}(X^k)$, $P_k(x)$ is a Legendre polynomial of degree k and λ_k is function of $\{\mu_X(k - 2i)\}_{i=0}^{Floor[k/2]}$.

Data $\{c_i\}_{i=1}^{n'}$ is a sample of the multiplicative noise C, where $n' \gg n$. This sample is not the same one used to mask the original data $\{x_i\}_{i=1}^{n}$. Thus, the values of $\{x_i\}_{i=1}^{n}$ cannot be obtained from $\{x_i^*\}_{i=1}^{n}$ and $\{c_i\}_{i=1}^{n'}$ by dividing. [4] introduced the sample-moment-based density approximate (the L2014 Approach in short) for estimating the density function of the original data $\{x_i\}_{i=1}^{n}$ based on noise-multiplied data $\{x_i^*\}_{i=1}^{n}$. Under the L2014 Approach, the estimated density function of X has the following expression

$$f_{X,K|\{\{x_i^*\}_1^n, \{c_i\}_1^{n'}\}}(x) = \sum_{k=0}^{K} a_k(x, a, b) \frac{\overline{(X^*)^k}}{\overline{C^k}}, \qquad x \in R, \tag{4}$$

where $\overline{(X^*)^k} = \sum_{i=1}^{n}(x_i^*)^k/n$; $\overline{C^k} = \sum_{i=1}^{n'}(c_i)^k/n'$; a and b are $\min\{x_i\}$ and $\max\{x_i\}$, respectively; K is the optimal upper order of moments such that $f_{X,K|\{\{x_i^*\}_1^n,\{c_i\}_1^{n'}\}}$ is the best estimated density function of the original data, subject to the information of $\{x_i^*\}$ and $\{c_i\}$.

The L2014 Approach is a theoretical approach for estimating the density function of the original data. To implement the approach in practice, without the original density function as the reference, determining the value of K is a critical issue. It is not the case that, the larger the value of K is, the closer the function $f_{X,K|\{\{x_i^*\}_1^n,\{c_i\}_1^n\}}$ to the original density function. Therefore, developing an algorithm for implementing the approach in practice is a challenge.

R package MaskDensity14. [5] introduced an algorithm for determining the value of K in the L2014 Approach. They adopted the algorithm to an R package MaskDensity14. For convenience, we refer this computational approach as the MaskDensity14 Approach.

The algorithm for determining the value of K consists of the following steps:

1. Set an initial upper order of moment, $K = 1$ and a maximum upper order of moment to be tested. The maximum upper order of moment set in MaskDensity14 is 100.
2. Independently simulate a sample $\{c_i\}_{i=1}^n$ from C and obtain the smoothing function $f_{X,K|\{x_i^*,c_i\}_{i=1}^n}$.
3. Simulate a sample $\{x_j'\}_{j=1}^n$ from $f_{X,K|\{x_i^*,c_i\}_{i=1}^n}$.
4. Independently simulate a second sample $\{c_j'\}_{i=1}^n$ from C. Mask $\{x_j'\}_{j=1}^n$ by using this new sample of noise and yield a new masked dataset $\{x_j'^*\}_{j=1}^n$.
5. Sort $\{x_j'^*\}_{j=1}^n$ and $\{x_i^*\}_{i=1}^n$, respectively. Evaluate the correlation $Cor(K)$ between the two sorted datasets. Keep track of the optimum upper order of moment such that $Cor(K_{\mathrm{opt}}) = \max_{k \leq K} Cor(k)$.
6. Update K to $K + 1$ and return to (2) if $K + 1 \leq 100$. Stop when $Cor(K)$ drops below a threshold taken as $Cor(K) < 1 - 10[1 - Cor(K_{\mathrm{opt}})]$ or $K + 1 > 100$.
7. Report K_{opt} as the optimum upper order of moment used.

Examples in [5], as well as our experience of using MaskDensity14, indicate that the software performans well in practice, but can be further improved. We note that Step 4 in the algorithm involves a sampling process. The randomness of the sampling will cause the randomness of the estimated density function. Therefore, repeatedly applying MaskDensity14 to the same set of noise-multiplied data will yield different estimated density functions, though most of them are mimic each other. [4] showed that $f_{X,K|\{x_i^*,c_i\}_{i=1}^n}(x)$ is an approximately unbiased estimator of $f_{X,K}(x)$ for each $x \in$ the range of X. If the randomness becomes an issue, to reduce the impact of the randomness in estimating the density function, it suggests that repeatedly apply MaskDensity14 to the same set masked data and obtain a sequence of estimated density functions of the original data. Then, use the functional mean of the sequence of estimated density functions as the final estimated density function of the original data. The suggestion works well in practice, but sometimes the process might be cumbersome.

The B-M L2014 Approach. [6,16] proposed a different algorithm, the computational Bayesian approach, for determining the optimal upper order moment K in the L2014 Approach. We call the computational Bayesian approach for implementing the L2014 Approach the Bayes-Moment L2014 Approach (the B-M L2014 Approach for short).

We assume that the multiplicative noise C is a positive random variable and bounded by the real numbers c_0 and c_1, where $0 < c_0 < c_1$.

Adopt the notation in the L2014 Approach above. Denote

$$f_{X^*,K}(x^*) = \int_{c_0}^{c_1} \frac{1}{c} f_C(c) f_{X,K}(\frac{x^*}{c}) dc, \quad x^* \in \text{the range of } X^*, \tag{5}$$

and

$$f_{X^*,K|\{x_i^*,\tilde{c}\}_1^n}(x^*) = \int_{c_0}^{c_1} \frac{1}{c} f_C(c) f_{X,K|\{x_i^*,\tilde{c}\}_1^n}(\frac{x^*}{c}) dc, \tag{6}$$

where $x^* \in$ the range of X^*.

[4] proved and demonstrated that $f_{X,K|\{x_i^*,\tilde{c}_i\}}$ can be a good estimation of f_X subject to K is appropriate. From (5) and (6), it concludes that $f_{X^*,K|\{x_i^*,\tilde{c}_i\}}$ can be a good estimation of f_{X^*}, subject to K is appropriate. The necessary condition that $f_{X,K|\{x_i^*,c_i\}}$ is a good approximation of f_X is that $f_{X^*,K|\{x_i^*,\tilde{c}_i\}}$ is a good approximation of f_{X^*}. Based on this logic, [6] proposed a new algorithm for determining the appropriate value K in the L2014 Approach. The algorithm consists of the steps described below.

1. Set an initial upper order of moment, $K = 1$ and a maximum upper order of moment to be tested. The maximum upper order of moment set is set by 100;
2. Decide N_S, the number of positions in the interval $[\min\{x_i^*\}, \max\{x_i^*\}]$. Denote positions as $z_0 = \min\{x_i^*\} < z_1 < \cdots < z_{N_S} < z_{N_S+1} = \max\{x_i^*\}$ such that $z_i - z_{i-1} = \triangle z_i$ are all equal, $i = 1, 2, \cdots, (N_S + 1)$.
3. Evaluate

$$S(K) = \sum_{j=1}^{N_S+1} \left[f_{X^*}(z_j) - f_{X^*,K|\{x_i^*,\tilde{c}_i\}_1^n}(z_j) \right]^2 \triangle z_j$$

at the positions $\{z_j\}_{j=0}^{N_S+1}$. Keep the track of the optimum upper order of moment K_{opt} such that $S(K_{\text{opt}}) = \min_{k \leq K} S(k)$.
4. Update K to $K + 1$ and return to (3) if $K + 1 \leq 100$. Stop when $S(K)$ jumps beyond a threshold taken as $S(K) > 100 \times S(K_{\text{opt}})$ or $K + 1 > 100$. To save the computational time, based on our experience, we use this criterion as a threshold for determining the optimal value of K.
5. Report K_{opt} as the optimum upper order of moment used.

No randomisation process is involved in the process of determining the optimum upper order of moment in the B-M L2014 Approach. The closeness of f_{X^*} and $f_{X^*,K|\{x_i^*,\tilde{c}_i\}}$ is evaluated at the positions $\{z_j\}_{j=0}^{N_S+1}$. The estimated density function yielded by the B-M L2014 Approach is unique, subject to the noise-multiplied data, the distribution of the multiplicative noise, the number N_S and the threshold set in the algorithm. The larger the number N_S is, the more costly the approach will be.

Simulation Studies for the MaskDensity14 and the B-M L2014 Approach. In this subsection, we apply MaskDensity14 and the B-M L2014 Approach to real-life data. We use two ways to evaluate if a set of noise-multiplied data can provide a reasonable level of protection on the set of the original data. The first way is to visually check the scatter plot of the original data X against the noise-multiplied data X^* and to evaluate the correlation coefficient of the original data and the noise-multiplied data. No visually observed function relationship between X and X^* and having a lower value of correlation coefficient between X and X^* (for instance, the values <0.9) are the necessary conditions to ensure that the noise-multiplied data X^* provides a reasonable level of protection on data X [17]. The second way is to check whether the value of SNR $= \mathrm{Var}(\log(X))/\mathrm{Var}(\log(C))$ is much less than 1^2.

Example 4. The United States Census dataset [18] contains 54 numerical variables arising from the extraction process. In this example, we consider the variable **AFNLWGT**. The number of observations of **AFNLWGT** is 1080. The values of observations of **AFNLWGT** are very large. If the noise-multiplied masking scheme is applied to **AFNLWGT**, the values of masked observations will become too large to be analysed. Therefore, we consider log(**AFNLWGT**) in this study. The smoothed density function of log(**AFNLWGT**) presents in Fig. 4.

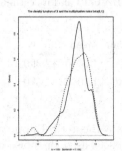 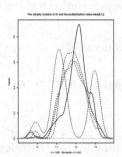

Fig. 4. On the left panel, the plots of density function of log(**AFNLWGT**) (black line) and its estimated density function (dash line)given by the B-M L2014 Approach. On the right panel, the plots of density function of log(**AFNLWGT**) (black line) and the estimated density functions given by MaskDensity14.

The distribution of log(**AFNLWGT**) is skewed to the left. Our experience shows that a skewed multiplicative noise might provide more protection for data

---

^{2} See the discussion of the KEtal2003 Approach.

with skewed distribution. In this example, we use the multiplicative noise with Beta$(6, 1)$ distribution to mask log($\mathbf{AFNLWGT}$)^3.

The scatter plot (not presented in the paper) shows that overall the observations of log($\mathbf{AFNLWGT}$) can be well protected by the noise-multiplied data. This fact is also conformed by other two measurements, the correlation coefficient 0.2727083 of the set of the observations log($\mathbf{AFNLWGT}$) and its noise-multiplied data, and the value of SNR = Var(log($\mathbf{AFNLWGT}$))$/$Var(log(Beta$(6, 1)$))) = 0.08357918.

We applied the B-M L2014 Approach to the noise-multiplied data. The range of the noise multiplied data is $[3.462, 12.89]$. For such short range, the N_S used is $N_S = 50$. The plot of the estimated density function was presented in Fig. 4. We also independently applied `MaskDensity14` to the same set of noise-multiplied data. The plots of the estimated density functions were also presented in Fig. 4. As we expected, the outputs of `MaskDensity14` are not stable, though most of the plots of the estimated density functions are close to each other.

In the following study, we only focus on the estimated density function given by the B-M L2014 Approach. We can see the estimated density function is not entirely close to the density function of log($\mathbf{AFNLWGT}$). We are interested whether the estimated density function provides the data user with the basic statistical information of log($\mathbf{AFNLWGT}$). We simulated a sample of size 1000 from the estimated density function. The summary statistics of log($\mathbf{AFNLWGT}$) and the simulation are reported in Table 2, respectively. The outputs of the summary statistics are close to each other and indicate that the B-M L2014 Approach can successfully retrieve the basic statistical information of the original data from the masked data.

Table 2. The summary statistics of log($\mathbf{AFNLWGT}$) and its estimated density function.

Distribution	Min	1st Qu	Median	Mean	3rd Qu	Max
True	9.515	11.750	12.100	12.040	12.390	13.440
Estimated	9.577	11.690	12.140	12.070	12.510	13.150

In summary, the L2014 Approach gives a theoretical approach for estimating the density function of the original data based on noise-multiplied data. `MaskDensity14` and the B-M L2014 Approach are computational methods for implementing the L2014 Approach.

To identify the the optimal upper order moment K, the B-M Approach requires the process of evaluating the difference between $f_{X^*,K|\{x_i^*,\tilde{c}\}}$ and f_{X^*} at positions $\{z_j\}$. The more the positions are used in the evaluation, the higher the

---

^{3} Other multiplicative noise distributions might be considered. Identifying a best multiplicative noise for masking the underlying data in terms of minimising the level of values disclosure risk and minimising the original data utility loss subject for future work.

computational cost the process will be. Fortunately, it is not necessary to use too many positions in the evaluation when the function f_{X^*} is not volatile. The B-M L2014 Approach is more computationally costed than MaskDensity14, but provides a stable inference outcome.

MaskDensity14 has its advantages. Running MaskDensity14 is less computationally expensive. The data user can apply the MaskDensity14 to the underlying masked data and quickly obtain preliminary information on the estimated density function of the original data. If the randomness of the outputs of MaskDensity14 is an issue, the B-M L2014 Approach is an alternative approach. The data user can also use the output of the MaskDensity14 as a reference for determining the number of positions of $\{z_i\}$ used in the B-M L2014 Approach.

3   Comparing Techniques for Noise-Multiplied Data and the Techniques for Noise-Added Data

Statistical information of the original data can be recovered from the noise-multiplied data in one of two ways. One way is to apply MaskDensity14 or the B-M L2014 Approach to noise-multiplied data directly. Another is to apply the AS2000 Approach or the AA2001 Approach to the log-transformation of the data if it is well defined. It is of interest whether the first manner is better in terms of retrieving more statistical information from the original data.

In this section, we use an example to demonstrate that MaskDensity14 and the B-M L2014 Approach can sometimes outperform the AS2000 Approach and the AA2001 Approach in retrieving the statistical information in the original data. MaskDensity14 and the B-M L2014 Approach are the same technique but with a different algorithm. Therefore, we only consider the B-M L2014 Approach in this section.

Example 5. Consider the same $\{x_i\}$ and $\{c_i\}$ as that in Example 2. In this example, we denote $\text{Data}_{\text{orig}} = \{x_i\}$ the original data and $\text{Data}_{\text{mult}} = \{x_i^*\} = \{x_i c_i\}$ the noise-multiplied data of $\text{Data}_{\text{orig}}$. The correlation coefficient of X and $X^* = XC$ is 0.451025. The plot of X vs. X^* is presented in Fig. 5.

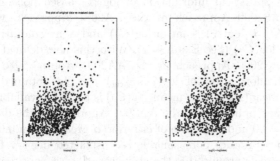

Fig. 5. On the left panel, the plot X vs X^*, where X is masked by the noise-multiplied data masking scheme. On the right panel, the plot $log(X)$ vs $log(X^*)$.

The plot of X against X^* and the value of the correlation coefficient indicate that the original data are well protected by the multiplicative noise C at a reasonable level. By taking the log-transformation on the noise-multiplied data, we can obtain the type of noise-added data for the data $\log(\text{Data}_{\text{orig}})$ of $\log(X)$. The value of SNR $= \text{Var}(\log(X))/\text{Var}(\log(C))$ given by the type of noise-added data is 0.2632817. It means that the values of the $\log(X)$ cannot be accurately estimated by using the KEtal2003 Approach, so do the values of X. In addition, the plot of $\log(\text{Data}_{\text{mult}})$ vs $\log(\text{Data}_{\text{orig}})$ (Fig. 5) also shows that $\log(X)$ are also well protected by the additive noise $\log(C)$.

When we apply the B-M L2014 Approach to the noise-multiplied data $\text{Data}_{\text{mult}}$, the number N_S used is 50. Figure 6 gives the plot of the density function of X and the plot of the estimated density function provided by the B-M L2014 Approach. The curve of the estimated density function mimics the curve of the density function of the original data.

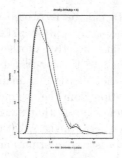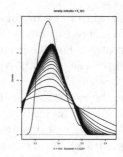

Fig. 6. On left panel, the plots of the density function of X (in black line) and the estimated density function (in dash line) given by the B-M L2014 Approach. On the right panel, the plots of the density function of the X' (solid line) and the estimated density function (dash line) given by the AS2000 Approach.

Recall that both the AS2000 Approach and the AA2001 Approach are designed for noise-added data where the associate additive noise has mean 0. To retrieve the statistical information of $\log(X)$ based noise-multiplied data by using the AS2000 Approach or the AA2001 Approach, we introduce a new variable $X' = \log(X) + \text{mean}(\log(C))$ and a new additive noise $C' = \log(C) - \text{mean}(\log(C))$ (See Example 2). With this new defined additive noise, we apply the AS2000 Approach or the AA2001 Approach to $\log(\text{Data}_{\text{mult}})$. The statistical information of $\log(\text{Data}_{\text{orig}})$ can be obtained from that of $\log(\text{Data}_{\text{orig}}) + \text{mean}(\log(C))$ as $\text{mean}(\log(C))$ is publicly available. The detailed discussion on the application of the AA2001 Approach to the data is in Example 2. Using $K = 20$ and the value of criterion α greater than 0.01, the AA2001 Approach cannot produce a reasonable reconstructed density function. Using $K = 20$ and the α is around 0.1, the skewness of X' is captured by the reconstructed density function. Choosing a correct pair (K, α) is an issue in the approach. The estimated density function of X' given by the AS2000 Approach is presented in Fig. 6. Visually speaking, the reconstructed density function is not a reasonable estimation of the density function of X'.

Comparing the outputs of the estimated density functions given by three approaches, respectively, the B-M L2014 Approach performs better.

4   Summary, Conclusion and Future Work

We have reviewed five existing approaches for estimating the density function of the original data based on masked data. Among them, the KEtal2003 Approach is not practical in terms of predictive disclosure risk. However, the concept of SNR is useful for evaluating the disclosure of noise-added data and the noise-multiplied data.

All of the algorithms, the AS2000, the AA2001, MaskDensity14, and the B-M L2014 Approaches alike are useful for big data mining with confidentiality. They give the public an opportunity to explore the statistical information of confidential microdata without exposing confidential information. Our studies show that the mathematical theories and computational algorithms used to support the algorithms are different and have different limitations. Even though this paper gives an example where the B-M L2014 Approach can more successfully estimate the density function of the original data than the AS2000 Approach and the AA2001 Approach, none of the methods dominates the others.

All techniques reviewed here are designed for univariate numerical data. Methods for estimating the mass function for categorical data and the joint density function for multivariate data is an important direction for future work.

Acknowledgments. Part of R code for implementing the AS2000 Approach was developed by Miss A. Fernando supported by the Winter Project Scholarship 2016, School of Mathematics and Applied Statistics, UoW.

References

1. Agrawal, R., Srikant, R.: Privacy-preserving data mining. ACM SIGMOD Rec. **29**, 439–450 (2000)
2. Agrawal, D., Aggarwal, C.C.: On the design and quantification of privacy preserving data mining algorithms. In: Proceedings of the Twentieth ACM SIGMOD-SIGACT-SIGART Symposium on Principles of Database Systems, pp. 247–255. ACM (2001)
3. Kargupta, H., Datta, S., Wang, Q., Sivakumar, K.: On the privacy preserving properties of random data perturbation techniques. In: 2003 Third IEEE International Conference on Data Mining, ICDM 2003, pp. 99–106. IEEE (2003)
4. Lin, Y.-X.: Density approximant based on noise multiplied data. In: Domingo-Ferrer, J. (ed.) PSD 2014. LNCS, vol. 8744, pp. 89–104. Springer, Cham (2014). https://doi.org/10.1007/978-3-319-11257-2_8
5. Lin, Y.X., Fielding, M.J.: MaskDensity14: an R package for the density approximant of a univariate based on noise multiplied data. SoftwareX **3**, 37–43 (2015)

6. Lin, Y.X.: Mining the statistical information of confidential data from noise-multiplied data. In: Proceedings of the 3rd IEEE International Conference on Big Data Intelligence and Computing (2017)
7. Domingo-Ferrer, J., Sebé, F., Castellà-Roca, J.: On the security of noise addition for privacy in statistical databases. In: Domingo-Ferrer, J., Torra, V. (eds.) PSD 2004. LNCS, vol. 3050, pp. 149–161. Springer, Heidelberg (2004). https://doi.org/10.1007/978-3-540-25955-8_12
8. Lin, Y.X., Mazur, L., Sarathy, R., Muralidhar, K.: Statistical information recovery from multivariate noise-multiplied data, a computational approach. Trans. Data Priv. **11**, 23–45 (2018)
9. Kim, J.J.: A method for limiting disclosure in microdata based on random noise and transformation. In: Proceedings of the Section on Survey Research Methods, pp. 303–308. American Statistical Association (1986)
10. Kim, J., Winkler, W.: Multiplicative noise for masking continuous data. Statistics 2003-01 (2003)
11. Mivule, K.: Utilizing noise addition for data privacy, an overview. In: Proceedings of the International Conference on Information and Knowledge Engineering (IKE), The Steering Committee of The World Congress in Computer Science, Computer Engineering and Applied Computing (WorldComp), p. 1 (2012)
12. Torra, V.: Data Privacy: Foundations, New Developments and the Big Data Challenge. SBD, vol. 28. Springer, Cham (2017). https://doi.org/10.1007/978-3-319-57358-8
13. Nayak, T.K., Sinha, B., Zayatz, L.: Statistical properties of multiplicative noise masking for confidentiality protection. J. Off. Stat. **27**(3), 527–544 (2011)
14. Muralidhar, K., Batra, D., Kirs, P.J.: Accessibility, security, and accuracy in statistical databases: the case for the multiplicative fixed data perturbation approach. Manag. Sci. **41**(9), 1549–1564 (1995)
15. Provost, S.B.: Moment-based density approximants. Math. J. **9**(4), 727–756 (2005)
16. Lin, Y.X.: A computational Bayesian approach for estimating density functions based on noise-multiplied data. Int. J. Big Data Intell. (2018). (in press)
17. Ma, Y., Lin, Y.X., Sarathy, R.: The vulnerability of multiplicative noise protection to correlational attacks on continuous microdata. Technical report, National Institute for Applied Statistics Research Australia, School of Mathematics and Applied Statistics, University of Wollongong, Australia (2017)
18. United States Census Bureau: United states census dataset (2000). Accessed 27 July 2000

Protecting Values Close to Zero Under the Multiplicative Noise Method

Yan-Xia Lin[✉]

School of Mathematics and Applied Statistics,
National Institute for Applied Statistics Research Australia,
University of Wollongong, Wollongong, Australia
yanxia@uow.edu.au

Abstract. Perturbing sensitive data is one of the standard ways of protecting confidential data. The multiplicative noise method is one of these data perturbation methods and this method has attracted researchers' attention in the recent decade. However, values close to zero in datasets cannot be well protected by using the multiplicative noise method directly. This paper proposes a method for safeguarding the values close to zero through noise-multiplied shifted data. We demonstrate that those values can be reasonably protected through noise-multiplied data by following the approach proposed in this paper. This paper also indicates that the density function of the original data can be reasonably reconstructed from the noise-multiplied shifted data by using the software *MaskDensity14* or *MaskDensityBM*.

Keywords: Confidential data · Masked data
Multiplicative noise method

1 Introduction

In order to make business microdata available to the public for data analysis, whilst still preserving adequate data privacy, releasing anonymized data to the public is commonly used in practice. In the recent decade, the multiplicative noise method, one of the anonymizing data techniques, has attracted researchers' attention [1–8]. More references can be found from review articles and books [9–11].

The multiplicative noise method offers uniform protection, in terms of the coefficient of variance of noise, to all observations [2,12,13]. The method is suitable for perturbing micro business data and economic modelling of income data [2,13]. However, when the literature says the method offers uniform protection to all observations, obviously, zeros and values close to zero are not under consideration. Zeros cannot be protected if the multiplicative noise is applied to the zeros directly. Data collected from real-life activities commonly contain zeros or values close to zero. Protecting those values without compromising too much data utility becomes an issue for the multiplicative noise method.

© Springer Nature Switzerland AG 2018
J. Domingo-Ferrer and F. Montes (Eds.): PSD 2018, LNCS 11126, pp. 247–262, 2018.
https://doi.org/10.1007/978-3-319-99771-1_17

This paper proposes the data-shifting approach for protecting zeros and values close to zero under the multiplicative noise method. We have described and implemented this idea of applying the data-shifting approach with our research in recent years [6,8,14]. However, we as well as other literature have not systematically presented any discussion on this issue. Therefore, we would like to raise this issue in this paper.

Under the data-shifting approach, the sensitive micro data (including zeros) are protected through the noise-multiplied shifted data. With the software developed recently for reconstructing the density function of the original data based on noise-multiplied data, we demonstrate that the statistical information of the original data can be retrieved from the noise-multiplied shifted data and an appropriate balance between data protection and data utility can be achieved.

The rest of this paper is constructed as follows. We propose the data-shifting approach in Sect. 2. In Sect. 2, we also introduce the commonly used measurement of value disclosure risk for noise-multiplied data. In principle, any random variable with any probability distribution can work as a multiplicative noise. Our experience shows that applying multiplicative noise with mixture distributions have been more advantageous in protecting the original data. In Sect. 3, we focus on the multiplicative noise with mixture uniform distributions. We discuss how to identify a multiplicative noise from the family of mixture uniform distributions by taking into account the commonly used measurements on disclosure risk. In Sect. 4, we briefly introduce the techniques for retrieving the statistical information of the original data from noise-multiplied shifted data. Using simulation studies, we demonstrate in Sect. 5 that the data-shifting approach can be beneficial in protecting values close to zero and the data-shifting approach does not significantly affect the output of retrieving the statistical information of the original data from noise-multiplied shifted data. The last section gives a summary of our paper.

2   Protecting Values Close to Zero and the Measurements of Value Disclosure Risk

Denote X a univariate continuous sensitive random variable and $\{x_i\}_{i=1}^n$ a set of realisations of X. The random variable C, namely the multiplicative noise, is used to protect the values of X through the multiplicative noise method. C is independent of X. The noise-multiplied data of $\{x_i\}$ are generated in the following way. Simulate a sample $\{c_i\}_{i=1}^n$ from C and multiply x_i by c_i, $i = 1, \ldots, n$, respectively. The dataset $\{x_i^* = x_i c_i\}_{i=1}^n$ is called the noise-multiplied dataset of $\{x_i\}_{i=1}^n$. To ensure x_i^* to be an unbiased estimator of x_i, $i = 1, \ldots, n$, it requires that $E(C) = 1$. It usually also requires that C be a positive random variable.

The values close to zero in $\{x_i\}_{i=1}^n$ might be in the following two scenarios.

Case 1. All the values appear near one of the tails of the distribution of $\{x_i\}$.
Case 2. All the values do not appear near the tails of the distribution of $\{x_i\}$.

We propose a data-shifting approach for protecting the values close to zero in $\{x_i\}$ as follows.

Step 1. Construct a new data set $\{\tilde{x}_i\}$ from $\{x_i\}$ by shifting $\{x_i\}$ by a constant a_0, i.e. $\{\tilde{x}_i\} = \{x_i + a_0\}$.

Step 2. Apply the multiplicative noise C to $\{\tilde{x}_i\}$, instead of $\{x_i\}$, and obtain the noise-multiplied shifted data $\{\tilde{x}_i^*\}$ of $\{x_i\}$.

For Case 1, the values close to zero appear near one of the tails of the distribution of $\{x_i\}$. The data provider can quickly identify a shifting parameter a_0 for the data-shifting approach such that all the data in the new dataset $\{\tilde{x}_i\} = \{x_i + a_0\}$ are significantly different from zero. For Case 1, the absolute value of a_0 is less likely to be large. The issue of protecting the values close to zero then becomes the issue of protecting the values significantly different from zero.

The data-shifting approach can also be applied to Case 2, but more issues need to be considered. Without loss of generality, we can assume that the values of $\{x_i\}$ are bounded within a finite interval $[-a, b]$, with $a, b > 0$. Thus, there is a real number $a_0 > a$ such that the shifted data $\{x_i + a_0\}$ contain no zeros. Since the zeros of the dataset $\{x_i\}$ are not on or near the tails of the distribution of $\{x_i\}$, the shifting parameter a_0 could be big if a is big. Protecting the values of the original data is not the only consideration in the process of data perturbation. The data provider also wishes the statistical information of the original data can be retrieved from the masked data. Our experience shows that, for $\{x_i\}$ and C with different probability distributions, the value of a_0 will affect the level of data protection in different ways. Data shifted by a big a_0 might lead to a certain level of data information loss after the multiplicative noise C masks the shifted data. If in the dataset $\{x_i\}$, there is a significant gap between the subset with values significantly different from zero and the subset with values close to zero, it is not necessary to shift all the values in $\{x_i\}$ by $a_0 > a$. Therefore, the way of deciding a shifting parameter is case depended. Generally speaking, it is difficult to give a general regulation for determining the shifting parameter in Case 2.

In this paper, we explore the impact of the variance of C and the value of a_0 on disclosure risk. In simulation studies, we only give examples where the value of the shifting parameter is not extremely big. The study carried out in this paper can be used as a reference for complex cases.

There is not a universal quantitative criterion for evaluating the value disclosure risk under the multiplicative noise method. Different contents of data require different levels of protection on the values of the data. Furthermore, the data provider usually does not know what external information is available for intruders in attacking the underlying data. It is impossible to set a universal quantitative criterion for evaluating the value disclosure risk. Therefore, in this paper, we only consider the following commonly used measurements in assessing the level of protection on the original data provided by the noise-multiplied data.

Without loss of generality, we always assume that $E(C) = 1$. Denote $X^* = XC$ where X is a sensitive variable and protected by C through the multiplicative noise method. In this paper, we consider the following three measurements for evaluating if a multiplicative noise C can provide reasonable level protection to X.

Measurement 1. For a small value $\delta > 0$, the noise C should ensure that

$$P\left(\left|\frac{X - X^*}{X}\right| < \delta \,\middle|\, X = x\right) = P(|C - 1| < \delta)$$

cannot be too close to 1. The smaller the probability, the more protection the noise C will provide to the original data (See [6, 15]).

Measurement 2. To protect the value of the original data through X^*, we require that

$$\frac{Var(X - X^*)}{Var(X)}$$

cannot be too close to 0. The larger the ratio is, the more protection X^* will provide to X ([10, 16–18]).

Measurement 3. The multiplicative noise C needs to be such as to ensure that the correlation coefficient of X and X^* is less than 0.9. The details discussion see [19]. The plot $\{x_i\}$ vs. $\{x_i^*\}$ can visually show whether the linear correlation between X and X^* is significant.

There is no standard to evaluate how small the probability in *Measurement 1.* should be. However, if C is mixture distributed as shown in Sect. 4, the probability defined in *Measurement 1* can be equal to zero. There is no standard to evaluate how large the ratio in *Measurement 2* should be. There is also no standard to assess how small the $cor(X, X^*)$ in *Measurement 3* should be. In one word, there are no quantitative standards in practice for the criteria above. The data provider needs to combine the information of the measurements and the nature of the underlying data to make his own decision. In this paper, for *Measurement 1*, we let $\delta = 0.05$ and we require

$$P\left(\left|\frac{X - X^*}{X}\right| < \delta \,\middle|\, X = x\right) = P(|C - 1| < \delta) = 0.$$

For *Measurement 3*, we require $cor(X, X^*) < 0.8$. We will show *Measurement 3* and *Measurement 2* are related. Thereby, once the standard for *Measurement 3* is given, the standard for *Measurement 2* will be determined.

Apparently, choosing a multiplicative noise for an underlying dataset also needs to be done in conjunction with the issue of utility loss control. However, we only focus on the value disclosure risk in this section.

Denote $\tilde{X} = X + a$ and $\tilde{X}^* = \tilde{X}C = (X + a)C$. We study the impact of the shifting parameter a and the variance of C on disclosure risk based on the measurements described above.

Since

$$P\left(\left|\frac{\tilde{X} - \tilde{X}C}{\tilde{X}}\right| < \delta \,\middle|\, \tilde{X} = \tilde{x}\right) = P(|C - 1| < \delta),$$

the shifting parameter gives no impact on the probability defined in *Measurement 1*. The measurement is only related to the distribution of C. To ensure the probability is as small as possible, the multiplicative noise C should not have mass within the interval $[1 - \delta, 1 + \delta]$.

In *Measurement 2*,

$$Var(\tilde{X} - \tilde{X}^*) = Var[\tilde{X}(1 - C)] = E[\tilde{X}^2(C - 1)^2] - [E(\tilde{X})]^2[E(1 - C)]^2$$
$$= Var(C)\{Var(X) + [E(X) + a]^2\}.$$

Therefore,

$$\frac{Var(\tilde{X} - \tilde{X}^*)}{Var(\tilde{X})} = \frac{Var(C)\{Var(X) + [E(X) + a]^2\}}{Var(X)}$$

$$= Var(C)\left(1 + \frac{[E(X) + a]^2}{Var(X)}\right). \tag{1}$$

Equation (1) gives the ratio of the variance of the difference between the original variable and its masked variable to the variance of the original variable. The larger the ratio is, the better the average level of protection the original data will receive. The ratio is an increasing function of $Var(C)$ and a. The larger the values of $Var(C)$ and a are, the more the noise is added to the original data. Logically, to control information loss, the values of $Var(C)$ and a cannot be unlimitedly increased.

By noting that

$$Cov(\tilde{X}, \tilde{X}^*) = E(\tilde{X}\tilde{X}^*) - E(\tilde{X})E(\tilde{X}^*)$$
$$= E[(X + a)^2]E(C) - [E(X + a)]^2(EC)^2 = Var(X + a) = Var(X)$$

and

$$Var(\tilde{X}^*) = E[(X + a)^2]Var(C) + Var(X + a) = E[(X + a)^2]Var(C) + Var(X),$$

the correlation coefficient of $\tilde{X}$ and $\tilde{X}^*$ can be calculated as follows

$$\rho(\tilde{X}, \tilde{X}^*) = \frac{Cov(\tilde{X}, \tilde{X}^*)}{\sqrt{Var(\tilde{X})Var(\tilde{X}^*)}}$$

$$= \frac{Var(X)}{\sqrt{Var(X)[E[(X + a)^2]Var(C) + Var(X)]}} = \frac{1}{\sqrt{1 + Var(C)\frac{E[(X + a)^2]}{Var(X)}}} \tag{2}$$

$$= \frac{1}{\sqrt{1 + \frac{Var(\tilde{X} - \tilde{X}^*)}{Var(\tilde{X})}}}.$$

The left-hand-side of (2) is the output of *Measurement 3* and the denominator of the right-hand-side of (2) is related to the output of *Measurement 2*. It turns out that *Measurement 2* and *Measurement 3* are related. [19] empirically showed that the level of the value disclosure risk is unacceptable if $\rho(\tilde{X}, \tilde{X}^*) \geq 0.9$. It is therefore equivalent to claim that it is unacceptable if $\frac{Var(\tilde{X} - \tilde{X}^*)}{Var(\tilde{X})} < (1/0.9)^2 - 1 = 0.2345679$.

3   Mixture Uniformly Distributed Multiplicative Noise

Any multiplicative noise C with mean equal to 1 and no mass in the interval $[1 - \delta, 1 + \delta]$ can ensure that the probability in *Measurement 1* is 0. However, to ensure that the level of disclosure risk, evaluated by *Measurement 2* and *Measurement 3* are acceptable, the data provider needs to search for an appropriate multiplicative noise from a wide range of random variables space through a tedious testing process. Our experience shows that a multiplicative noise with probability distribution from the family of mixture uniform distributions is more likely to be an appropriate multiplicative noise in practice. To simplify the discussion in this paper, we restrict the probability distribution of C to the family of mixture uniform distributions,

$$f_C(.; b_1, b_2, b_{01}, b_{02}, p) = pUnif(1 - b_1, 1 - b_{01}) + (1 - p)Unif(1 + b_{02}, 1 + b_2) \quad (3)$$

where $0 < p < 1$, $\delta \leq b_{01}, b_{02} < 1$, both b_1 and b_2 are positive, and $b_1 < 1$. To make sure $E(C) = 1$, the values of the parameters $b_1, b_2, b_{01}, b_{02}, p$ cannot be independently determined.

To further simplify the study in this paper, we only consider the following scenario where $b_{01} = b_{02}$, $p = 0.5$ and $b_1 = b_2$, i.e.

$$f_C(.; b_1, b_2, b_{01}, b_{02}, p) = 0.5Unif(1 - b_1, 1 - b_{01}) + 0.5Unif(1 + b_{01}, 1 + b_1). \quad (4)$$

Regarding the level of protection of the original data and the level of control of the loss of the utility of the original data, the shifting parameter a and $Var(C)$ play essential roles in the multiplicative noise method. When $\rho(\tilde{X}, \tilde{X}^*)$ is fixed, a and $Var(C)$ cannot be independently chosen. We first derive $\rho(\tilde{X}, \tilde{X}^*)$ and $Var(C)$ in terms of the shifting parameter a.

From (2), we have

$$\rho = \rho(\tilde{X}, \tilde{X}^*) = \frac{1}{\sqrt{1 + Var(C)\frac{E[(X+a)^2]}{Var(X)}}}$$

therefore,

$$\frac{1}{\rho^2} = 1 + Var(C)\frac{E(X^2) + 2aE(X) + a^2}{Var(X)}$$

$$Var(C) = \frac{1/\rho^2 - 1}{1 + \frac{[E(X)+a]^2}{Var(X)}}. \quad (5)$$

For the multiplicative noise C, we have $E(C) = 1$ and

$$\begin{aligned}
Var(C) &= E(C^2) - [E(C)]^2 = 0.5E(U_1^2) + 0.5E(U_2^2) - 1 \\
&= 0.5[Var(U_1) + (EU_1)^2 + Var(U_2) + (EU_2)^2] - 1 \\
&= \tfrac{1}{3}(b_1^2 + b_1 b_{01} + b_{01}^2)
\end{aligned} \quad (6)$$

If we choose $b_{01} = 0.05$ or slightly greater than 0.05, say 0.055, the multiplicative noise C will ensure the probability in *Measurement 1*, $P(|C - 1| < 0.05) = 0$. From (6), we can use the value $Var(C)$ to determine the value of b_1, i.e.

$$b_1 = \left\{ -b_{01} + \sqrt{b_{01}^2 - 4[b_{01}^2 - 3Var(C)]} \right\} / 2,$$

as b_1 needs to be positive. Thus, once the value $Var(C)$ is determined, the distribution of C is determined.

The values of $E(X)$ and $Var(X)$ are determined by the underlying original data, while the data provider determines the value of ρ. Given $E(X)$, $Var(X)$ and ρ, Eq. (5) shows the relationship between $Var(C)$ and a. Based on the noise-multiplied data only, [19] proved that the data user could establish a linear regression model for modeling the relationship between the values of the original data and the values of the masked data. Given the value of the masked data, the data user can use the model to predict the value of the original data. [19] showed that the higher the value of ρ is, the more accurate the prediction will be. Therefore, the data provider needs to keep the value of ρ as low as possible. Our experience shows that ρ should be under 0.9 [19].

Thereby, we can follow the steps below to identify an appropriate multiplicative noise C for the underlying data $\{x_i\}$:

(1) Identify a real number a_0 such that there are no zeros in the set of shifted data $\{x_i + a_0\}$.
(2) Given $E(X)$, $Var(X)$ and ρ, identify the value $Var(C)$ from (5).
(3) Decide $b_{01} \geq 0.05$, and obtain b_1 from (6).

Then, a probability distribution of C can be identified from the distribution family (4).

4   Retrieving the Statistical Information from Noise-Multiplied Shifted Data

Once the set of the noise-multiplied shifted data $\{\tilde{x}_i^* = (x_i + a_0)c_i\}$ is formed, the remaining issue is about retrieving the statistical information of $\{x_i\}$ from the data $\{\tilde{x}_i^* = (x_i + a_0)c_i\}$. The discussion on evaluating utility preservation or utility loss can be found from the literature, [2, 20–22] and references therein.

Suggested by the literature, one manner used to evaluate the utility of the original data, retrieved from the masked data is to check whether the moments of the original data can either be preserved or can be accurately estimated based on the masked data. People pay more attention to the k-moments in the evaluation, $k \leq 4$. The relationship between the moments of X and the moments of $\tilde{X} = X + a_0$ can be derived as follows:

$$E(X) = E(\tilde{X}) - a_0$$
$$E(X^k) = E(\tilde{X}^k) - \sum_{i=0}^{k-1} \binom{k}{i} a_0^{k-i} E(X^i)$$

$k = 1, 2, \cdots$. If the shifting parameter a_0 and the moments information of the multiplicative noise C are publicly available, the kth moment of $\tilde{X}$ can be evaluated as follows

$$E(\tilde{X}^k) = \frac{E[(\tilde{X}^*)^k]}{E(C^k)},$$

through the moments of $\tilde{X}^*$ and the moments of C, so is the kth moment of X, $k = 1, 2, \cdots$. Theoretically, the processes of data shifting and data masking would not cause loss to the information of the moments of the original data.

Consequently, replacing the theoretical value of the moments with the associated sample moments, the moments of the original data can be reasonably estimated if the sample size of the data is big. Therefore, retrieving the information of the moments of the original data from the noise-multiplied shifted data is feasible.

Another method, the sample-moment-based density approximant [8], can be beneficial in retrieving the statistical information of the original data. The technique is employed for reconstructing the density function of the original data based noise-multiplied data. The summary statistics of the original data can be estimated through the summary statistics of the sample drawn from the reconstructed density function. Two software *MaskDensity14* and *MaskDensityBM*[1] are developed for implementing the sample-moment-based density approximant [23–25]. Both softwares have the same power in implementing the method of the sample-moment-based density approximant. The inference results produced by *MaskDensityBM* are more stable. For details see [24,25].

In the next section, we use *MaskDensityBM* to retrieve the statistical information of the original data based on noise-multiplied shifted data. The process of reconstructing the density function of the original data consists of two steps. (i) Apply *MaskDensityBM* to the noise-multiplied shifted data and obtain the reconstructed density function $\hat{f}_{\tilde{X}}$ for $\{\tilde{x}_i\} = \{x_i + a_0\}$. (ii) Shift $\hat{f}_{\tilde{X}}$ by a_0 and obtaining the reconstructed density function $\hat{f}_X$. We evaluate the level of utility loss by visually comparing the actual density function of the original data and its reconstructed density function. Also, we evaluate the level of utility loss by comparing the absolute difference of the summary statistics of the sample from the reconstructed density function with the summary statistics of the original data.

From [8], the density function of $\tilde{X}$ can be reasonably approached by the sample-moment-based density approximant $f_{\tilde{X}, K | \{\tilde{x}_i^*\}, \{c_i'\}}$, subject to the information of $\{\tilde{x}_i^*\}$, $\{c_i'\}$ and K, where $\{\tilde{x}_i^*\}$ is the noise-multiplied data of $\{\tilde{x}_i\}$; $\{c_i'\}$ is a sample from the multiplicative noise C and $\{c_i'\}$ is different from the sample which was used to mask $\{\tilde{x}_i\}$; and K is the optimal upper order of moments used in the sample-moment-based density approximant. Identifying the value of K without using the information of the original data directly is a challenge.

---

[1] Software *MaskDensityBM* is available on request.

For reading convenience, we briefly introduce the logic of the algorithm used in *MaskDensityBM* in determining the K and the computational treatment adopted in the software.

Assume that the multiplicative noise C is a positive random variable and bounded by real numbers c_{00} and c_{01}, where $0 < c_{00} < c_{01}$. Based on the Bayes' theorem,

$$f_{\tilde{X}^*}(\tilde{x}^*) = \int_{c_{00}}^{c_{01}} \frac{1}{c} f_C(c) f_{\tilde{X}}(\frac{\tilde{x}^*}{c}) dc \qquad \tilde{x}^* \in \text{ the range of } \tilde{X}^* \qquad (7)$$

Denote

$$f_{\tilde{X}^*,K|\{\tilde{x}_i^*\},\{c_i'\}}(x^*) = \int_{c_{00}}^{c_{01}} \frac{1}{c} f_C(c) f_{\tilde{X},K|\{x_i^*\},\{c_i'\}}(\frac{x^*}{c}) dc$$

$$\approx \sum_{j=0}^{NC} \frac{1}{c_j} f_C(c_j) f_{\tilde{X},K|\{x_i^*\},\{c_i'\}}(\frac{x^*}{c_j}) \triangle c_j \qquad (8)$$

where $c_0 = c_{00}$ and $c_{NC} = c_{01}$. From (7) and (8), we can conclude that the necessary condition that $f_{\tilde{X},K|\{x_i^*\},\{c_i'\}}$ is a good approximation of $f_{\tilde{X}}$ is that $f_{\tilde{X}^*,K|\{\tilde{x}_i^*\},\{c_i'\}}$ is a good approximation of $f_{\tilde{X}^*}$. Therefore, the optimal upper order K should be the K such that

$$S(K) = \sum_{j=1}^{N_S+1} \left[f_{\tilde{X}^*}(z_j) - f_{\tilde{X}^*,K|\{\tilde{x}_i^*\},\{c_i'\}}(z_j) \right]^2 \triangle z_j$$

takes a small value, where $z_0 = \min\{\tilde{x}_i^*\} < z_1 < \cdots < z_{N_S} < z_{N_S+1} = \max\{\tilde{x}_i^*\}$. The details of the algorithm used in *MaskDensityBM* can be found from [24, 25].

Therefore, applying *MaskDensityBM* to the noise-multiplied shifted data, the data user needs to decide two integers N_s and NC for the evaluation process. In the simulation studies of next section, we set $N_s = 50$ and $NC = 100$.

5   Simulation Studies

In this section, we use two examples to show the application of the data-shifting method in protecting the values of the original data and display the fact that the statistical information of the original data can still be reasonably retrieved from the masked shifted data.

Example 1. The original data is a sample $\{x_i\}_{i=1}^{1000}$ drawn from $N(0,2)$. The multiplicative noise C has density function

$$f_C = 0.5 Unif(1 - b_1, 0.95) + 0.5 Unif(1.05, 1 + b_1).$$

We need to identify a multiplicative noise C with an appropriate b_1 in terms of what the level of values protection and the amount of utility loss are under control. Figure 1 gives the plot of $Var(C)$ against a for each ρ fixed. Those pairs

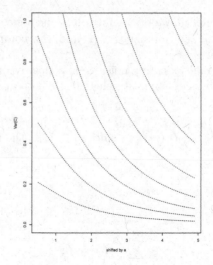

Fig. 1. The plot of $Var(C)$ vs a. The lower curve corresponding to $\rho = 0.9$. The subsequent curves move up corresponding to the ρ with increment by 0.1.

$(a, Var(C))$, which lead to ρ less than 0.9, can be visually identified from the plot. The larger the value $Var(C)$ is, the more perturbation the multiplicative noise C will add to the original data. Regarding the protection of the values of the original data, we should not consider a noise with small variance.

Table 1 gives the summary statistics of $\{x_i\}$. Assume that all the values close to zero are sensitive. Given $\min\{x_i\} = -3.974$, we shift all the original data by $a_0 = 5$. Thus, the new data dataset $\{x_i + 5\}$ does not contain any zeros. If we require that $\rho = 0.5$, it means we wish to control the correlation coefficient of the original shifted data and their masked data is at most 0.5. With the decision of $a_0 = 5$, $\rho = 0.5$ and $b_{01} = 0.055$, we use (5) to obtain $Var(C) = 0.2189947$. This is a theoretical value. Then, we use (6) to determine b_1, which is 0.7876061. Thus, the distribution of the multiplicative noise C is determined.

We, then simulate a sample $\{c_i\}_{i=1}^{1000}$ from the multiplicative noise C and use this sample to mask the shifted original data $\{x_i + a_0\}$. The noise-multiplied shifted data can be generated by *MaskDensityBM*. The software produces a set of masked data for $\{x_i + a_0\}$, whilst creating a binary file containing the information of the multiplicative noise for data users (details see [23]).

Figure 2 shows the plot of the original data vs. their masked data. Comparing the plot of the original data vs. the noise-multiplied data and the plot of the original data vs. the noise-multiplied shifted data, we find that the data-shifting approach does improve the level of protection on the values. Under the condition that the data provider accepted $\rho = 0.5$, this multiplicative noise C meets the requirement of *Measurement 1–Measurement 3*. The sample correlation coefficient of $\{x_i + 5, (x_i + 5)c_i\}$ and the sample variance $\{c_i\}$ are 0.49774 and 0.2189947, respectively. They are not the same as their theoretical values, but close enough.

So far we only check the level of value disclosure risk. We also need to evaluate utility loss. We can obtain the reconstructed density function of the original data from the noise-multiplied shifted data by using *MaskDensityBM*. Then, assess information loss by comparing the difference between the plot of the density function of the original data and the plot of the constructed density function.

The plots of the density function of $\{x_i\}$ and reconstructed density function are presented in Fig. 3. The reconstructed density function mimics the density function of the original data. We also report the summary statistic of the original data and the summary statistics of a sample simulated from the reconstructed density function in Table 1. They are not the same in values but reasonably close to each other. We also independently used C to mask the set of original data 50 times and obtained 50 sets of noise-multiplied shifted data. Then, used *MaskDensityBM* to obtain the estimated density functions from the sets of noise-multiplied shifted data, respectively. Consequently, we obtained 50 outputs of the summary statistics from the reconstructed density functions. Table 1 reports the results across the 50 independent simulations. The result is impressive.

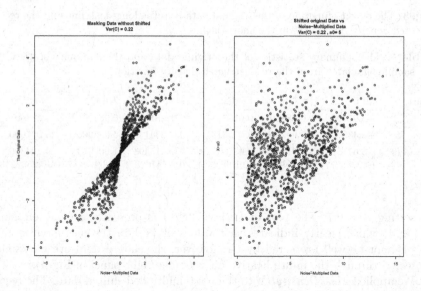

Fig. 2. The left panel is the plot of $\{x_i\}$ vs $\{x_i c_i\}$, where $Var(C) = 0.22$. The right panel is the plot of $\{x_i + a_0\}$ vs $\{(x_i + a_0)c_i\}$, where $a_0 = 5$ and $Var(C) = 0.22$.

Example 2. The original data $\{x_i\}_{i=1}^{1000}$ were simulated from the mixture $0.3Normal(2, 1^2) + 0.7Normal(6, 1.5^2)$.

The summary statistics of $\{x_i\}$ are listed in Table 2. The minimum value of this sample is -0.6609. In this example, we shifted $\{x_i\}$ by $a_0 = 2$ and the multiplicative noise C has mixture uniform distributions with $b_{01} = b_{02} = 0.055$, $b_1 = b_2 = 0.5076463$. The parameters of b_1 and b_2 are determined from (6)

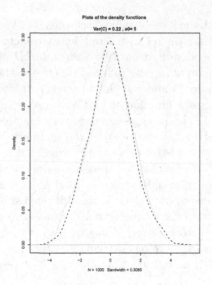

Fig. 3. The density function of the original data is in the long-dash line and the reconstructed density function is in the short-dash line.

Table 1. The summary statistics of the original data and the summary statistics of the sample simulated from the reconstructed density function

Data	Min.	1st Qu.	Median	Mean	3rd Qu.	Max.
X	−3.97400	−0.88860	0.01302	0.02281	0.93990	4.58400
New X	−3.438000	−0.992900	0.070510	0.003341	0.953900	3.997000
Mean (sd.)	−3.671526 (0.2738277)	−0.9170799 (0.1189605)	0.03433718 (0.1097246)	0.05415493 (0.07476287)	1.007943 (0.09784875)	4.204396 (0.2562243)

by setting $\rho = 0.7$. The plot of $\{x_i\}$ vs $\{x_i c_i\}$ is presented at the left panel in Fig. 4, which clearly indicates that values of $\{x_i\}$ in the neighbourhood of zero cannot be well protected by C. However, the shifted data are reasonably protected through the multiplicative noise C (the right panel in Fig. 4).

We applied *MaskDensityBM* to the noise-multiplied shifted data. The reconstructed density function of the original data is presented in Fig. 5, and it mimics the density function of the original data. We simulated a sample, denoted *SimuX* with the size of 1000 from the reconstructed density function. The summary statistics of *SimuX* is listed in Table 2, the values of the statistics are close to those given by the original data, respectively.

Both Examples 1 and 2 demonstrate that we can identify a multiplicative noise C and a shifting-parameter a_0 for protecting values close to zero in the original data by using the data-shifting approach. The existing techniques for estimating the density function of the original data is beneficial in retrieving the statistical information of the original data from their shifted masked data.

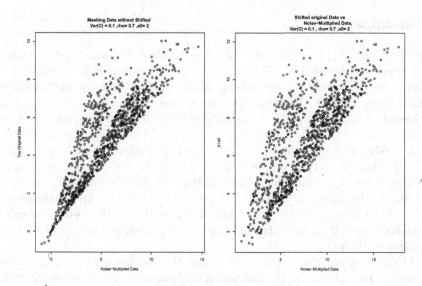

Fig. 4. The left panel is the plot of $\{x_i\}$ vs $\{x_ic_i\}$. The right panel is the plot of $\{x_i + a_0\}$ vs $\{(x_i + a_0)c_i\}$.

Table 2. The summary statistics of the original data and the summary statistics of the sample simulated from the reconstructed density function.

Data	Min.	1st Qu.	Median	Mean	3rd Qu.	Max.
X	-0.6609	2.9100	5.1380	4.7850	6.4890	10.0400
$SimuX$	-0.5981	3.1910	5.2850	4.9010	6.5410	9.5540

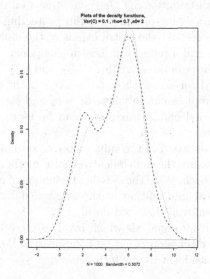

Fig. 5. The plots of the density function of the original data (in bold dash line) and reconstructed density function of the original data (in thin dash line).

6 Summary

Directly employing the multiplicative noise method to protect values close to zero might not be efficient. This paper proposes a solution for protecting the values close to zero under the scheme of the multiplicative noise, namely the data-shifting method. The principle idea of the method is to generate noise-multiplied shifted data and use the noise-multiplied shifted data to protect the original data.

To allow data users to retrieve the statistical information of the original data by using the method proposed in this paper, the shifting-parameter a_0 and the knowledge of the sample moments of the multiplicative noise C, need to be available to the public. However, if C is selected by using the three measurements addressed in Sect. 2, the level of identification/prediction disclosure risk can be controlled, even if a_0 and the information of the sample moments of C are available to the public.

The two values, the variance of C and the shifting-parameter a_0, have a significant impact on the level of protection provided by the noise-multiplied shifted data to the original data. The values cannot be independently determined and are associated with $\rho(\tilde{X}, \tilde{X}^*)$ (See (2)). To ensure the statistical information of the original data can be reasonably retrieved from the noise-multiplied shifted data, the value of $\rho(\tilde{X}, \tilde{X}^*)$ cannot be set as small as we wish. Therefore, it needs a balance between the level of value protection and utility loss control and the balance can be achieved through choosing the appropriate values of the variance of C and the shifting-parameter a_0.

There is no general regulation for choosing the variance of C and the shifting-parameter a_0 in the data-shifting method. The probability distribution of the original data and the distribution of C have an impact on the performance of the variance of C and the shifting-parameter a_0 in protecting the original data. To simplify the study, we restrict the distributions of the multiplicative noise C to the family of mixture uniform distributions in this paper. The mixture uniform distributions involves four parameters $(b_1, b_2, b_{01}$ and $b_{02})$. There are at least two degrees of freedom in determining these parameters in the data-shifting method. We only work on a simple scenario where $b_1 = b_2$ and $b_{01} = b_{02} = 0.055$ in this paper. The study carried out in this paper can be used as a reference for the study of complex cases.

This paper demonstrates that the values close to zero in the original data can be well protected by using the multiplicative noise method in conjunction with the data-shifting approach. With the existing technique of reconstructing density function, the statistical information of the original data could be reasonably retrieved from noise-multiplied shifted data.

The utility loss for data analysis which involves more than one variable is a significant issue. The discussion could be complicated. We do not address it in this paper but will consider it in our future work.

References

1. Hwang, J.T.: Multiplicative errors-in-variables models with applications to recent data released by the U.S. department of energy. J. Am. Stat. Assoc. **81**, 680–688 (1986)
2. Kim, J., Winkler, W.: Multiplicative noise for masking continuous data. Technical report, Statistical Research Division, U.S. Bureau of the Census, Washinton D.C. 20233 (2003)
3. Kim, J., Jeong, D.: Truncated triangular distribution for multiplicative noise and domain estimation. In: Government Statistics-JSM, pp. 1023–1030 (2008)
4. Oganian, A.: Multiplicative noise for masking numerical microdata with constraints. "SORT", Special Issue, pp. 99–112 (2011). http://hdl.handle.net/2099/11378. ISSN 1696-2281
5. Ruiz, N.: A multiplicative masking method for preserving the skewness of the original micro-records. J. Off. Stat. **28**, 107–120 (2012)
6. Lin, Y.X., Wise, P.: Estimation of regression parameters from noise multiplied data. J. Priv. Confid. **4**, 55–88 (2012)
7. Klein, M., Mathew, T., Sinha, B.: Noise multiplication for statistical disclosure control of extreme values in log-normal regressopm samples. J. Priv. Confid. **6**, 77–125 (2014)
8. Lin, Y.-X.: Density approximant based on noise multiplied data. In: Domingo-Ferrer, J. (ed.) PSD 2014. LNCS, vol. 8744, pp. 89–104. Springer, Cham (2014). https://doi.org/10.1007/978-3-319-11257-2_8
9. Mivule, K.: Utilizing noise addition for data privacy, an overview. In: Proceedings of the International Conference on Information and Knowledge Engineering (IKE2012), The Steering Committee of The World Congress in Computer Science, Computer Engineering and Applied Computing (WorldComp), pp. 65–71 (2012)
10. Mendes, R., Vilela, J.P.: Privacy-preserving data mining: methods, metrics, and applications. IEEE Access **5**, 1–21 (2017)
11. Torra, V.: Data Privacy: Foundations, New Developments and the Big Data Challenge. SBD, vol. 28. Springer, Cham (2017). https://doi.org/10.1007/978-3-319-57358-8
12. Nayak, T.K., Sinha, B., Zayatz, L.: Statistical properties of multiplicative noise masking for confidentiality protection. J. Off. Stat. **27**(3), 527–544 (2011)
13. Muralidhar, K., Batra, D., Kirs, P.J.: Accessibility, security, and accuracy in statistical databases: the case for the multiplicative fixed data perturbation approach. Manag. Sci. **41**(9), 1549–1564 (1995)
14. Lin, Y.X., Mazur, L., Sarathy, R., Muralidhar, K.: Statistical information recovery from multivariate noise-multiplied data, a computational approach. Trans. Data Priv. **11**, 23–45 (2018)
15. Sinha, B., Nayak, T., Zayatz, L.: Privacy protection and quantile estimation from noise multiplied data. Sankhya B **73**, 297–315 (2012)
16. Oliveira, S.R.M., Zaiane, O.R.: Privacy preserving clustering by data transformation. J. Inf. Data Manag. **1**, 37–51 (2010)
17. Adam, N.R., Worthmann, J.C.: Security-control methods for statistical databases: a comparative study. ACM Comput. Surv. **21**, 515–556 (1989)
18. Muralidhar, K., Parsa, R., Sarathy, R.: A general additive data perturbation method for database security. Manag. Sci. **45**(10), 1399–1415 (1999)

19. Ma, Y., Lin, Y.X., Sarathy, R.: The vulnerability of multiplicative noise protection to correlational attacks on continuous microdata. Technical report, National Institute for Applied Statistics Research Australia, School of Mathematics and Applied Statistics, University of Wollongong, Australia (2017)
20. Agrawal, D., Aggarwal, C.C.: On the design and quantification of privacy preserving data mining algorithms. In: Proceedings of the Twentieth ACM SIGMOD-SIGACT-SIGART Symposium on Principles of Database Systems, pp. 247–255. ACM (2001)
21. Burridge, J.: Information preserving statistical obfuscation. Stat. Comput. **13**, 321–327 (2003)
22. Domingo-Ferrer, J., Sebé, F., Castellà-Roca, J.: On the security of noise addition for privacy in statistical databases. In: Domingo-Ferrer, J., Torra, V. (eds.) PSD 2004. LNCS, vol. 3050, pp. 149–161. Springer, Heidelberg (2004). https://doi.org/10.1007/978-3-540-25955-8_12
23. Lin, Y.X., Fielding, M.J.: Maskdensity14: An R package for the density approximant of a univariate based on noise multiplied data. SoftwareX **3**, 37–43 (2015)
24. Lin, Y.X.: Mining the statistical information of confidential data from noise-multiplied data. In: Proceedings of the 3rd IEEE International Conference on Big Data Intelligence and Computing (2017)
25. Lin, Y.X.: A computational Bayesian approach for estimating density functions based on noise-multiplied data. Int. J. Big Data Intell. (2018). (in press)

Efficiency and Sample Size Determination of Protected Data

Bradley Wakefield[✉] and Yan-Xia Lin

National Institute for Applied Statistics Research Australia,
School of Mathematics and Applied Statistics, University of Wollongong,
Wollongong, Australia
bnw722@uowmail.edu.au

Abstract. This paper assesses the usefulness of a proposed multiplicative perturbation method by contrasting the statistical efficiency achieved in point hypothesis testing of simple proportions with that of the differentially private aggregated Laplace mechanism. This efficiency is evaluated by obtaining an analytical expression that determines the sample size required for protected data to retain a given significance level and power.

Keywords: Noise multiplied data · Hypothesis test
Confidential data · Data protection · Multiplicative perturbation
Proportions

1 Introduction

Information sharing and data transparency is an essential component of modern day research. Especially within the area of health and medicine, clinical trial transparency through clinical trial registries and public sharing of results and information has become ever more prevalent. ClinicalTrials.gov is currently the largest registry in the world and grants public access to information provided by more than 78,900 trials sponsored by government agencies and privacy industries [26]. This proliferation of information, although likely to foster new advances through collaboration of research, may lead to the trust and privacy of participants being compromised. Individuals often only agree to actively participate in trials and surveys after receiving assurances that the information they provide will not be made available in a form in which their identity can be ascertained [7,8,29]. Consequently, the recent demand for statistical protection methodologies has considerably increased.

A range of data protection methods have been proposed and applied in practice [3,6,10,14,19,20,24,28,29] (and references therein). These methodologies usually relate to the protection of all or part of the private data and involve masking either the observations or statistics that are produced within the data set. Applying conventional statistical analysis to protected data, information

© Springer Nature Switzerland AG 2018
J. Domingo-Ferrer and F. Montes (Eds.): PSD 2018, LNCS 11126, pp. 263–278, 2018.
https://doi.org/10.1007/978-3-319-99771-1_18

pertaining to individual observations of the data should be unobtainable, however general statistical inference must be possible and maintain a reasonable level of accuracy. The balance between protection and utility is a decision ultimately made by the data agency and the way in which this tradeoff is performed is highly dependent on the data protection method implemented. Two types of data protection methods are considered in this paper: the (ϵ-)differential privacy method by means of the aggregated Laplace mechanism [6,26] and the multiplicative noise method i.e. applying a multiplicative perturbation to each observation of the data [9,10,14].

Differential privacy refers to a privacy protection standard that ensures no single observation is identifiable based on a response to any single query [6]. The privacy preservation standard of differential privacy can be achieved through varying statistical mechanisms [4,6,18,25,27] (and references therein). These mechanisms are usually categorised as methods which aggregate data and release a perturbed output or methods that alter individual observations and release protected **micro data**. The advantage of achieving protected **micro data** is that it enables the data user to perform more extensive analysis than that envisaged by the data issuer whereas releasing only aggregations of data, provides limitations to the extent in which the data can be analysed. Within the differential privacy framework for categorical variables, mechanisms which output **micro data** include random swapping, randomised response and other post randomisation probability mechanisms [23]. However, these methods are not commonly used in practiced as the standard of privacy required is at too great a cost to data utility. Perturbed aggregated output mechanisms include the Laplace noise mechanism and the exponential noise mechanism [4,6,18,26].

The multiplicative noise method is a data masking scheme that is used for data protection through observational perturbation and has attracted statisticians' attention in the recent decade. Protecting the value of the individual observation are necessary in many scenarios. For instance, in clinical decision making and hypothesis testing, data needs to be collected from patients/volunteers who participate in clinic trials. If the data collector is an untrusted third party, the patients/volunteers may not willingly let the data collector access the true values of their information. Another example of the importance of protected micro data can be found when considering a smart metering system, which as discussed in [1], requires perturbation from the individual household smart meter reading before the data is sent to the supplier for the data end user.

A description of the traditional multiplicative noise method is provided by [10]. The definition of the multiplicative noise method applied in this paper is the general notion of masking some Bernoulli distributed observation by multiplying the observation by some independent random noise (see Sect. 3).

The noise multiplied **micro data** can be analysed in order to recover statistical information of various attributes [10,11] and can also be utilised in the estimation of parameters in linear regression [9,14]. Currently, many non-conventional statistical data analysis methods, based on noise multiplied data, have been developed [9,11,13–15,19,22]. Through the use of noise multiplied **micro data**,

data users have more opportunities to perform meaningful statistical inference on the published information without the need to send various aggregated data requests to the data agency, as may be the case with the Laplace noise mechanism.

Differential privacy is a privacy model that offers a priori levels of privacy protection on the data (according to the ϵ parameter), whereas the privacy achieved by the multiplicative noise is statistically evaluated a posteriori. Furthermore, the multiplicative noise method belongs to an approach of statistical disclosure control (SDC) related to analytical validity, whereas as discussed in [2], differential privacy focuses mainly on formal privacy guarantees.

The example of [26] applying the ϵ-differential privacy to statistical hypothesis testing inspires our curiosity in comparing these two methods. This paper investigates whether there are any advantages in applying the multiplicative noise method to the same type of statistical problem concerned in [26].

Hypothesis testing is a valued statistical tool by which scientists and researchers apply statistical evidence to test and support hypotheses. Typically, this process involves determining whether or not there is sufficient evidence to reject a null hypothesis. Whilst designing an experiment in order to conduct an appropriate hypothesis test, it is critical to ensure the test performed achieves statistical significance and sufficient statistical power. A considerable amount of literature has been developed for determining the sample size required to achieve the required significance level and statistical power when considering unprotected data. However, these methods fail to properly account for the variation introduced into protected data that has in some way been perturbed or masked. Furthermore, the sample size required of data protected by various privacy preservation techniques to maintain sufficient statistical significance and power, also provides a metric by which comparison of the statistical efficiency of these methods can be performed.

Considering only simple proportion data in a simple hypothesis test, this paper assesses the extent to which required sample size is altered across the two privacy protection techniques. Applying the concept and result derived by [26] for the ϵ-differentially private aggregated Laplace mechanism (**referred to in this paper as the differential privacy method**), to the multiplicative noise masking scheme, further inference can then be made about the comparative efficiencies of these methods. This paper acknowledges the differences in the inherit nature of the privacy protection offered by these techniques and refers to the limitations of each, whilst referencing protection measures that evaluate certain aspects of protection strength. Ultimately, this study provides statisticians and data agencies with advice and examples for informed assessment of the balance between information loss, disclosure risk and the costs of data collection offered by these techniques, given that a decision can be made about the preferred data analysis practice.

This paper is constructed as follows. Section 2 briefly reviews the ϵ-differential privacy method and gives the sample size adjustment under the ϵ-differential privacy method. The multiplicative noise method is introduced in Sect. 3. The pre-knowledge on the multiplicative noise data masking scheme is briefly described in this section. The sample size adjustment based on noise multiplied data is derived in Sect. 3. A study of comparison between the multiplicative noise method and the ϵ-differential privacy method is presented in Sect. 3. The final section gives the conclusion of this paper.

2   Sample Size Determination Under Differential Privacy Framework: Classical Hypothesis Testing with a Single Proportion

For reading convenience, we briefly introduce the formula of sample size determination under the differential privacy approach [26].

Let X be Bernoulli(p) distributed and consider the following hypothesis test:

$$H_o : p = p_o \quad \text{versus} \quad H_a : p = p_o + \delta, \quad \delta > 0. \tag{1}$$

Denote $\{x_1, x_2, \cdots, x_N\}$ as a random sample of X, i.e. a set of original data. The sample size N required to ensure the detection of a significant difference, subject to a certain level of significance α and power $1 - \beta$ of the hypothesis test is given by:

$$N_o = (z_{1-\alpha/2} + z_{1-\beta})^2 \sigma^2 / \delta^2. \tag{2}$$

where $\sigma^2 = \bar{p}(1 - \bar{p})$; $\bar{p} = p_o + \delta/2$; $z_{1-\alpha/2}$ and $z_{1-\beta}$ are critical values from the standard normal distribution, and the statistic $\hat{p} = \frac{1}{N} \sum_{i=1}^{N} x_i$ is a sufficient statistic of p (see [26]). That is, the necessary condition to ensure the efficiency of the test is $N \geq N_0$.

If the observations $\{x_i\}_{i=1}^{N}$ are confidential, the data set cannot be issued to the public. In some circumstances, the data agency might also feel uncomfortable in releasing the value of $\hat{p}$.

[26] suggested using the differentially private aggregated Laplace mechanism to protect the original data. Under the differential privacy framework, the original data is concealed from the data user. The value of $\hat{p}$ is perturbed by adding a Laplace noise $L(\sqrt{2}/(\epsilon N_p))$ (i.e. an observation from the Laplace distribution with mean 0 and standard deviation $\sqrt{2}/(\epsilon N_p)$). Only the perturbed value is issued to the public and hence the data user must apply the perturbed $\hat{p}$ in the hypothesis testing. The parameter $\epsilon > 0$ in the added noise $L(\sqrt{2}/(\epsilon N_p))$ is a measure of the information leakage. The smaller the ϵ is, the more the noise will be added to $\hat{p}$. The $N_p > N_o$ is used to determine the sufficient size of the sample required by the differential privacy approach. It will ensure that the hypothesis test has the same level of efficiency produced by the sample $\{x_1, \cdots, x_{N_o}\}$ [26]. The size N_p is determined as follows:

$$N_p = N_o K = N_o \left(\frac{1}{2} + \frac{1}{2} \sqrt{1 + \frac{8\delta^2}{\epsilon^2 (z_{1-\alpha/2} + z_{1-\beta})^2 \sigma^4}} \right). \tag{3}$$

That is, the size of the sample should have at least N_p if the original data is going to be protected through the differential privacy approach with the same level of statistical efficiency with respect to the hypothesis test (1).

3   The Multiplicative Noise Method and Sample Size Determination

The multiplicative noise data masking scheme which requires a random noise C to protect the confidential data set $\{x_1, x_2, \cdots, x_N\}$. The random noise C is independent of X, with mean $E(C) \neq 0$ and known variance $Var(C)$. The protected data set is denoted as $\{x_1^*, x_2^*, \ldots, x_N^*\}$ where $x_i^* = x_i c_i$ and $c_1, \cdots, c_N$ is a random sample of C. The noise multiplied data (or masked data) can be considered as a sample of $X^* = XC$. The noise multiplied data set can then be released to the public for external analysis as sensitive observations are obscured by the random noise. It is not necessary to force $E(C) = 1$. If the values of $E(C)$ is released to the public or somehow available to be used in the analysis, using C to mask X is equivalent to using $C/E(C)$ to mask X. In this study, we always force $E(C) = 1$.

It can be shown that, given $X = x$, $\hat{X} = X^*/E(C)$ is an unbiased estimator of x. The original data is protected through the noise multiplied data. The probability distribution of the noise C has a direct impact on the level of the protection. Lin and Wise [14] introduce a measurement $R(\delta_e)$ for evaluating the level of data protection ([12] proposed a similar measurement.). The measure $R(\delta_e)$ is defined as follows, for any $x \neq 0$,

$$Pr\left(\left|\frac{\hat{X} - X}{X}\right| < \delta_e | X = x\right) = F_C(E(C)(1 + \delta_e)) - F_C(E(C)(1 - \delta_e))$$

where $F_C(\cdot)$ is the cumulative distribution function of C. The measurement depicts the probability of relative difference between X and it's corresponding unbiased estimator $\hat{X}$ less than δ_e. The probability given by a smaller value δ_e, say 0.05 is of particular interest. Checking the correlation coefficient of X and X^* is also necessary [16]. These measurements enable us to determine which distributions of the multiplicative noise may offer better protection on the original data and how the mean and variance of the noise variable affects the protection.

With an appropriate multiplicative noise C, the multiplicative noise data masking scheme will provide a reasonable level protection on the original data. However, the standard masking scheme described above will not provide value protection for $x = 0$. The issue of protecting the values "0" will be considered in next subsection.

3.1   Sample Size Determination Under the Multiplicative Noise Framework: Classical Hypothesis Testing with Single Proportion

Consider the sample $x_1, x_2, \ldots, x_N$, the values of the data given by N independent subjects, and the same hypothesis test described in Sect. 2. Recall

$\hat{p} = \bar{x} = \frac{1}{N} \sum_{i=1}^{N} x_i$ is the estimator of $\mu_X = p$. We consider the following scenario where each subject is required to send noise-multiplied datum to a data collector. In a general situation, firstly, the data collector sends the information of the probability distribution of the multiplicative noise C to each subject. After receiving the information of the probability distribution of C, each subject independently simulates a value c from C and yields noise-multiplied value xc to the data collector. However, for the $x_1, x_2, \ldots, x_N$ considered in this paper, the situation is slightly different. The observation x_i, $i = 1, \cdots, N$, takes values "0" or "1", and the multiplicative noise method does not protect "0". Therefore, for the type of data considered in this paper, the process of generating noise-multiplied data need to be modified. Here is the suggestion for the modification. Instead of only sending the information of the probability distribution of the multiplicative noise C to each subjects, the data collector also needs to send out a positive real number a. After received the information, each subject can generate the noise-multiplied data as follows: $x^* = (x + a)c$ and send the masked data to the data collector.

In this paper, we will consider the noise-multiplied data in the form $x^* = (x + a)c$, where a is a shift parameter and c a sample value from the underlying multiplicative noise. We show how the sample size is determined when the noise-multiplied data are in the format $x^* = (x + a)c$. In Sect. 3.2.2, we use examples to explain how to determine the shifting value a in terms of reducing the value of disclosure risk.

A transformation of the random sample $\{x_1, x_2, \cdots, x_N\}$ to $\{y_1, y_2, \ldots, y_N\}$ given by $y_i = x_i + a$ for some fixed constant $a > 0$ is performed. Consequently the hypothesis test can be expressed in terms of Y as:

$$H_0 : \mu_Y = p_0 + a \quad vs \quad H_a : \mu_Y = p_0 + a + \delta. \tag{4}$$

It is now Y which is protected with the multiplicative noise method with $Y^* = CY$. Using the central limit theorem, we know that as N gets larger,

$$\frac{\overline{Y^*} - E(Y^*)}{\sqrt{Var(Y^*)/N}} \to N(0, 1),$$

where $\overline{Y^*} = \sum_{i=1}^{N} y_i^*/N = \sum_{i=1}^{N}(y_i c_i)/N$. We use $(\overline{Y^*} - E(Y^*))/\sqrt{Var(Y^*)/N}$ as a test statistic for determining the significance and power of the above test. Following the same argument used by [26] in deriving N_p under the differential privacy approach, we can determine the sample size N_m required for the noise multiplied data such that the hypothesis testing based on the noise multiplied data maintains the significance level α and power $1 - \beta$. The size N_m can be identified and solved from the following equation

$$E(C)(p_0 + a) + z_{1-\alpha/2}\sigma_{Y^*}/\sqrt{N_m} = E(C)(p_0 + a + \delta) - z_{1-\beta}\sigma_{Y^*}/\sqrt{N_m} \tag{5}$$

where $\sigma_{Y^*}^2 = Var(Y^*) = E(C^2)Var(Y) + [E(Y)]^2 Var(C)$. Following the same argument in [26], the $E(Y)$ and $Var(Y)$ in (5) are replaced by $\bar{p} + a$ and $\bar{p}(1 - \bar{p})$, where $\bar{p} = p_0 + \delta/2$. Hence,

$$N_m = \frac{\left\{Var(C)\left[\bar{p}^2 + 2a\bar{p} + a^2 + \bar{p}(1-\bar{p})\right] + [E(C)]^2\bar{p}(1-\bar{p})\right\}(z_{1-\alpha/2} + z_{1-\beta})^2}{[E(C)]^2\delta^2}$$

$$= N_o\frac{Var(C)(\bar{p} + 2a\bar{p} + a^2) + [E(C)]^2\bar{p}(1-\bar{p})}{\bar{p}(1-\bar{p})[E(C)]^2} > N_o. \tag{6}$$

3.2   The Multiplicative Noise Method vs. The Differential Privacy Method

The multiplicative noise method and the differential privacy method under consideration are two different data protection approaches. The differential privacy method discussed in this paper protects the original data by issuing perturbed sufficient statistics or estimates whereas the multiplicative noise method releases the masked data (**micro data**).

In this section, we investigate the multiplicative noise method vs. the differential privacy approach (proposed in [26]) as well as addressing necessary matters pertaining to the effective use of the multiplicative noise method. This investigation is performed by considering four key aspects: (i) the necessary condition for ensuring a smaller size of sample required for the hypothesis; (ii) determination of the shift parameter a; (iii) the protection strategy for sensitive observations and (iv) the protection level on individual entries of the original data set.

The Necessary Condition for Ensuring a Smaller Size of Sample Required for the Hypothesis. The size of the sample required for effective hypothesis testing has a direct impact on the cost and outcome of the research. It is an area of great concern for the data agency in identifying an appropriate multiplicative noise which maintains a consistent level statistical efficiency with respect to hypothesis testing. In this subsection, it is shown that with appropriate restrictions on the variance of the multiplicative noise, greater efficiency (smaller sample size required) can be obtained compared to that of the differentially privacy approach.

Comparing (3) and (6), we obtain $N_m < N_p$ if the variance of the multiplicative noise C meets the following condition

$$Var(C) < \frac{\bar{p}(1-\bar{p})[E(C)]^2}{\bar{p} + 2a\bar{p} + a^2}\left(-1 + \frac{1}{2} + \frac{1}{2}\sqrt{1 + \frac{8\delta^2}{\epsilon^2\bar{p}^2(1-\bar{p})^2(z_{1-\alpha/2} + z_{1-\beta})^2}}\right)$$

$$= \frac{\bar{p}(1-\bar{p})[E(C)]^2}{\bar{p} + 2a\bar{p} + a^2}\left(-1 + \frac{N_p}{N_o}\right)$$

$$= \frac{\bar{p}(1-\bar{p})[E(C)]^2}{N_o(\bar{p} + 2a\bar{p} + a^2)}(N_p - N_o) = UBV, \tag{7}$$

where UBV is the abbreviation of "the Upper Boundary of Variance". That is, with the same levels of significance and power of the test, the size of the sample required by the multiplicative noise approach can be less than that required by the differential privacy approach if $Var(C)$ meets the condition in (7).

Determination of the Shift Parameter a. Without loss of generality, we assume the mean of the noise C is 1. From (7), the upper boundary of $Var(C)$ is associated with the shift parameter a. As the value of a increases, the value of UBV will decrease. Given $E(C) = 1$, if $Var(C)$ is very close to 0, then the value of $y^* = c(x + a)$ will be close to the value of $y = x + a$. Given the fact that a is publicly accessible, the original value x will receive a high level of disclosure risk. As the value of a decreases toward 0, the value of UBV will increase. The range of $[0, UBV]$ will tend to be wider. This will provide the data agency with a wider range of choice in selecting the possible value of $Var(C)$ during the construction of the multiplicative noise C to meet (7). However, if the value of a is too small and close to 0, then the difference of x and $x + a$ is negligible. Especially, when $x = 0$, the masked value $(0 + a)c$ might be very close to 0 and the masking scheme could fail to provide sufficient protection for the x.

To investigate the relationship between a, N_m and $Var(C)$, we carry out the following simulation study and show how to determine a and $Var(C)$ in practice such that N_m is less than N_p and the disclosure risk is under control.

Example 1. Consider the hypothesis test

$$H_0: \quad \mu = p_0 = 0.65 \quad vs \quad H_a: \quad \mu = p_0 + \delta = p_0 + 0.05 = 0.85, \qquad (8)$$

and the level of significance and power of the test are chosen to be $\alpha = 0.05$ and $1 - \beta = 0.9$ respectively. Let $\{x_i\}$ be a sample drawn from a population under H_0 and let $\{y_i = x_i + a\}$ be the shifted data. The multiplicative noise C is constructed as $I_{(W=0)}X_1 + I_{(W=1)}X_2$, where I is an indicator function; $W \sim Bernoulli(0.5)$; $X_1 \sim N(1 - \sqrt{0.7 \times \sigma_C^2}, 0.3 \times \sigma_C^2)$ and $X_2 \sim N(1 + \sqrt{0.7 \times \sigma_C^2}, 0.3 \times \sigma_C^2)$, where X_1 and X_2 are independent. Thus, $E(C) = 1$ and $Var(C) = \sigma_C^2$.

The value UBV given by (7) is varied as a increases from 0.1 to 2. For each a fixed, to ensure $N_m < N_p = 255.15$, we can assign any value to σ_C^2 as long as the value is between 0 and UBV. To simplify our study, we set $Var(C) = \sigma_C^2 = UBV - 0.02$ in this example. For each a, the values of $Var(C)$, N_m, $R(0.05)$ and the sample correlation coefficient of the shifted original data and their unbiased masked data are reported in Table 1.

From Table 1, all measures of $R(0.05)$ are reasonably small whereas the sample correlation coefficients are around 0.5. Both these measures are key tools used to try and assess the level of protection offered by the multiplicative noise method and in practice, it is preferable that this correlation coefficient is closer to zero as this lessons the likelihood that a data intruder can predict the individual entries in the original data set. However, since X is a categorical variable and the value of $Var(C)$ is bounded by UBV, the correlation coefficient cannot be as small as perhaps desired. Although considering this study only has two different values a and $1 + a$, the information provided by the correlation coefficient might be of little help in the estimation of the individual entries of the shifted data based on their masked data. Consequently it is imperative that the plot of the masked data vs. the shifted data is reviewed to ensure adequate protection has

Table 1. The report of $Var(C)$, N_m, $R(0.05)$ and the sample correlation coefficient of the shifted original data and their unbiased masked data where $N_p = 255.15$.

a	σ_C^2	N_m	$R(0.05)$	Cor.	a	σ_C^2	N_m	$R(0.05)$	Cor.
0.1	0.8412	250.37	0.02478	0.508	1.1	0.1971	236.19	0.05156	0.516
0.2	0.6991	249.43	0.02720	0.515	1.2	0.1764	234.19	0.05456	0.492
0.3	0.5876	248.38	0.02968	0.522	1.3	0.1586	232.09	0.05762	0.520
0.4	0.4991	247.22	0.03222	0.492	1.4	0.1430	229.89	0.06076	0.477
0.5	0.4279	245.96	0.03482	0.471	1.5	0.1293	227.57	0.06397	0.513
0.6	0.3700	244.60	0.03748	0.495	1.6	0.1173	225.15	0.06726	0.516
0.7	0.3223	243.12	0.04018	0.476	1.7	0.1066	222.64	0.07065	0.462
0.8	0.2826	241.55	0.04294	0.549	1.8	0.0972	220.01	0.07413	0.491
0.9	0.2494	239.87	0.04576	0.552	1.9	0.0887	217.28	0.07771	0.548
1.0	0.2212	238.08	0.04863	0.527	2.0	0.0811	214.44	0.08140	0.534

been provided. This plot, demonstrated by Fig. 1, portrays the mapping between the masked and original data and provides a clear indication as to the level of protection offered.

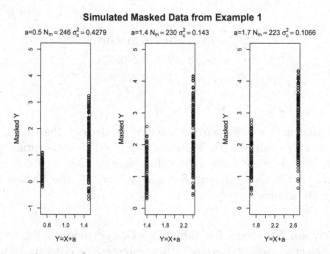

Fig. 1. The plots of shifted masked data vs the original shifted data i.e. Y^* vs Y applying the shift and variances values outlined in Example 1 and Table 1. The original data has been randomly sampled from the Bernoulli distribution with $p = 0.65$.

Considering the outputs given by $a = 0.5, 1.4$ and 1.7 from Fig. 1 as an example. In both cases, $a = 0.5$ and $a = 1.4$ achieve correlation coefficients around 0.47. However, Fig. 1 clearly shows that, when $a = 0.5$, all the masked

$Y = 0 + a$ appear in a close neighborhood of 0.5 and the masked $Y = 1 + a$ widely spread across 0 to 4 providing a relatively narrow overlap. The wider the overlap interval is, the more the individuals of masked $Y = 1 + a$ and $Y = 0 + a$ take similar values within the same range. Thus, the data intruder will have less chance of obtaining a correct estimate when choosing between $X = 0$ and $X = 1$. The plots indicate that using $a = 1.4$ or $a = 1.7$ to shift the values of the original data, will provide a greater level of protection to the original data than that obtained using $a = 0.5$.

In this simulation study, Fig. 1 shows that Y^* will take values less than 1.1, 2.5 and 3 roughly corresponding to $a = 0.5$, $a = 1.4$ and $a = 1.7$, respectively, if the original entry takes value 0. For convenience, we call 1.1, 2.5 and 3 the *upper critical values* (denoted by UC_a) of the masked shifted data corresponding to the values of $a = 0.5, 1.4$ and 1.7, respectively. The value of UC_a is varied and related to the original data and the noise sample. Once the original data were masked and issued to the public, the value of UC_a is fixed. Considering the worst case scenario, if the data intruder knows the real probability $P(X = 1) = 0.65$, the probability distribution of C and the value of UC_a, the data user should be able to work out the probability of Y^* greater than UC_a. These calculations have been reported the probability in Table 2.

Table 2. The Probability of Y^* greater than the critical value UC_a.

	$P(Y^* > UC_a)$
$UC_a = 1.1, a = 0.5$	0.3976808
$UC_a = 2.5, a = 1.4$	0.3106577
$UC_a = 3, a = 1.7$	0.2713907

The original data entries with masked value greater than UC_a receive no protection in this example. Table 2 shows, only 60% of the original data are protected under different levels if the data agency uses $a = 0.5$ to shift the original data. If the data agency uses $a = 1.4$ or 1.7, the percentage will be increased to 70 or more.

The Protection Strategy for Sensitive Observations. Example 1 clearly shows that selecting an appropriate shift parameter a is critical. An appropriate a can enable more data to be protected. However, even using the shift parameter $a = 1.4$ or 1.7, around 30% of the total number of the original data could not be protected if the data intruder was able to access the information of $P(X = 1)$, probability distribution C and UC_a. This notion raises a considerable problem when assessing the effectiveness of the multiplicative noise method in protecting categorical data. If all the entries with $X = 1$ are sensitive, the failure to protect a proportion of the data set would be a serious problem even if the percentage of the number of $X = 1$ having masked values greater than UC_a is only 5%.

The differentially private approach being considered does not encounter this issue as it releases perturbed aggregated data. Even if the data system is an enquirer system, the aggregated differential privacy approach can ensure that a minimum standard of privacy is maintained on all observations. However, in the event that not all observations are sensitive, this enforced standard could be considered excessive and offers considerable utility sacrifices for the purpose of unnecessary privacy protection. Thereby, the important issue is about the protection strategy for sensitive observations.

To protect the participants, all the ID information in a confidential data set has to be removed before the data set is released to the public. If an issued data set contains only one variable X and this variable is a categorical variable (taking values 1 or 0), the data set has no confidentiality issue if no other extra information can be used to link the values of X with the removed ID information.

An entry of the categorical variable X becomes sensitive when a relationship between the value of the entry and other additional information can be established, as the ID of the entry can be identified through relationships with other variables. In this scenario, a published data set usually involves X and other variables, and it is through the risk of linkage that the values of X can be considered as sensitive. In this situation the risk of linkage on a particular entry of X would vary from observation to observation and hence it may not be necessary to impose a set level of protection on all observations. It is through this framework lens that one may suggest that the multiplicative noise method may be a possible alternative to current differential privacy methods.

Due to the randomness of noise simulation, strategies must be developed that ensure that the entries with less protection are not sensitive. Assume that only a few entries with $X = 1$ are sensitive in the original data. Consider the scenario that, after shifted and masked, some of the sensitive entries have the masked values greater than the critical value, the value UC_a discussed in previous subsection. We suggest the following additional strategy to protect those entries. The data agency can swap the masked values of these sensitive entries with other masked values of $X = 1$. Those entries of $X = 1$ need to meet at least two conditions (i) the original entries are not sensitive and (ii) the masked values are much less than UC_a. With such swapping treatment, the conclusion of hypothesis testing will not be affected, as the original values of the swapped data are the same and hence calculations of the statistics would remain unaffected. This treatment also does not affect any data analysis pertaining to the noise multiplied data given by other variables if the original data set involves other attributes. It is because that, under the multiplicative noise method, X and other attributes (if there are any) are independently masked by independent multiplicative noises which usually have different probability distributions. Due to page limit, we do not provide further details discussion in this paper.

The Protection Level on Individual Entries in the Original Data Set. Although differentially private **micro data** mechanisms exist, the statistical usefulness of these methods is relatively small compared to that of the approaches considered in this paper. For example considering the Warner randomised response mechanism [29] which has been shown to be differentially private, in order to obtain **micro data** with an information leakage parameter $\epsilon = 0.1$ the randomisation parameter must be approximately 0.475 which implies that nearly every second observation needs to be randomised in some form or another. Applying the same hypothesis test as previously outlined, the required sample size for the randomised response mechanism is given by $N_0 \left[1 + \frac{e^\epsilon}{(1-e^\epsilon)p(1-p)} \right]$ and hence under the same parameters outlined in Example 1 would require a sample size of approximately 26,296 around 103 times larger than that of the aggregated Laplace mechanism used by [26] with the same information leakage parameter.

Whereas for data masked by the multiplicative noise method, the statistical utility of this method can be set to rival that of the Laplace mechanism and apply a reasonable level of protection. Although each original observation may face a certain level of disclosure risk, by choosing an appropriate multiplicative noise to mask the underlying data and applying additional appropriate data protection strategies to the masked data, the agency might be able to control the level of disclosure risk. The key advantage of the multiplicative noise method is that the issued noise multiplied data has more application in data analysis [9,14,15,22].

The differential privacy method and the multiplicative noise method are two very different data protection approaches and hence it is not particularly fair to determine a winner when comparing the two methods purely based on the probability of identifying the values of the original individual observations or on the limitations of applying the protected the data. However, when assessing which method may be more appropriate to use in a particular setting it is important to compare the protection the two methods offers to individual entries of the original data set, if the comparison is possible.

As described, the differential privacy approach considered in this paper does not release any information related to the individual entries of the original data set. However, if the data agency allows the data user queries for the sample means of any subsets of the original data set, the data user might be able to catch some information on each individual entries of the original data set. Assume that this scenario is held. Following the design of the differential privacy, the data agency will allow to provide the data user with disturbed sample means of the subsets of entries. The sample mean of a subset entries is perturbed by the additive Laplace$(\sqrt{2}/(\epsilon N_p))$ noise.

Assume that the data user receives the following protected information[1]

$$\frac{1}{N_p} \sum_{i=1}^{N_p} x_i + e_0 \quad \text{and} \quad \frac{1}{N_p} \sum_{i \neq j} x_i + e_j,$$

$j = 1, 2, \cdots, N_p$, where $\{e_0, e_1, \cdots, e_{N_p}\}$ is a random sample from the Laplace$(\sqrt{2}/(\epsilon N_p))$ noise. After algebra, the data user can obtain the perturbed individual entry x_j^* of the original data and express it in the following way

---

[1] When N_p is larger, there is no significant difference between $1/(N_p - 1)$ and $1/N_p$. To simplify the calculation, we use $1/N_p$ instead of $1/(N_p - 1)$.

$$x_j^* = N_p\left(\frac{1}{N_p}\sum_{i=1}^{N_p}x_i + e_0 - \frac{1}{N_p}\sum_{i\neq j}x_i - e_j\right) = x_j + N_p(e_0 - e_j),$$

$j = 1, 2, \cdots, N_p$. The x_j^* is the unbiased estimator of x_j, given x_j. Denote $L_X^* = X + N_p(e1 - e2)$, i.e. X masked by the additive noise $N_p(e1 - e2)$, where $e1$ and $e2$ are i.i.d random variables with Laplace($\sqrt{2}/(\epsilon N_p)$) distribution. In terms of attacking the value of the original data, the data intruder might be interested in the following probability

$$P(X = 0|L_X^* \text{ is observed}) \text{ and } P(X = 1|L_X^* \text{ is observed}).$$

Using the data in Example 1, we found that knowing the noise added data gives no advantage in predicting any particular individual entries in the original data set. This is the key advantage of the differential privacy approach.

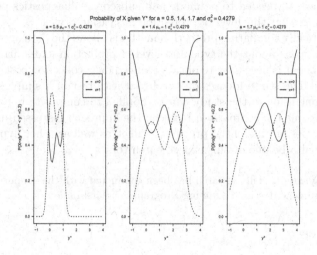

Fig. 2. The plots of $P(X = 0|y^* < Y^* < y^* + 0.2)$ (in dash line) and $P(X = 1|y^* < Y^* < y^* + 0.2)$ (in solid line), with shift parameters a = 0.5, 1.4, 1.7 and fixed variance $\sigma_c^2 = 0.4279$ and $p = 0.65$. The probability distribution of the multiplicative noise C is the same as that defined in Example 1.

The protection level the multiplicative noise method provided to each entry is varied. The variance can be controlled if the data agency so chooses by forsaking some of the statistical efficiency. The plots in Fig. 2 show the variation of $P(X = 0|y^* < Y^* < y^* + 0.2)$ and $P(X = 1|y^* < Y^* < y^* + 0.2)$ is bounded between 0.65 and 0.45, respectively, when the variance of the noise is fixed at $\sigma_c^2 = 0.4279$ and the shift values are altered by 0.5, 1.4, and 1.7. The variation of the protection is minimised comparing to the variation when $\sigma_c^2 = 0.143$ or 0.1066. (The plots for these case are omitted from this paper.)

Therefore the multiplicative noise method, although not achieving the standard level of protection offered by the differential privacy approach, is tailorable to the needs of the data agency and is much more accessible for analysis. Moreover, by adopting various strategies to protect sensitive observations, the multiplicative noise method can still provide the individual entries of the original data set with a reasonable level of protection.

4 Conclusion

In this paper, we have looked at two differing schemes that can be used to provide privacy protection for participants in clinical trials, surveys or scientific studies.

This paper ultimately deduces that it is possible to obtain a **micro data** based protection method (the multiplicative noise method) that provides a reasonable level of protection when compared to that of the aggregated Laplace mechanism. However, an appropriate choice of the distribution of the multiplicative noise, as well as appropriate choices for the mean, variance and shift parameters are all needed to be made and subsequent diagnostics performed in order to ensure that this level of protection is satisfactory.

Systematically comparing protection methods allows the data agency to make informed decisions about the type and level of protection and statistical utility especially when it is necessary to obtain reduced data collection costs i.e. smaller sample sizes. The study focuses on the hypothesis test of a simple proportion. The same conclusion of the study can be applied to composite hypothesis testing. Carrying out the comparison between the approach discussed in this paper with other types of differential privacy settings, as well as other hypothesis test settings will also be one of our next research topics.

Acknowledgement. This research has been conducted with the support of the Australian Government Research Training Program Scholarship.

References

1. Ács, G., Castelluccia, C.: I have a DREAM! (DiffeRentially privatE smArt Metering). In: Filler, T., Pevný, T., Craver, S., Ker, A. (eds.) IH 2011. LNCS, vol. 6958, pp. 118–132. Springer, Heidelberg (2011). https://doi.org/10.1007/978-3-642-24178-9_9
2. Drechsler, J.: My understanding of the differences between the CS and the statistical approach to data confidentiality. In: IFE Research (ed.) 4th IAB Workshop on Confidentiality and Disclosure (2011). http://doku.iab.de/veranstaltungen/2011/ws_data2011_drechsler.pdf
3. Duncan, G.T., Lambert, D.: Disclosure-limited data dissemination. J. Am. Stat. Assoc. **81**, 10–18 (1986)
4. Dwork, C., Smith, A.: Differential privacy for statistics: what we know and what we want to learn. J. Priv. Confid. **2**, 135–154 (2010)
5. Dwork, C., Roth, A.: The algorithmic foundations of differential privacy. Found. Trends Theor. Comput. Sci. **9**, 211–407 (2013)
6. Dwork, C.: Differential privacy. In: Bugliesi, M., Preneel, B., Sassone, V., Wegener, I. (eds.) ICALP 2006. LNCS, vol. 4052, pp. 1–12. Springer, Heidelberg (2006). https://doi.org/10.1007/11787006_1

7. Gostin, L.O.: Privacy and security of personal information in a new health care system. J. Am. Med. Assoc. **270**, 2487–2493 (1993)
8. Green, A.K., et al.: The project data sphere initiative: accelerating cancer research by sharing data. Oncologist **20**, 464–471 (2015)
9. Hwang, J.T.: Multiplicative errors-in-variables models with applications to recent data released by the U.S. Department of Energy. J. Am. Stat. Assoc. **81**, 680–688 (1986)
10. Kim, J.J., Winkler, W.E.: Multiplicative Noise for Masking Continuous Data, Research Report Series (Statistics #2003-01), Statistical Research Division, US Bureau of the Census, Washington D.C., pp. 1–17 (2003)
11. Kim, J.J., Jeong, D.M.: Truncated triangular distribution for multiplicative noise and domain estimation. Sect. Gov. Stat. - JSM **2008**, 1023–1030 (2008)
12. Klein, M., Mathew, T., Sinha, B.: Noise multiplicative for statistical disclosure control of extreme values in log-normal regression samples. J. Priv. Confid. **6**, 77–125 (2014)
13. Lin, Y.-X., Fielding, M.J.: MaskDensity14: an R package for the density approximant of a univariate based on noise multiplied data. SoftwareX **3–4**, 37–43 (2015). https://doi.org/10.1016/j.softx.2015.11.002
14. Lin, Y.-X., Wise, P.: Estimation of regression parameters from noise multiplied data. J. Priv. Confid. 61–94 (2012)
15. Lin, Y.-X.: Density approximant based on noise multiplied data. In: Domingo-Ferrer, J. (ed.) PSD 2014. LNCS, vol. 8744, pp. 89–104. Springer, Cham (2014). https://doi.org/10.1007/978-3-319-11257-2_8
16. Ma, Y., Lin, Y.-X., Sarathy, R.: The vulnerability of multiplicative noise protection to correlational attacks on continuous microdata. In: 2016 Working Paper, School of Mathematics and Applied Statistics, National Institute for Applied Statistics Research Australia, University of Wollongong, Australia (2016)
17. McSherry, F.D.: Privacy integrated queries: an extensible platform for privacy-preserving data analysis. In: Proceedings of the 2009 ACM SIGMOD International Conference on Management of Data, Providence, Rhode Island, USA, pp. 19–30, https://doi.org/10.1145/1559845.1559850 (2009)
18. McSherry, F., Talwar, K.: Mechanism design via differential privacy. In: Proceedings of the 48th Annual IEEE Symposium on Foundations of Computer Science, Washington, DC, USA, pp. 94–103 (2007). https://doi.org/10.1109/FOCS.2007.41
19. Oganian, A.: Multiplicative noise protocols. In: Domingo-Ferrer, J., Magkos, E. (eds.) PSD 2010. LNCS, vol. 6344, pp. 107–117. Springer, Heidelberg (2010). https://doi.org/10.1007/978-3-642-15838-4_10
20. Oganian, A.: Multiplicative noise for masking numerical microdata data with constraints. SORT - Stat. Oper. Res. Trans. (Special Issue), 99–112 (2011)
21. Sarathy, R., Muralidhar, K.: Evaluating laplace noise addition to satisfy differential privacy for numeric data. Trans. Data Priv. **4**, 1–17 (2011)
22. Sinha, B., Nayak, T.K., Zayatz, L.: Privacy protection and quantile estimation from noise multiplied data. Sankhya B **73**, 297–315 (2011)
23. Shlomo, N., Skinner, C.J.: Privacy protection from sampling and perturbation in survey microdata. J. Priv. Confid. **4**, 155–169 (2012)
24. Torra, V.: Data Privacy: Foundations, New Developments and the Big Data Challenge. Springer, Cham (2017). https://doi.org/10.1007/978-3-319-57358-8
25. Wang, Y., Lee, J., Kifer, D.: Differentially private hypothesis testing (2015). Revisited, CoRR, arXiv: 1511.03376

26. Vu, D., Slavkovic, A.: Differential privacy for clinical trial data: preliminary evaluations. In: Proceedings of the 2009 IEEE International Conference on Data Mining Workshops, Washington, DC, USA, pp. 138–143 (2009). https://doi.org/10.1109/ICDMW.2009.52

27. Wang, Y., Wu, X., Hu, D.: Using randomized response for differential privacy preserving data collection. In: Proceedings of the Workshops of the (EDBT/ICDT) 2016 Joint Conference, (EDBT/ICDT) Workshops 2016, Bordeaux, France, 15 March 2016 (2016). http://ceur-ws.org/Vol-1558/paper35.pdf

28. Willenborg, L., De Waal, T.: Elements of Statistical Disclosure Control. LNS, vol. 155. Springer, New York (2012). https://doi.org/10.1007/978-1-4613-0121-9

29. Warner, S.L.: Randomized response: a survey technique for eliminating evasive answer bias. J. Am. Stat. Assoc. **60**, 63–69 (1965)

Quantifying the Protection Level of a Noise Candidate for Noise Multiplication Masking Scheme

Yue Ma[✉], Yan-Xia Lin, Pavel N. Krivitsky, and Bradley Wakefield

National Institute for Applied Statistics Research Australia,
School of Mathematics and Applied Statistics, University of Wollongong,
Wollongong, NSW 2500, Australia
ym894@uowmail.edu.au

Abstract. When multiplicative noises are used to perturb a set of original data, the data provider needs to ensure that the original values are not likely to be learned by data intruders from the noise-multiplied data. Different attacking strategies for unveiling the original values have been recognised in the literature, and the data provider needs to ensure that the noise-multiplied data is protected against these attacking strategies by selecting an appropriate noise generating variable. However, there are many potential attacking strategies, which makes the quantification of the protection level of a noise candidate difficult. In this paper, we argue that, to quantify the protection level a noise candidate offers to the original data against an attacking strategy, the data provider might look at the average value disclosure risk it produces. Correspondingly, we propose an optimal estimator which maximizes the average value disclosure risk. As a result, the data provider could use the maximized average value disclosure risk as a single measure for quantifying the protection level a noise candidate offers to the original data. The measure could help the data provider with the process of noise generating variable selection in practice.

Keywords: Multiplicative noise masking · Value disclosure risk
Attacking strategy · Statistical disclosure control

1 Introduction

Data can be released to the public either as microdata or as tabular summaries. Microdata contains information of data respondents across several attributes, such as personal income and property tax. Releasing microdata provides data users with a wider range of statistical analysis than tabular data, but has higher risks of identity and value disclosure. To balance these two conflicting aspects, methods of statistical disclosure control (SDC) are introduced and studied. Many SDC manners, including data perturbation, rank swapping, data shuffling, etc., are employed to produce protected microdata for public use, ensuring that the

© Springer Nature Switzerland AG 2018
J. Domingo-Ferrer and F. Montes (Eds.): PSD 2018, LNCS 11126, pp. 279–293, 2018.
https://doi.org/10.1007/978-3-319-99771-1_19

released microdata preserves particular statistical properties of the original data as well as meet the requirement of disclosure risk control set by the data agency (see Fuller (1993), Burridge (2003), Muralidhar and Sarathy (2006), Oganian (2011), Ruiz (2011) and reference therein). This paper regards evaluating value disclosure risk of applying noise multiplication masking scheme to a set of positive and continuous data.

The noise multiplication masking scheme perturbs each original observation by multiplying it with a random noise term generated from a noise generating variable C. The data provider releases the noise-multiplied data together with the density function of the noise generating variable f_C to the public. Methodologies for analysing noise-multiplied data have been developed by taking into account the information of f_C such that population parameter estimates could be recovered from the noise-multiplied data (Nayak et al. 2011; Sinha et al. 2011; Lin and Wise 2012; Klein et al. 2014). Using multiplicative noises has been advocated by many researchers. Multiplicative noises provide uniform protections, in terms of the coefficient of variation of the noises, to all sensitive observations (Nayak et al. 2011). The masking mechanism is easy to implement in practice and a balanced utility-risk tradeoff is achieved by selecting an appropriate noise generating variable from a pool of noise candidates, which is referred to as "tuning mechanism" in Klein et al. (2014). The masking method has been used in practice by the U.S. Energy Information Administration and the U.S. Bureau of Census (Kim and Jeong 2008).

Disclosure risk could be classified as identification disclosure risk and value disclosure risk (Melville and McQuaid 2012). Value disclosure occurs if a target original value is reasonably inferred by a data intruder using an attacking strategy, leading to data confidentiality breach. To understand value disclosure risks associated with the masking scheme, some intrusion attacking strategies have been modelled in the literature with the assumption that data intruders have no prior knowledge about the original data. For instance, Klein et al. (2014) considered the scenario where in a microdata, the response variable of a generalised linear model is masked by multiplicative noises while explanatory variables are unmasked. The authors showed that a data intruder may use the predicted value based on the generalised regression model to estimate a target original value. Another two attacking estimators, namely "unbiased attacking estimator" (Nayak et al. 2011; Lin and Wise 2012) and "correlation-attack estimator" (Ma et al. 2018), will be reviewed in Sect. 5. For an original observation to be masked by a noise candidate C, its value disclosure risk against an attacking strategy could either be quantified in terms of a confidence interval (Nayak et al. 2011; Agrawal and Srikant 2000), or in terms of a probability. Under probabilistic value disclosure risk measure, the value disclosure risk is the probability that its value is disclosed by the attacking strategy. Probabilistic disclosure risk measure has been used in practice by the Australian Bureau of Statistics (Chipperfield et al. 2018). In this paper, we use probabilistic disclosure risk measure for quantifying value disclosure risks.

Because the data provider has no control over which attacking strategy a data intruder might use, an ideal noise generating variable should provide enough protection to each original observation against all attacking strategies. This is the requirement of the worst-case differential privacy (DP), which means that no private information would be revealed even if an adversary knows all information about the data except the private information. The data provider might say that a noise candidate provides enough protection to the original data if for any original observation, the maximum value disclosure risk against any attacking strategy is below a threshold value. However, it is difficult to find such a noise candidate. We will show in the simulation study that, even if assuming adversaries has no prior knowledge, a noise candidate which protects one observation well against an attacking strategy might not protect another observation against the attacking strategy adequately. Alternatively, the data provider might look at the average value disclosure risk for quantifying the protection level of a noise candidate offers to the original data against an attacking strategy, which is the mean of the value disclosure risks of all original observations against the attacking strategy. For large-sized data, the average value disclosure risk provides a rough idea about the proportion of original observations which could be disclosed by the attacking strategy in a set of noise-multiplied data. The data provider might want the proportion to be below an acceptable level, especially if all observations might be identified or targeted. Under this setting, the data provider might say that a noise candidate offers enough protection to the original data if the average value disclosure risks against several attacking strategies are all below a threshold level simultaneously.

We consider the following setting in this paper: To attack an original observation y_i, the data intruder uses an attacking estimator $g(Y_i^*)$, where $Y_i^* = y_i C$ and $g(Y_i^*)$ is a function of Y_i^*. We will show in Sect. 5, there are many attacking estimators, either discussed in the literature or not, satisfy this setting. We also assume the data provider uses the average value disclosure risk to quantify the protection level of a noise candidate offers to the original data against an attacking strategy. We will illustrate in the simulation study that a noise candidate which protects the original data well against one attacking estimator might not protect against another attacking estimator well. Therefore, in practice, to evaluate the protection level of a noise candidate, the data provider might need to evaluate value disclosure risks against several attacking estimators. This is a complicated job for the data provider. In this paper, we derive an attacking estimator, namely the optimal estimator, which maximizes the average value disclosure risk. Therefore, for a noise candidate, instead of measuring its value disclosure risks against several attacking estimators which are functions of Y_i^*, the data provider could simply measure the average value disclosure risk against the optimal estimator to understand the protection level the noise candidate yields to the original data. The data provider could use this single measure for noise generating variable selection in practice. We note that other attacking estimators which are not functions of Y_i^* exist, such as the attacking estimator introduced in Klein et al. (2014). These cases are not considered in this paper and could be considered in the future.

In the literature, some distributions have been proposed as good noisy distributions because they offer strong protections against the unbiased estimator for any set of original data. Those distributions include bi-modal normal distribution (Lin and Wise 2012), truncated triangular distribution (Kim and Jeong 2008) and mixture of uniforms distribution (Klein et al. 2014). We will show in the simulation study that these distributions might not provide good protections to the original data according to our protection level quantification measure.

This paper is organised as follows: Sect. 2 overviews methods for analysing noise-multiplied data and introduces an overall utility loss measure we adopt in this paper. Section 3 introduces the probabilistic value disclosure risk measure and the optimal estimator. Section 4 introduces methods for evaluating value disclosure risks against the optimal estimator. Section 5 provides three discussions. Section 6 presents a simulation study. Section 7 concludes the paper.

2   Analysing Noise-Multiplied Data and Data Utility Loss

For univariate data, the noise multiplication masking method works as follow:

Suppose a set of original data $y = \{y_i\}_{i=1}^n$ are independent realizations from Y. To mask y, the data provider chooses a noise generating variable C with $E(C) = 1$. Y and C are independent. A set of noise terms $c = \{c_i\}_{i=1}^n$ are independently drawn from C and multiplied with y to produce the noise-multiplied data $y^* = \{y_i^*\}_{i=1}^n = \{y_i c_i\}_{i=1}^n$. $\{y_i^*\}_{i=1}^n$ could be treated as independent realizations from Y^*, where $Y^* = YC$. The data provider releases y^* together with the density function of noise f_C to the public. We assume both Y and C are positive and continuous, and n is large.

For convenience, in this paper we discuss data utility loss in terms of parameter estimates of Y only. Methods for recovering parameter estimates of Y from y^* have been developed in the literature. Nayak et al. (2011) showed that, unbiased moments estimates of Y could be easily recovered from y^*. Sinha et al. (2011) proposed a Bayesian approach for estimating quantiles of Y from y^*. However, the method requires a data user to have prior knowledge about the distribution of Y. Another way of recovering quantiles is to estimate f_Y using a reconstruction algorithm (Agrawal and Aggarwal 2001; Lin 2014). Lin (2014) showed that a sample-moment-based density reconstruction algorithm could be used for estimating f_Y based on y^* and f_C. Consequently, data users could obtain estimates of the quantiles of Y from the reconstructed density estimate, which is denoted as $\hat{f}_Y$. Numerical examples in Lin (2014) and Lin and Fielding (2015) showed that accurate quantile estimates could be obtained in this way. In summary, estimates of moments and quantiles of Y could be recovered from y^*, given f_C. For obtaining regression coefficients estimates from noise-multiplied datasets, see Lin and Wise (2012) and Klein et al. (2014).

When recovering a parameter estimate from noise-multiplied data, the recovered estimate is less accurate than the one data users would obtain by analysing the original data. Overall data utility loss is an aggregate measure of loss of accuracies across several parameters. There is no unique way to measure overall data utility loss. In the literature, the way of measuring overall data utility loss varies according to different data masking scenarios as well as which parameters estimates could be recovered from the masked data. See Domingo-Ferrer and Torra (2001); Shlomo (2010) and Yancey et al. (2002) for several overall utility loss measures.

In this paper, we note the following: 1. for large-sized sample, moments estimates of Y are normally recovered accurately regardless of the distribution of C; 2. under the reconstruction algorithm approach, the accuracies of the quantile estimates of Y depend on the accuracy of the reconstructed density estimate $\hat{f}_Y$. Unlike moments estimates, the accuracy of $\hat{f}_Y$ is largely affected by the distribution of C. (See results in Agrawal and Aggarwal (2001)). Therefore, in this paper we adopt the overall data utility loss measure proposed in Agrawal and Aggarwal (2001). The authors defined the overall data utility loss measure according to how much do the reconstructed density estimates and the original density function overlap. Mathematically, the overall data utility loss is defined as:

$$UL(f_Y, C) = \frac{1}{2}E[\int_{-\infty}^{\infty} |f_Y(y) - \hat{f}_{Y,C}(y)|dy],$$

where $\hat{f}_{Y,C}(y)$ is the reconstructed density function of f_Y based on the data masked by C. $UL(f_Y, C)$ is bounded between 0 and 1, with 0 indicates no utility loss. This utility loss measure is similar to the pMSE which could be seen as a loss measure for synthetic data (Snoke et al. 2018). A low $UL(f_Y, C)$ value means that many parameter estimates of Y could be accurately recovered from the noise-multiplied data if C is the noise generating variable. We note that functionals of the distribution might be estimated directly better than by going through a density estimator (Bickel and Ritov 2003). In this paper we do not pursue this discussion.

3   Probabilistic Disclosure Risk Measure and the Optimal Estimator

In this section we introduce the probabilistic value disclosure risk measure against a particular attacking estimator. We also introduce the optimal estimator which maximizes the average value disclosure risk.

For the purpose of recovering the value of a continuous original datum from its noise-multiplied datum, it is tough to guess the exact value of the original datum, and it is not necessary to achieve it. Using an Acceptance Rule to decide if an estimate $\tilde{y}_i$ can be accepted as a correct guessing of the original value y_i is sufficient. Using acceptance rule to define value disclosure is adopted by the Australian Bureau of Statistics for tabulated business data (Chipperfield et al. 2018). Following Lin and Wise (2012) and Klein et al. (2014), we adopt the following definition of value disclosure:

Suppose a data intruder uses $\tilde{y}_i$ as an estimate of the original value y_i. The expression of $\tilde{y}_i$ is determined by his own attacking strategy and in this paper we assume $\tilde{y}_i = g(y_i^*)$. To classify the $\tilde{y}_i$ as a valid estimate of y_i, it is sufficient for $\tilde{y}_i$ to be reasonably close to y_i. **Disclosure** *of y_i occurs if $\tilde{y}_i$ satisfies $|\frac{\tilde{y}_i - y_i}{y_i}| \le \delta$, where δ is the* **acceptance rule** *defined in* Lin and Wise (2012) *and is a small positive number. For instance, for a positive observation y_i, if we set $\delta = 0.05$, we say that $\tilde{y}_i$ discloses y_i if $0.95 y_i \le \tilde{y}_i \le 1.05 y_i$. The value of δ is determined by the data provider.*

In this paper we assume the original data is only subject to attacking estimators which are functions of Y_i^*. Following Klein et al. (2014); Lin and Wise (2012), the probabilistic value disclosure risk measure for evaluating the disclosure risk of y_i against an attacking estimator $\tilde{Y}_i$ takes the following form:

$$R(\tilde{Y}_i, \delta | Y_i = y_i) = P(\frac{|\tilde{Y}_i - Y_i|}{Y_i} < \delta | Y_i = y_i),$$

where $\tilde{Y}_i = g(Y_i^*)$, $Y_i^* = C Y_i$.

The data provider balances utility-risk tradeoff by selecting an appropriate noise generating variable from a set of noise candidates. As we argued in the introduction, in this paper we assume the data provider uses the average value disclosure risk $\overline{\{R(\tilde{Y}_i, \delta | Y_i = y_i)\}_{i=1}^n} = \frac{\sum_{i=1}^n R(\tilde{Y}_i, \delta | Y_i = y_i)}{n}$ to quantify the protection level a noise candidate C offers to the original data against the attacking estimator $\tilde{Y}_i$. As the data provider has no control over which attacking strategy a data intruder might use to attack the original data, therefore, to assess the protection level of a noise candidate, the data provider might need to evaluate the average value disclosure risks against many potential attacking estimators. A noise candidate is qualified for masking the original data if the average value disclosure risks are below a certain level simultaneously. To facilitate the process, we propose an "optimal estimator" which maximizes the average probabilistic value disclosure risk. The optimal estimator takes the following form:

$$Z_i^{opt} = argmax_{g(Y_i^*)} \int_{\frac{g(Y_i^*)}{(1+\delta)}}^{\frac{g(Y_i^*)}{(1-\delta)}} f_{Y_i | Y_i^*}(y) dy, \tag{1}$$

where f_Y is the density function of Y, $Y_i^* = Y_i C$ and $f_{Y_i | Y_i^*}(y) = \frac{1}{y} f_C(\frac{Y_i^*}{y}) f_{Y_i}(y)$. The derivation of Z_i^{opt} is given in Appendix. Correspondingly, the value disclosure risk of y_i against the optimal estimator Z_i^{opt} is defined as

$$R_{opt}(y_i, C, \delta) = P(|\frac{Z_i^{opt} - Y_i}{Y_i}| < \delta | Y_i = y_i)$$

The average value disclosure risk of using C to mask the original data against the optimal estimator is $R_{overall}(C)$. That is:

$$R_{overall}(C) = \frac{\sum_{i=1}^n R_{opt}(y_i, C, \delta)}{n}$$

We propose that the data provider measures $R_{overall}(C)$ to quantify the level of protection a noise candidate C offers to the original data. If $R_{overall}(C)$ is below an acceptable level, then C offers enough protection to the original data against any other estimators which are functions of Y_i^*. We note that when y_i is subject to other forms of attacks which are not sole functions of Y_i^*, such as the one proposed in Klein et al. (2014), the data provider might need to use $R_{overall}(C)$ in conjunction with other corresponding average value disclosure risk measures to quantify the protection level of C. In this paper we assume the original data is only subject to attacking estimators which are functions of Y_i^*.

4   Finding $R_{opt}(y_i, \delta)$

In this section we discuss how the data provider could find $R_{opt}(y_i, C, \delta)$ in practice.

Case 1: f_Y is known. If f_Y is known to the data provider then the mathematical expression of Z_i^{opt} might be derived analytically. Suppose $Q(y)$ is the antiderivative of $\frac{1}{y} f_C(\frac{Y_i^*}{y}) f_{Y_i}(y)$, then

$$H(Z_i^{opt}) = \int_{\frac{Z_i^{opt}}{\delta+1}}^{\frac{Z_i^{opt}}{1-\delta}} \frac{1}{y} f_C(\frac{Y_i^*}{y}) f_{Y_i}(y) dy = Q(\frac{Z_i^{opt}}{1-\delta}) - Q(\frac{Z_i^{opt}}{\delta+1})$$

To find the argument Z_i^{opt} where $H(Z_i^{opt})$ is maximized, we take derivative of $H(Z_i^{opt})$ to yield:

$$\frac{dH(Z_i^{opt})}{dZ_i^{opt}} = \frac{1}{Z_i^{opt}} f_C(\frac{Y_i^*(1-\delta)}{Z_i^{opt}}) f_Y(\frac{Z_i^{opt}}{1-\delta}) - \frac{1}{Z_i^{opt}} f_C(\frac{Y_i^*(\delta+1)}{Z_i^{opt}}) f_Y(\frac{Z_i^{opt}}{1+\delta})$$

We set $\frac{dH(Z_i^{opt})}{dZ_i^{opt}} = 0$ to find Z_i^{opt}. As an example, if $Y \sim LN(\mu_1, \sigma_1^2)$, $C \sim LN$ (μ_2, σ_2^2), we can show that $Z_i^{opt} = exp(\frac{(\sigma_1^2+\sigma_2^2)ln[(1+\delta)(1-\delta)]+2\sigma_1^2 ln(y_i C)-2\sigma_1^2\mu_2+2\sigma_2^2\mu_1}{2(\sigma_1^2+\sigma_2^2)})$. Correspondingly, we could obtain $R_{opt}(y_i, C, \delta)$ for each y_i if the expression of Z_i^{opt} is available.

Case 2: f_Y is unknown. In practice, the data provider may only have a set of original data $\{y_i\}_{i=1}^n$ available without knowing the true underlying distribution of Y. To find $R_{opt}(y_i, C, \delta)$ in this case, the data provider might need to estimate the density function f_Y. For instance, the data provider might obtain a kernel density estimate K_Y based on $\{y_i\}_{i=1}^n$, and replace $f_{Y|Y_i^*}$ by $\frac{1}{y} f_C(\frac{Y_i^*}{y}) K_Y(y)$ in Eq. (1) to approximate Z_i^{opt}, and then approximate $R_{opt}(y_i, C, \delta)$ for each y_i.

The data provider might have different strategies to approximate $R_{opt}(y_i, C, \delta)$. In the simulation study, we assume Y is bounded between $[a, b]$ and C is bounded between $[c, d]$, where a and b are the minimum and maximum of the original data. Therefore, we note that Y^* is bounded between $[ac, bd]$. We adopt the following strategy to approximate $R_{opt}(y_i, C, \delta)$: 1. We generating n_1 multiple copies of y_i^*, where $y_i^* \sim y_i C$. Denote them as $\{y_{il}^*\}_{l=1}^{n_1}$; 2. We note that the support of $f_{Y|Y^*}$ is $[max(a, \frac{Y^*}{d}), min(b, \frac{Y^*}{c})]$. Therefore, for each y_{il}^*, we take n_2 points between $[max(a, \frac{y_{il}^*}{d}), min(b, \frac{y_{il}^*}{c})]$ with equal increment Q. Denote the points as $\{h_v\}_{v=1}^{n_2}$; 3. For each h_v, we attempt to estimate $\int_{\frac{h_v}{(1+\delta)}}^{\frac{h_v}{(1-\delta)}} f_{Y|Y_i^*}(y) dy$. Because Y is bounded between $[a, b]$, therefore $\frac{h_v}{(1+\delta)}$ cannot be lower than a and $\frac{h_v}{(1-\delta)}$ cannot be greater than b. Therefore $\int_{\frac{h_v}{(1+\delta)}}^{\frac{h_v}{(1-\delta)}} f_{Y|Y_i^*}(y) dy = \int_{max(a, \frac{h_v}{(1+\delta)})}^{min(b, \frac{h_v}{(1-\delta)})} f_{Y|Y_i^*}(y) dy$. To estimate this integral, we take n_3 points between $max(a, \frac{h_v}{(1+\delta)})$ and $min(b, \frac{h_v}{(1-\delta)})$ with equal increment Δ. Let $m_j = max(a, \frac{h_v}{(1+\delta)}) + (j-1)\Delta$. We approximate the integral by $I_{h_v} = \sum_{j=1}^{n_3} \frac{1}{m_j} f_C(\frac{y_{il}^*}{m_j}) K_Y(m_j) \Delta$. As a result, for $\{h_v\}_{v=1}^{n_2}$, we obtain a corresponding set $\{I_{h_v}\}_{v=1}^{n_2}$; 4. We let $z_{i,y_{il}^*}^{opt} = argmax_{h_v}\{I_{h_v}\}_{v=1}^{n_2}$. Then $z_{i,y_{il}^*}^{opt}$ is the optimal estimate of y_i given y_{il}^*; 5. We repeat the above process for each y_{il}^*, $l = 1, 2, \cdots, n_1$. As a result, we obtain $\{z_{i,y_{il}^*}^{opt}\}_{l=1}^{n_1}$; 6. We count the number of $z_{i,y_{il}^*}^{opt}$ such that $z_{i,y_{il}^*}^{opt} \in [y_i(1-\delta), y_i(1+\delta)]$. Denote the number as n_4; 7. We estimate $R_{opt}(y_i, C, \delta)$ by n_4/n_1.

We use the above steps for estimating $R_{opt}(y_i, C, \delta)$ in the simulation study. We note that the above steps is a naive and might not be the most efficient way for estimating $R_{opt}(y_i, C, \delta)$. However, they are sufficient for illustration purpose of this paper.

5   Discussion

In this section we discuss the optimal estimator and other possible attacking estimators mainly from data intruders' point of view. For illustration, we introduce another three attacking estimators for estimating an original value y_i. The three attacking estimators will also be used for illustration purpose in the simulation:

Unbiased estimator: Nayak et al. (2011) showed that $Y_i^* = CY_i$ is an unbiased estimator of y_i as $E(Y_i^*|Y_i = y_i) = y_i$. Therefore, Y_i^* is the unbiased estimator for y_i.

Correlation-attack estimator: Ma et al. (2018) showed that when the population correlation between Y and Y^* is high, then a simple linear model might be adequate to explain the relationship between the two variables. As a result, the correlation-attack estimator $\hat{Y}_i$ takes the following form: $\hat{Y}_i = (1 - \rho_{YY^*}^2)\mu_Y + \rho_{YY^*}^2 Y_i^*$, where $\rho_{YY^*}^2$ is the population correlation between Y and Y^*, μ_Y is the population mean of Y. These two quantities could be estimated directly from the noise-multiplied data y^*.

Maximum a posteriori (MAP) estimator: This estimator is not discussed in the literature. The MAP estimator is simply the mode of $f_{Y_i|Y_i^*}(y)$, where $f_{Y_i|Y_i^*}(y) = \frac{1}{y} f_C(\frac{Y_i^*}{y}) f_{Y_i}(y)$. Denote the estimator as Z_i, then $Z_i = argmax_y f_{Y_i|Y_i^*}(y)$.

Discussion 1: Interpretation of z_i^{opt} from the data intruder's point of view

Given y_i^*, a realization of Z_i^{opt} takes the following form:

$$z_i^{opt} = argmax_{g(y_i^*)} \int_{\frac{g(y_i^*)}{(1+\delta)}}^{\frac{g(y_i^*)}{(1-\delta)}} f_{Y|Y_i^*=y_i^*}(y) dy = argmax_{g(y_i^*)} \int_{\frac{g(y_i^*)}{(1+\delta)}}^{\frac{g(y_i^*)}{(1-\delta)}} \frac{1}{y} f_C(\frac{y_i^*}{y}) f_Y(y) dy.$$

From the data intruder's point of view, suppose the data intruder has no prior knowledge of y_i and assumes that $y_i \sim Y_i$. Given y_i^*, suppose the data intruder uses an estimate $g(y_i^*)$ to attack y_i, then the probability that $g(y_i^*)$ discloses y_i could be expressed as

$$P(|\frac{g(y_i^*) - Y_i}{Y_i}| < \delta | Y_i^* = y_i^*) = P(\frac{g(y_i^*)}{1+\delta} < Y_i < \frac{g(y_i^*)}{1-\delta} | Y_i^* = y_i^*).$$

The expression of z_i^{opt} means that z_i^{opt} maximizes the above posterior probability of disclosing y_i. Therefore, if the data intruder wants to maximize the posterior probability of disclosing y_i, the data intruder may use z_i^{opt} to attack y_i.

To find z_i^{opt}, the data intruder needs information of $f_C(\frac{y_i^*}{y})$ and $f_Y(y)$. Because f_C and y_i^* are public knowledge, the data intruder knows $f_C(\frac{y_i^*}{y})$. However, $f_Y(y)$ is unknown and needs to be estimated. The data intruder could estimate f_Y by using the reconstruction algorithm we mentioned in Sect. 2. As a result, the data intruder could obtain an approximated value of z_i^{opt} by replacing $f_Y(y)$ by $\hat{f}_Y(y)$ in the expression of z_i^{opt}.

Discussion 2: Estimator selection from the data intruder's point of view

To attack y_i, the data intruder might use any of the four estimators Y_i^*, $\hat{Y}_i$, Z_i^{opt} and Z_i. For the data provider, it could use the corresponding value disclosure risk measures to tell which attacking estimator is more effective for attacking y_i. For instance, in the simulation study, we can show that if y_i is around 185000, then it has a value disclosure risk of 1 against the optimal estimator if C_1 is used to mask the original data. Therefore, the data provider knows that if the optimal estimator is used to attack the corresponding noise-multiplied value y_i^*, it will lead to value disclosure. However, the data intruder might not know that the optimal estimator will surely lead to value disclosure of y_i^*.

To attack y_i^*, the data intruder might come up with his own rule for determining which attacking estimator to use to disclose y_i. For instance, the data intruder might use the correlation-attack estimator if the estimated sample correlation between the noise-multiplied data and the original data is very high. Alternatively, as we argued in Discussion 1, Z_i^{opt} maximizes the data intruder's posterior probability of disclosing y_i. If the probability is used as a decision rule, then the data intruder will use the optimal estimator to attack y_i. Discussion on how to choose between the correlation-attack estimator and the unbiased estimator based on mean squared errors can be found in Ma et al. (2018). Regardless, the decision rule used by a data intruder does not necessarily lead to the best choice of estimator for attacking y_i. Therefore, the data provider might be less worried about those original values which suffer very high disclosure risks against an attacking estimator when evaluating the protection level of a noise candidate C, especially if the corresponding records have low identity disclosure risks.

Discussion 3: Value disclosure risk measure and the worst-case DP
Ideally when releasing noise-multiplied data to the public, the worst-case differential privacy (DP) could be achieved. In that way, no private information could be revealed even if an adversary knows all information about the data except the private information. This requirement could be achieved in some statistical limitation methods, such as ϵ-differential private synthetic data (Snoke et al. 2018). Our value disclosure risk measure aims to ensure a sufficient level of uncertainty when a data intruder with no prior knowledge attacks the noise-multiplied data, but it does not guarantee the worst-case DP. Releasing data which satisfy the worst-case DP might render a high loss of data utility and therefore might not be considered by a statistical agency. How to generate noise-multiplied data which achieves the worst-case DP while maintaining enough data utility requires future study.

6   Simulation Study

In this section we present a simulation study using real-life data. We consider several noise candidates. The main purpose of the simulation study is to illustrate several points which we mentioned in this paper. Following Klein et al. (2014), we use the public use data from the 2000 Current Population Survey (CPS) March supplement. The entire data set contains household, family, and individual records. As extreme income values are normally considered sensitive and need to be protected, we consider **the top 5% household income values** under household income attribute as the original data. The original data contains 2533 positive observations ranging from 140000 to 768742. We denote the original data as $\{y_i\}_{i=1}^{2533}$. We set the acceptance rule $\delta = 0.1$ throughout this section. We consider the following four noise candidates: $C_1 \sim I_1U_1 + (1-I_1)U_2$, where $P(I_1 = 0) = P(I_1 = 1) = 0.5$, $U_1 \sim U(0.5, 0.9)$ and $U_2 \sim U(1.1, 1.5)$; $C_2 \sim I_2U_3 + (1-I_2)U_4$, where $P(I_2 = 0) = P(I_2 = 1) = 0.5$, $U_3 \sim U(0.3, 0.9)$ and $U_4 \sim U(1.1, 1.7)$; $C_3 \sim U(1 - 0.5\sqrt{93/75}, 1 + 0.5\sqrt{93/75})$; $C_4 \sim U(0.245, 1.755)$. Among the four noise candidates, C_1 and C_3 have the same variance. Similarly for C_2 and C_4. C_1 and C_2 represent those types of noise candidate which have been advocated by researchers as they offer strong protections against the unbiased estimator.

We first measured the value disclosure risks of each y_i against each of the four attacking estimators. To find $R_{opt}(y_i, C, 0.1)$, we obtained a kernel density estimate K_Y based on the original data by using the function 'density()' with default parameters in R. We computed the average value disclosure risks against each attacking estimator. The results are given in Table 1. To comment on the table, we see that C_1 and C_2 provide very good protection to the original data against the unbiased estimator because the corresponding average value disclosure risks are 0. However, comparing C_1 with C_3 (with the same variance), we see that C_3 protects the original data better against the other three estimators than C_1. Similarly story can be seen when we compare C_2 and C_4. The result shows that a noise candidate which protects the original data well against one attacking estimator may not provides sufficient protection against other estimators. The result may also suggest that, those noise candidates represented by C_1 and C_2 which have been advocated by some researchers might not provide better protections than other noise candidates.

Denote $R_{cor}(y_i, C, 0.1)$ as the value disclosure risk of y_i against the correlation-attack estimator if C is used to mask y_i. We provide value disclosure risks $\{R_{cor}(y_i, C_1, 0.1)\}_{i=1}^{2533}$ against the original data $\{y_i\}_{i=1}^{2533}$ plot in Fig. 1(a), and we provide $\{R_{opt}(y_i, C_1, 0.1)\}_{i=1}^{2533}$ against $\{y_i\}_{i=1}^{2533}$ plot in Fig. 1(b). We also provide similar plots under the case of C_3 in Fig. 2. We see that, for an attacking strategy, a noise candidate might not be able to protect all observations. For instance, we see in Fig. 1(b) that some observations have value disclosure risks of 1 against the optimal estimator. We also see that, a noise candidate which protects an original observation well against one attacking estimator might not protect it well against another estimator. For instance, we see in Fig. 2(b) that observations around 200000 only have value disclosure risks of around 0.2 against the optimal estimator. However, the corresponding value disclosure risks against the correlation-attack estimator are above 0.3. The story is reversed for observations greater than 500000. While an ideal noise candidate should protects all observations against all attacking strategies, these two figures showed that it is difficult to find such a noise candidate. Therefore, using average value disclosure risk to quantify the level of protection of a noise candidate to the original data against an attacking estimator, and ensure that the average disclosure risk is below a certain level seems to be a reasonable criteria for noise generating variable selection.

For each of the noise candidates, we computed the level of overall utility loss. To compute this value, take C_1 for instance, we firstly produced q copies of noise-multiplied data using C_1. For the i-th copy of noise-multiplied data, we used an R-package MaskDensity14 (Lin and Fielding 2015) to implement the sample-moment-based density reconstruction algorithm in R, and we obtained a copy of the reconstructed density function $\hat{f}_{Y,C_1}^i$. Therefore, we obtained

Table 1. The average value disclosure risks against each attacking estimator

Noise	Unbiased estimator	Correlation-attack estimator	MAP estimator	Optimal estimator
C_1	0	0.357	0.543	0.628
C_2	0	0.307	0.508	0.567
C_3	0.179	0.327	0.501	0.547
C_4	0.124	0.301	0.484	0.519

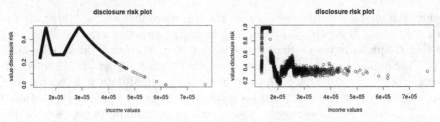

(a) $\{R_{cor}(y_i, C_1, 0.1)\}_{i=1}^{2533}$ against income values plot

(b) $\{R_{opt}(y_i, C_1, 0.1)\}_{i=1}^{2533}$ against income values plot

Fig. 1. Value disclosure risks against income values plots for C_1.

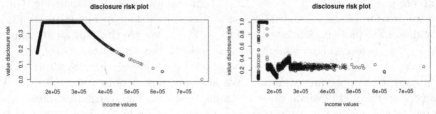

(a) $\{R_{cor}(y_i, C_3, 0.1)\}_{i=1}^{2533}$ against income values plot

(b) $\{R_{opt}(y_i, C_3, 0.1)\}_{i=1}^{2533}$ against income values plot

Fig. 2. Value disclosure risks against income values plots for C_3.

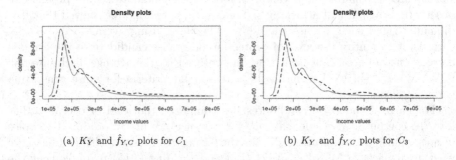

(a) K_Y and $\hat{f}_{Y,C}$ plots for C_1

(b) K_Y and $\hat{f}_{Y,C}$ plots for C_3

Fig. 3. Kernel density estimate against the reconstructed density estimates for C_1 and C_3. The solid lines are density plots of the original data. The dashed lines are reconstructed density plots.

q copies of the reconstructed density function. We estimated $UL(f_Y, C_1)$ by $\frac{1}{2q}\sum_{i=1}^{q}\int_{-\infty}^{\infty}|K_Y(y) - \hat{f}_{Y,C_1}^{i}(y)|dy$, where K_Y is the kernel density estimate of Y. Denote $\hat{f}_{Y,C}(y) = \sum_{i=1}^{q}\hat{f}_{Y,C}^{i}/q$. Figure 3 provides $\hat{f}_{Y,C}(y)$ and $K_Y(y)$ plots for illustrating the accuracies of the reconstructed density functions. The overall utility losses for the four noise candidates are 0.190, 0.209, 0.230 and 0.204 respectively. Except C_4, we see that there is a trade-off between $R_{overall}(C)$ (last column of Table 1) and the overall utility loss. We see that C_1 offers the lowest level of overall utility loss but the worst protection to the original data, while

C_3 offers a better protection at the expense of a higher amount of utility loss. The data provider could choose a noise candidate which achieves the desired utility-risk tradeoff for masking the original data.

In this simulation, we see from Table 1 that $R_{overall}(C_4) = 0.519$, meaning that it offers the highest level of protection to the original data among the four noise candidates. To find a noise candidate with better protection, it can be shown that if $C \sim LN(-0.144, 0.536^2)$ truncated between 0 and 5 is considered, then $R_{overall}(C)$ is only 0.424, which provides a better protection than C_4. In practice the data provider could decide its own threshold level such that $R_{overall}(C)$ needs to be below the threshold level in order for C to be considered.

7 Conclusion

In this paper we propose a measure for quantifying the protection level a noise candidate C for noise multiplication masking scheme. To attack the original data from its noise-multiplied version, the data intruder might adopt different attacking strategies. From the data provider's perspective, we argue that to quantify the protection level of a noise candidate against a particular attacking strategy, the data provider might look at the average value disclosure risk. We propose an optimal attacking estimator which maximizes the average value disclosure risk, and the corresponding maximized average value disclosure risk is the measure for quantifying the protection level a noise candidate C offers to the original data. As a result, the data provider could use this single measure instead of using multiple value disclosure risk measures for noise generating variable selection in practice. We note that in this paper we assume an attacking estimator is a function of the noise-multiplied variable. Relaxing this assumption for more generalised results requires future study.

Acknowledgements. We thank the anonymous reviewers for their constructive comments on the paper. This research has been conducted with the support of the Australian Government Research Training Program Scholarship.

Appendix

In this section we show how z_i^{opt} is derived. Suppose for a set of original data $\{y_i\}_{i=1}^n$, the following probabilistic disclosure risk measure to be used by the data provider.

$$P(\frac{|\tilde{Y}_i - Y_i|}{Y_i} < \delta) = P(\frac{|\tilde{Y} - Y|}{Y} < \delta), i = 1, \cdots, n$$

In the following we assume $Y > 0$, $C > 0$, $\tilde{Y} = g(Y^*)$, where $Y^* = CY$. We observe that the disclosure risk of an observation y is given as:

$$P(\frac{|g(Y^*) - Y|}{Y} < \delta | Y = y) = \int_0^\infty I(\frac{g(y^*)}{1+\delta} < y < \frac{g(y^*)}{1-\delta}) f_{Y^*|Y}(y^*|Y = y) dy^*$$

Therefore $P(\frac{|g(Y^*)-Y|}{Y} < \delta|Y)$ is a function of random variable Y.

The average disclosure risk for the original data is

$$R_{overall} = \frac{\sum_{i=1}^n P(\frac{|g(Y^*)-Y|}{Y} < \delta|Y = y_i)}{n}$$

Suppose $E_Y(P(\frac{|g(Y^*)-Y|}{Y} < \delta|Y))$ exists, therefore we have

$$R_{overall} \xrightarrow{P} E_Y(P(\frac{|g(Y^*)-Y|}{Y} < \delta|Y))$$

as $n \to \infty$.

The objective is to find an expression of $g(Y^*)$ which maximizes $R_{overall}$ as $n \to \infty$. Because $\{Y^*|Y = y\} = yC$, therefore $f_{Y^*|Y}(y^*|Y = y) = \frac{1}{y}f_C(\frac{y^*}{y})$. We observe the following:

$$\begin{aligned}
E_Y(P(\frac{|g(Y^*)-Y|}{Y} < \delta|Y)) &= \int_0^\infty \int_0^\infty I(\frac{g(y^*)}{1+\delta} < y < \frac{g(y^*)}{1-\delta}) f_{Y^*|Y=y}(y^*) dy^* f_Y(y) dy \\
&= \int_0^\infty \int_0^\infty I(\frac{g(y^*)}{1+\delta} < y < \frac{g(y^*)}{1-\delta}) \frac{1}{y} f_C(\frac{y^*}{y}) f_Y(y) dy dy^* \\
&= \int_0^\infty f_{Y^*}(y^*) \int_0^\infty I(\frac{g(y^*)}{1+\delta} < y < \frac{g(y^*)}{1-\delta}) f_{Y|Y^*=y^*}(y) dy dy^* \\
&= \int_0^\infty f_{Y^*}(y^*) \int_{\frac{g(y^*)}{(1+\delta)}}^{\frac{g(y^*)}{(1-\delta)}} f_{Y|Y^*=y^*}(y) dy dy^*
\end{aligned}$$

(2)

Therefore, $E_Y(P(\frac{|g(Y^*)-Y|}{Y} < \delta|Y))$ is maximized if $\int_{\frac{g(y^*)}{(1+\delta)}}^{\frac{g(y^*)}{(1-\delta)}} f_{Y|Y^*=y^*}(y) dy$ is maximized. The form of $g(y^*)$ which maximizes $\int_{\frac{g(y^*)}{(1+\delta)}}^{\frac{g(y^*)}{(1-\delta)}} f_{Y|Y^*=y^*}(y) dy$ is $z_{opt} = argmax_{g(y^*)} \int_{\frac{g(y^*)}{(1+\delta)}}^{\frac{g(y^*)}{(1-\delta)}} f_{Y|Y^*=y^*}(y) dy$. Therefore, the optimal estimator Z_{opt} takes the following form

$$Z_{opt} = argmax_{g(Y^*)} \int_{\frac{g(Y^*)}{(1+\delta)}}^{\frac{g(Y^*)}{(1-\delta)}} f_{Y|Y^*}(y) dy$$

References

Agrawal, R., Aggarwal, C.: On the design and quantification of privacy preserving data mining algorithms. In: Proceedings of the 20th Symposium on Principles of Database Systems, Santa Barbara, California, USA (2001)

Agrawal, R., Srikant, R.: Privacy preserving data mining. In: Proceedings of the ACM SIGMOD, pp. 439–450 (2000)

Bickel, P.J., Ritov, Y.: Nonparametric estimators which can be "plugged-in". Ann. Stat. **31**(4), 1033–1053 (2003)

Burridge, J.: Information presrving statistical obfuscation. Stat. Comput. **13**, 321–327 (2003)

Chipperfield, J., Newman, J., Thompson, G., Ma, Y., Lin, Y.X.: Prospects for protecting aggregate business microdata via a remote server. J. Off. Stat. (Major Revision) (2018)

Domingo-Ferrer, J., Torra, V.: Disclosure protection methods and information loss for microdata. In: Confidentiality, Disclosure and Data Access: Theory and Practical Applications for Statistical Agencies, pp. 91–110 (2001)

Fuller, W.A.: Masking procedures for microdata disclosure limitation. J. Off. Stat. **9**, 383–406 (1993)

Kim, J., Jeong, D.M.: Truncated triangular distribution for multiplicative noise and domain estimation. In: Section on Government Statistics-JSM 2008, pp. 1023–1030 (2008)

Klein, M., Mathew, T., Sinha, B.: Noise multiplication for statistical disclosure control of extreme values in log-normal regression samples. J. Priv. Confid. **6**, 77–125 (2014)

Lin, Y.-X.: Density Approximant based on noise multiplied data. In: Domingo-Ferrer, J. (ed.) PSD 2014. LNCS, vol. 8744, pp. 89–104. Springer, Cham (2014). https://doi.org/10.1007/978-3-319-11257-2_8

Lin, Y.X., Fielding, M.J.: MaskDensity14: a R package for the density approximant of a univariate based on noise multiplied data. SoftwareX **3–4**, 37–43 (2015)

Lin, Y.X., Wise, P.: Estimation of regression paremeters from noise multiplied data. J. Priv. Confid. **4**, 61–94 (2012)

Ma, Y., Lin, Y.X., Sarathy, R.: The vulnerability of multiplicative noise protection to correlation-attacks on continuous microdata. Sankhya B (Major Revision) (2018)

Melville, N., McQuaid, M.: Research note-generating shareable statistical databases for business value: multiple imputation with multimodal perturbation. Inf. Syst. Res. **23**(2), 559–574 (2012)

Muralidhar, K., Sarathy, R.: Data shuffling - a new masking approach for numerical data. Manag. Sci. **52**, 658–670 (2006)

Nayak, T.K., Sinha, B., Zayatz, L.: Statistical properties of multiplicative noise masking for confidentiality protection. J. Off. Stat. **27**(3), 527–544 (2011)

Oganian, A.: Multiplicative noise for masking numerical microdata with constraints. SORT Special Issue: Priv. Stat. Database 99–112 (2011)

Ruiz, N.: A multiplicative masking method for preserving the skewness of the original micro-records. J. Off. Stat. **28**, 107–120 (2011)

Shlomo, N.: Releasing microdata: disclosure risk estimation, data masking and assessing utility. J. Priv. Confid. **2**(1), 73–91 (2010)

Sinha, B., Nayak, T.K., Zayatz, L.: Privacy protection and quantile estimation from noise multiplied data. Sankhya B **73**(2), 297–315 (2011)

Snoke, J., Raab, G.M., Nowok, B., Dibben, C., Slavković, A.: General and specific utility for synthetic data. J. Roy. Stat. Soc. Ser. A: Stat. Soc. **181**, Part 3, 663–688 (2018)

Yancey, W.E., Winkler, W.E., Creecy, R.H.: Disclosure risk assessment in perturbative microdata protection. In: Domingo-Ferrer, J. (ed.) Inference Control in Statistical Databases. LNCS, vol. 2316, pp. 135–152. Springer, Heidelberg (2002). https://doi.org/10.1007/3-540-47804-3_11

Record Linkage

Generalized Bayesian Record Linkage and Regression with Exact Error Propagation

Rebecca C. Steorts[1(✉)], Andrea Tancredi[2], and Brunero Liseo[2]

[1] Department of Statistical Science, affiliated faculty, Computer Science, Biostatistics and Bioinformatics, the information initiative at Duke (iiD), and the Social Science Research Institute (SSRI) Duke University; Principal Researcher, Center for Statistical Research Methodology, Duke University and U.S. Census Bureau, Durham, USA
beka@stat.duke.edu
[2] Department of Methods and Models for Economics, Territory and Finance, La Sapienza, Rome, Italy
{andrea.tancredi,brunero.liseo}@uniroma.it

Abstract. Record linkage (de-duplication or entity resolution) is the process of merging noisy databases to remove duplicate entities. While record linkage removes duplicate entities from such databases, *the downstream task* is any inferential, predictive, or post-linkage task on the linked data. One goal of the downstream task is obtaining a larger reference data set, allowing one to perform more accurate statistical analyses. In addition, there is inherent record linkage uncertainty passed to the downstream task. Motivated by the above, we propose a generalized Bayesian record linkage method and consider multiple regression analysis as the downstream task. Records are linked via a random partition model, which allows for a wide class to be considered. In addition, we jointly model the record linkage and downstream task, which allows one to account for the record linkage uncertainty exactly. Moreover, one is able to generate a feedback propagation mechanism of the information from the proposed Bayesian record linkage model into the downstream task. This feedback effect is essential to eliminate potential biases that can jeopardize resulting downstream task. We apply our methodology to multiple linear regression, and illustrate empirically that the "feedback effect" is able to improve the performance of record linkage.

1 Introduction

Record linkage (de-duplication or entity resolution) is the process of merging noisy databases to remove duplicate entities. While record linkage removes duplicate entities from such databases, *the downstream task* is any inferential, predictive or post-linkage task on the linked data. In this paper, we propose a joint model for the record linkage and the downstream task of linear regression. Our proposed model can link records over an arbitrary number of databases

© Springer Nature Switzerland AG 2018
J. Domingo-Ferrer and F. Montes (Eds.): PSD 2018, LNCS 11126, pp. 297–313, 2018.
https://doi.org/10.1007/978-3-319-99771-1_20

(lists or files). We assume there is duplication within each database, known as "duplicate detection." Our record linkage model can be expressed as a random partition model, which leads to a large family of distributions. Next, we jointly model the record linkage task and the downstream task (linear regression), which allows for the exact propagation of the record linkage uncertainty into the downstream task. Crucially, this generates a feedback propagation mechanism from the proposed Bayesian record linkage model into the downstream task of linear regression. This feedback effect is essential to eliminate potential biases that can jeopardize resulting inference in the downstream task. We apply our methodology to multiple linear regression, and illustrate empirically that the "feedback effect" is able to improve performance of record linkage.

1.1 Prior Work

Our work builds off [14, 16–18], which all proposed Bayesian record linkage models well suited for categorical data. [18] modeled the fully observed records through the "hit-and-miss" measurement error model [2]. One natural way to handle record linkage uncertainty is via a joint model of the record linkage and downstream task. [10] introduced a record linkage model for continuous data based on a multivariate normal model with measurement error. Turning to just record linkage tasks, [16, 17] were the first to perform simultaneous record linkage and de-duplication on multiple files by using the fully observed records, creating a scalable record linkage algorithm. In similar work, de-duplication in a single database framework was tackled from a Bayesian perspective in [14] by using the information provided by the comparison data.

Related work regarding the record linkage and downstream task has been considered under specific assumptions. [9] assumed that the two databases represent a permutation of the same database of units and proposed an estimator (LL) of the regression coefficients which is unbiased, conditionally on the matching probabilities provided by the record linkage task. [7] extended this approach to handle more complex and realistic linkage scenarios and logistic regression problems. Generalizations of the LL estimator have been also provided by [8] using estimating equations. In addition, [4] proposed to consider the probabilities of being a match—provided by the record linkage algorithm—as an ingredient to be used within a multiple imputation scenario. Finally, [5] proposed a Bayesian method that jointly models the record linkage and the association between the overlapping features in two different databases. The authors consider somewhat simpler situation where the number of records to match in the two databases is relatively small and relies upon a specific blocking criteria. In addition, one potential limitation of the approach is the assumption of specific matching pattern. For each single block of comparisons, all cases in the smaller database will certainly appear in the other databases. We refer to [6] for details.

Section 2 introduces our Bayesian record linkage model, providing extensions to priors on random partitions. Section 3 generalizes our record linkage methodology to the downstream task of linear regression. Section 4 provides experiments for the record linkage task on synthetic data. We then provide three experiments on the joint record linkage and downstream task of linear regression on synthetic data. Section 5 provides a discussion and extensions to future work.

2   Bayesian Record Linkage and Priors on Partitions

In this section, we introduce notation used through the paper, our Bayesian record linkage model, and an alternative and more intuitive construction for the prior on co-referent records, known as the linkage structure λ.

2.1   Notation

Assume L databases (lists, data sets, or files) $F_1, F_2 \ldots, F_L$ that consist of either qualitative and/or categorical records, which are noisy due to the data collection process. Each record corresponds to an underlying latent entity (statistical unit) of partially overlapping samples (or populations). In addition, assume all databases have p overlapping features (fields). Assume L sets of records are collected from a given population of size N_{pop} where $1 \leq N_{pop} \leq \infty$ in the same framework as [15,17]. As such, assign a label j' ($j' = 1, \ldots, N_{pop}$) to each member of the population. Next, let $\tilde{v}_{j'} = (\tilde{v}_{j'1}, \ldots, \tilde{v}_{j'p})$ be the vector of the p categorical overlapping features for the population individual j'. Finally, denote the entire set of population records by $\tilde{v} = (\tilde{v}_1, \ldots, \tilde{v}_{N_{pop}})$.

2.2   Bayesian Record Linkage Model

Assume the set of population records $\tilde{v}$ is generated independently, for $j' = 1, \ldots, N_{pop}$, from a vector of independent categorical variables $\tilde{V} = (\tilde{V}_1, \ldots, \tilde{V}_\ell, \ldots, \tilde{V}_p)$ such that $\tilde{V}_l \in \{v_{\ell 1}, \ldots, v_{\ell M_\ell}\}$ and

$$P\left(\tilde{V}_\ell = v_{\ell s}\right) = \theta_{\ell v_{\ell s}} \quad s = 1, \ldots, M_\ell, \tag{1}$$

where M_ℓ is the number of categorical values for the ℓth feature. At the sample level, assume that one does not observe the "true" population values due to measurement errors. Thus, the observed records, which is a database of size N_i, $i = 1, \ldots, L$, consists of distorted versions of subsets of the vectors $\tilde{v}_{j'}$. Let $v_{ij} = (v_{ij1}, \ldots, v_{ijp})$ denote the observed values for the j-th record of the i-th database, where $i = 1, \ldots, L$ and $j = 1, \ldots, N_i$. Denote the observed records (across the L databases) by $v = (v_{11}, \ldots, v_{1N_1}, \ldots, v_{L1}, \ldots v_{LN_L})$. Next, let the set of latent indicator variables $\lambda_{ij} \in \{1, \ldots, N_{pop}\}$ denote the unknown co-reference (matching) pattern between the observed records v and the population records $\tilde{v}$, where $\lambda_{ij} = j'$ indicates that the population record j' generated the observed record v_{ij}.[1] In general, let $\lambda = (\lambda_{11}, \ldots, \lambda_{1N_1}, \ldots, \lambda_{L1}, \ldots, \lambda_{LN_L})$ denote the linkage structure.

---

[1] The relation $\lambda_{ij_1} = \lambda_{ij_2}$, with $j_1 \neq j_2$, implies that records j_1 and j_2 of the i-th database represents are co-referent to the same population record. This is an instance of duplicate-detection within the same database. When $\lambda_{i_1 j_1} = \lambda_{i_2 j_2}$, with $i_1 \neq i_2$, one has the usual record linkage framework with the same individual appearing in two different databases.

Next, we formalize the distortion mechanism when the population records are observed in the L databases using the *hit-and-miss model* [2]. Let $V_{ij\ell}$ be the random variable that generates observed record $v_{ij\ell}$. Assume that $V_{ijl} \in \{v_{l1}, \ldots, v_{lM_\ell}\}$, that is, $V_{ij\ell}$ has the same support of $\tilde{V}_\ell$. Let $\delta_{a,b} = 1$ if $a = b$ and $\delta_{a,b} = 0$ if $a \neq b$, which implies that

$$P(V_{ij\ell} = v_{\ell s} \mid \lambda_{ij}, \tilde{v}, \alpha_\ell) = (1 - \alpha_\ell)\delta_{\tilde{v}_{\lambda_{ij}\ell}, v_{\ell s}} + \alpha_\ell \theta_{\ell v_{\ell s}} \quad s = 1, \ldots, M_\ell \quad (2)$$

for $i = 1, \ldots, L$; $j = 1, \ldots, N_i$; $\ell = 1, \ldots, p$, where $\alpha_\ell \in [0, 1]$ represents the distortion probability for the ℓ-th overlapping feature. Here, the true population value is observed with probability $1 - \alpha_\ell$, and a different value is drawn from the random variable $\tilde{V}_\ell$ generating the population values with probability α_ℓ. Finally, assuming the conditional independence among all the overlapping features given their respective unobserved population counterparts, one obtains

$$p(v \mid \tilde{v}, \lambda, \alpha) = \prod_{i=1}^{L} \prod_{j=1}^{N_i} \prod_{\ell=1}^{p} P(v_{ijl} \mid \tilde{v}, \lambda, \alpha) = \prod_{i=1}^{L} \prod_{j=1}^{N_i} \prod_{\ell=1}^{p} [(1 - \alpha_\ell)\delta_{\tilde{v}_{\lambda_{ij}\ell}, v_{ij\ell}} + \alpha_\ell \theta_{\ell v_{ij\ell}}]. \quad (3)$$

We assume that the distortion probabilities are exchangeable, that is

$$\alpha_\ell \overset{iid}{\sim} \mathrm{Beta}(f, g), \, \ell = 1, \ldots, p,$$

and we assume the probabilities $\theta_{\ell 1} \ldots \theta_{\ell M_\ell}$ are considered known and equal to the corresponding population frequencies. The model summarized by Eqs. (1) and (3) can be viewed as a latent variable model where the unobserved population records $\tilde{v}$ generate the observed records v and $\alpha = (\alpha_1 \ldots, \alpha_p)$ can be viewed as the unknown model distortion parameter.

Remark: A convenient property of the hit-miss model is that one can integrate out the unknown population values $\tilde{v}$ to directly obtain the distribution $p(v|\alpha, \lambda)$. The resulting marginal distribution $p(v|\alpha, \lambda)$ is the product of within-cluster distributions. To improve mixing, we use a Metropolis within Gibbs algorithm to simulate from the joint posterior $p(\lambda, \alpha|v)$ (See Appendix A).

2.3   The Prior Distribution for λ

In this section, we propose a more intuitive and subjective construction of a prior distribution on λ. Let z denote the random partition of the observed records determined by λ and let $\mathcal{P}$ denote the set containing all the possible partitions of the N observed records. The distribution on the sample labels λ induces a distribution on $\mathcal{P}$. Furthermore, matches and duplicates are completely specified given the knowledge of the random partition z, which is invariant with respect to the labelings of the partition blocks. Given this construction, one can directly focus on the partition distribution of the observed records without linking the labels distribution to a sample design and to a population size N_{pop}, see for example, [14]. One can effectively consider the distribution of λ as a prior distribution for the latent linkage structure and concentrate only on its probabilistic properties.

Both the interpretations of the role of λ (either as a consequence of the sampling design or a model represented by partitions) may provide useful insights for a correct choice of its prior distribution. One difficult and related question in the record linkage literature has been the subjective specification on the space of partitions. A simple, alternative prior for the number of distinct entities $k(z)$ can be obtained looking at the following allocation rule for the record labels which is based on a generalization of the Chinese Restaurant Process, namely the Pitman-Yor process (PYP) ([3,13]). (See Appendix B for details).

3   The Downstream Task of Linear Regression

In this section, we propose record linkage methodology for the downstream task of linear regression. Consider the model $\tilde{Y} = \sum_{l=1}^{p} \tilde{X}_l \beta_l + \epsilon$ for the population units, where the goal is to estimate the regression coefficients $\beta = (\beta_1, \ldots, \beta_p)^t$. We observe Y and $X = (X_1, \ldots, X_p)$, where X represents a noisy measurement of the true covariates $\tilde{X} = (\tilde{X}_1, \ldots \tilde{X}_p)$ and Y is a random copy of the corresponding population variable $\tilde{Y}$.

To better illustrate our approach, we consider two scenarios. In the first scenario—*the complete regression scenario*—each database reports a set of overlapping features, the response variable, and the covariates. Let y_{ij} and $x_{ij} = (x_{ij1} \ldots, x_{ijp})$ denote the observed values for the j-th unit of the i-th database, where $i = 1, \ldots, L$ and $j = 1, \ldots, N_i$. In addition, let (y, x) denote the entire set of regression data observed across the L databases. In the *complete scenario*, there is not a bias problem concerning the estimation of the β coefficients. In the second scenario—*the broken regression scenario*—we assume that the overlapping features are observed in each database, the response variable is observed in only the first database, and specific subsets of covariates are observed in the other databases. In this situation, let (y, x) denote the observation y_{1j}, where $j = 1, \ldots, N_1$ and x_{ij}, where $i = 2, \ldots, L$ and $j = 1, \ldots, N_i$. Note that x_{ij} represents only a fixed subset of the values $x_{ij1} \ldots x_{ijp}$ for $j = 1, \ldots, N_i$. Here, there is a bias issue regarding estimating the β coefficients.[2]

3.1   Simple Linear Regression

In this section, we consider linear regression and the two scenarios mentioned above with a single covariate X. First, consider the *complete regression scenario*. Let $\tilde{X}_{j'}$ be the true value of observation X corresponding to the records of cluster $C_{j'}$. Now consider a cluster $C_{j'} = \{(i, j)\}$ with one record. Given the true value of $\tilde{X}_{j'} = \tilde{x}_{j'}$ and membership to cluster $C_{j'}$, we assume that the response variable Y_{ij} follows a standard normal regression model with covariate $\tilde{x}_{j'}$, where the observed value for the covariate X_{ij} is normal with mean $\tilde{x}_{j'}$ and Y_{ij} and X_{ij} are independent. That is,

---

[2] In both scenarios, we assume that the covariates have zero mean and the regression model does not have an intercept.

$$\begin{bmatrix} Y_{ij} \\ X_{ij} \end{bmatrix} \mid \tilde{X}_{j'} = \tilde{x}_{j'} \sim N_2 \left[\begin{pmatrix} \beta & 0 \\ 0 & 1 \end{pmatrix} \begin{bmatrix} \tilde{x}_{j'} \\ \tilde{x}_{j'} \end{bmatrix}, \begin{pmatrix} \sigma_{y|\tilde{x}}^2 & 0 \\ 0 & \sigma_{x|\tilde{x}}^2 \end{pmatrix} \right]. \tag{4}$$

We assume that $\tilde{X}_{j'} \sim N(0, \sigma_{\tilde{x}}^2)$, which allows one to integrate $X_{j'}$ via Eq. 4. In fact, setting $Z_{ij} = (Y_{ij}, X_{ij})'$, one can easily show that conditionally on the event $\{(i,j) \in C_{j'}\}$, it follows that

$$Z_{ij} \sim N_2 \left[\begin{pmatrix} 0 \\ 0 \end{pmatrix}, \sigma_{\tilde{x}}^2 \begin{pmatrix} \beta^2 & \beta \\ \beta & 1 \end{pmatrix} + \begin{pmatrix} \sigma_{y|\tilde{x}}^2 & 0 \\ 0 & \sigma_{x|\tilde{x}}^2 \end{pmatrix} \right]. \tag{5}$$

For ease of notation, let I_n denote the $n \times n$ identity matrix, 0_n denote the n-vector of zero; 1_n denote a vector of all 1's, and $J_n = 1_n 1_n'$. Next, set

$$B = \begin{pmatrix} \beta^2 & \beta \\ \beta & 1 \end{pmatrix} \quad \text{and} \quad \Sigma = \begin{pmatrix} \sigma_{y|\tilde{x}}^2 & 0 \\ 0 & \sigma_{x|\tilde{x}}^2 \end{pmatrix}.$$

Consider a cluster $C_{j'} = \{(i_1, j_1), (i_2, j_2)\}$ with two records. The two pairs $Z_{i_1 j_1}$ and $Z_{i_2 j_2}$ are random vectors, both depending on the same "true" value $\tilde{X}_{j'}$. Let $\otimes$ be the Kronecker product. Conditionally on $\tilde{X}_{j'} = \tilde{x}_{j'}$ and on the cluster membership, we replicate the model for a cluster with one record by assuming that $Z_{i_1 j_1}$ and $Z_{i_2 j_2}$ are two independent bivariate normal random variables with joint distribution

$$N_4 \left[\left(I_2 \otimes \begin{pmatrix} \beta & 0 \\ 0 & 1 \end{pmatrix} \right) (1_4 \tilde{x}_{j'}), I_2 \otimes \Sigma \right]. \tag{6}$$

Then the marginal distribution of $(Z_{i_1 j_1}, Z_{i_2 j_2})'$ is

$$\begin{pmatrix} Z_{i_1 j_1} \\ Z_{i_2 j_2} \end{pmatrix} \sim N_4 \left(0_4, I_2 \otimes \Sigma + \sigma_{\tilde{x}}^2 J_2 \otimes B \right).$$

This argument can be extended to any cluster size. When $\operatorname{card}(C_{j'}) = n$, the marginal distribution of $Z = (Z_{i_1 j_1}, \ldots, Z_{i_n j_n})$ is again multivariate normal: $Z \sim N_{2n} \left(0_{2n}, I_n \otimes \Sigma + \sigma_{\tilde{x}}^2 J_n \otimes B \right)$.

Next, consider the *broken regression scenario*. In this case, when some information is missing—either the covariate in the first database or the response variable in some of the other databases—one can easily marginalize over the missing variables by using standard properties of multivariate normal distribution. Let $(y, x)_{C_j'} = ((y_{ij}, x_{ij}) : \lambda_{ij} = j')$ denote the set of regression observations, which conditionally on λ, correspond to the j'-th population unit. For example, for a cluster $C_{j'} = \{(1, j)\}$ with one record in the first database, we denote this as $(y, x)_{C_j'} = y_{1j}$. Using the marginal density of Y_{ij} in Eq. 5, we can write the likelihood, conditional on λ, as $p((y, x)_{C_j'} \mid \lambda, \beta, \sigma_{y|\tilde{x}}^2, \sigma_{x|\tilde{x}}^2)$. Similarly, suppose $C_{j'} = \{(i, j)\}$ with $i > 1$, then $(y, x)_{C_j'} = x_{ij}$ and the likelihood is given by marginal density of X_{ij}. Next, consider a cluster $C_{j'} = \{(1, j_1), (i_2, j_2)\}$ with a record in the first database and the other record in a different database, i.e. $i_2 > 1$. It follows that $(y, x)_{C'_j} = (y_{1j_1}, x_{i_2j_2})$ and the corresponding likelihood is found by marginalizing over the missing values X_{1j_1}, Y_{2j_2} in Eq. 6, where we obtain the joint density in Eq. 5. Finally, it follows that the likelihood function (as a function of $\lambda, \beta, \sigma^2_{y|\tilde{x}}, \sigma^2_{x|\tilde{x}}$) for both the complete and broken regression scenarios can be generally written as $p(y, x|\lambda, \beta, \sigma^2_{x|\tilde{x}}, \sigma^2_{y|\tilde{x}}) = \prod_{j'=1}^{N_{pop}} p((y, x)_{C'_j}|\beta, \sigma^2_{x|\tilde{x}}, \sigma^2_{y|\tilde{x}}).$[3]

In order to handle the record linkage and downstream regression task simultaneously, we assume conditional independence on λ between the overlapping features in the record linkage model and the set of variables in the downstream task of linear regression. Assuming conditional independence, we find

$$p(\lambda, \beta, \alpha, \sigma^2_{y|\tilde{x}}, \sigma^2_{x|\tilde{x}}|v, x, y) \propto p(v|\lambda, \alpha)p(y, x|\lambda, \beta, \sigma^2_{y|\tilde{x}}, \sigma^2_{x|\tilde{x}})$$
$$\times p(\lambda)p(\alpha)p(\beta, \sigma^2_{y|\tilde{x}}, \sigma^2_{x|\tilde{x}}). \qquad (7)$$

The first factor is related to the record linkage process, and second factor is related to the downstream task of linear regression, and the other factors represent the prior distributions. We assume independent diffuse priors for $\beta, \sigma^2_{y|\tilde{x}}, \sigma^2_{x|\tilde{x}}$. To update the appropriate regression parameters $\beta, \sigma^2_{y|\tilde{x}}, \sigma^2_{x|\tilde{x}}$, we use the Metropolis-Hastings algorithm in Appendix A. Using the factorization of the posterior in Eq. (7), the proposed method can be generalized to any statistical model.

3.2 Multiple Linear Regression

We extend the downstream task to that of multiple regression, first considering the complete regression scenario. Let $C_{j'}$ denote a cluster of size n, $Y_{C_{j'}}$ denote a vector with n observations of the response variable in this cluster, and $X_{C_{j'}}$ denote the $n \times p$ matrix with the values of the p covariates observed in the cluster units. Let $[YX]_{C_{j'}}$ denote the vector of $n(p+1)$ elements with the n rows of the matrix $(Y_{C_{j'}}, X_{C_{j'}})$ vertically stacked and let $\tilde{X}_{j'}$ denote the vector containing the true values of the p covariates. Equation 4 can be generalized assuming that

$$[YX]_{C_{j'}} \mid \tilde{X}_{j'} \sim N_{n(p+1)} \left[\left(I_{n \times n} \otimes \begin{pmatrix} \beta^t & 0^t_p \\ 0_{p \times p} & I_{p \times p} \end{pmatrix} \right) (1_{2n} \otimes \tilde{X}), I_{n \times n} \otimes \begin{pmatrix} \sigma^2_{y|\tilde{x}} & 0 \\ 0 & \Sigma_{x|\tilde{x}} \end{pmatrix} \right],$$

where

$$1_{2n} \otimes \tilde{X} \sim N_{2np} \left(0_{2np}, (1_n 1^t_n) \otimes \begin{pmatrix} \Sigma_{\tilde{x}} & \Sigma_{\tilde{x}} \\ \Sigma_{\tilde{x}} & \Sigma_{\tilde{x}} \end{pmatrix} \right).$$

This way the marginal distribution of $[YX]_{C_{j'}}$ is $n(p + 1)$-variate normal with zero mean and covariance matrix

$$\left(I_{n \times n} \otimes \begin{pmatrix} \beta^t & 0^t_p \\ 0_{p \times p} & I_{p \times p} \end{pmatrix} \right) \left((1_n 1^t_n) \otimes \begin{pmatrix} \Sigma_{\tilde{x}} & \Sigma_{\tilde{x}} \\ \Sigma_{\tilde{x}} & \Sigma_{\tilde{x}} \end{pmatrix} \right) \left(I_{n \times n} \otimes \begin{pmatrix} \beta^t & 0^t_p \\ 0_{p \times p} & I_{p \times p} \end{pmatrix} \right)^t +$$
$$\left(I_{n \times n} \otimes \begin{pmatrix} \sigma^2_{y|\tilde{x}} & 0 \\ 0 & \Sigma_{x|\tilde{x}} \end{pmatrix} \right),$$

---

[3] We assume that population units j' that do not have an observed cluster size contribute to the likelihood with a factor equal to 1.

which simplifies into

$$(1_n 1_n^t) \otimes \begin{pmatrix} \beta^t \Sigma_{\tilde{x}} \beta & \beta^t \Sigma_{\tilde{x}} \\ \Sigma_{\tilde{x}} \beta^t & \Sigma_{\tilde{x}} \end{pmatrix} + I_{n \times n} \otimes \begin{pmatrix} \sigma_{y|\tilde{x}}^2 & 0 \\ 0 & \Sigma_{x|\tilde{x}} \end{pmatrix}.$$

The likelihood provided by the multiple regression model is the product of the factors $p([YX]_{C_{j'}} = [y, x]_{C_{j'}} | \beta, \sigma^2 y|\tilde{x}, \Sigma_{x|\tilde{x}})$ for the observed clusters. The same considerations from linear regression regarding modeling the prior and the computational aspects apply to multiple linear regression. Note the major difference is in the marginalization pattern in the broken regression scenario. In fact, for a cluster joining records across more than one database, we may need to integrate out the covariate values missing in the databases that share a cluster.

4   Experiments

To investigate the performance of our proposed methodology we consider the RLdata500 data set from the RecordLinkage package in R. This synthetic data set consists of 500 records, each comprising first and last name and full date of birth. We modify this data set to consider two databases, where each database contains 250 records, respectively, with duplicates in and across the two databases. To consider the case without duplicate detection, we modify the original RLdata500 such that it has no duplicate records within each of the two databases. Without duplicate detection is a special case of our general methodology (see Appendix C). We provide experiments for both record linkage and the downstream task.

4.1   Record Linkage with and Without Duplicate-Detection

We provide two record linkage experiments—one with duplicate detection and one without duplicate detection. In Figs. 1 and 2, we report the prior and the posterior for $k(z)$ and the performance of the record linkage procedure measured in terms of the posteriors of the false negative rates (FNR) and the false discovery rates (FDR). (For a review of FNR and FDR, see [1, 15]).

Figure 1 (with duplicate detection) illustrates that the resulting posteriors of $k(z)$ appears robust to the choices of θ and σ (first row). We observe similar behavior for the posteriors of FNR and FDR (second and third rows). Figure 2 illustrates that as we vary the PYP parameters, the posterior of T is weakly dependent on their values. The two database framework without duplicate detection leads, a posteriori, to similar FNR (second row) and lower FDR (third row) compared to the previous case. (See Appendix D for the PYP parameter settings).

4.2   Regression Experiments

We consider three regression experiments on the RLdata500 data set. In Experiment I, we consider the complete regression scenario in a single database framework with duplicate detection. In Experiment II, we consider the broken regression scenario with record linkage and duplicate detection. In Experiment III, we

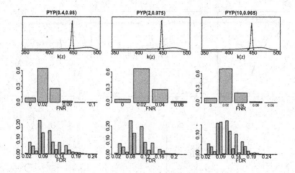

Fig. 1. Prior and posteriors for $k(z)$ (first row), FNR posteriors (second row), FDR posteriors (third row) for the `RLdata500` data set.

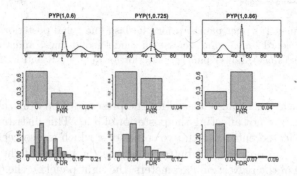

Fig. 2. Prior and posteriors for t (first row), FNR posteriors (second row), FDR posteriors (third row) for the `RLdata500` data set assuming a two database record linkage framework without duplicate-detection.

consider the broken multiple regression scenario in a two database framework without duplicate detection. (See Appendix E for details).

Figure 3 gives the results of Experiment I. The posteriors of $(\beta, \sigma_{y|\tilde{x}}, \sigma_{x|\tilde{x}})$ from our joint modeling approach (first row, solid lines) do not show remarkable differences when compared to their true counterpart (first row, dotted lines), which were obtained by fitting the regression model conditional on the true value of λ. The similarity between the posteriors is mainly due to the large concentration of λ around the true pattern of duplications. The mode of the posterior of the number of distinct entities is exactly the true value (450), where the FNR and FDR are considerably smaller with respect to case without the y and x columns. Hence, the effect of considering the information provided by the regression model has improved the record linkage process.

Figure 4 gives the results of Experiment II. The posteriors (first row, solid lines) of $(\beta, \sigma_{y|\tilde{x}}, \sigma_{x|\tilde{x}})$ are similar to the corresponding true posteriors (first row, dashed lines). We report the posteriors obtained by fixing λ equal to the point estimate provided by the hit-and-miss model applied to the categorical variables alone (first row, dotted lines). The posteriors of β and $\sigma^2_{y|\tilde{x}}$ obtained with the

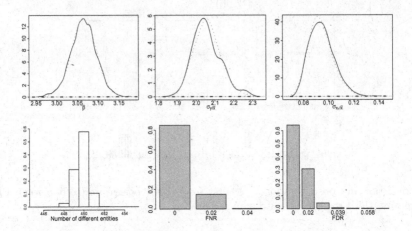

Fig. 3. Experiment I. Upper panels: prior (dotdash lines) and posterior of $\beta, \sigma_{y|\tilde{x}}, \sigma_{x|\tilde{x}}$ with the joint record linkage and regression model (solid lines) and the true linkage structure (dotted lines). Lower panels: posterior for $k(z)$, FNR and FDR.

plug-in approach are strongly biased for the presence of false matches which, on the other hand, are not affecting the posterior of $\sigma_{x|\tilde{x}}$. This distribution depends on the 13 duplicated entities with two copies of x which are correctly accounted for in the plug-in approach. To better illustrate the causes of the distortion in the estimation of the regression parameters, the right panel on the top row shows all the (x, y) pairs resulting from the plug-in approach. The solid black circles represent the true matches, and the empty red circles represent the false matches, with independent y and x values. We report the corresponding regression lines, where the three false matches are lowering the β estimate and increasing the $\sigma_{y|\tilde{x}}$ estimate. Further analysis reveals that the posterior for $k(z)$ (second row) with the integrated hit-miss and regression model is less concentrated with respect to the first experiment but it is more concentrated with respect to the single hit-miss model. We reduce the FDR, leaving the FNR almost unchanged. We coin this the *feedback effect* of the regression from the downstream task. For example, if we consider a false link, the posterior probability of being a match will typically be down-weighted by the low likelihood arising from the regression part of the model. Hence, in addition to centering the estimates of the regression coefficient β, the joint regression-hit miss model improves record linkage performance.

Figure 5 gives the results of the Experiment III. The joint model gives posteriors similar to the true ones while the plug-in approach gives biased estimates and larger variability (first row, left upper panels). The presence of false matches in the plug-in approach gives a positive bias in estimating the variance $\sigma_{y|\tilde{x}}$ and affects the posterior of the measurement error parameters (first row, right upper panels). The posteriors of $\sigma_{x_1|\tilde{x}}$ and $\sigma_{x_2|\tilde{x}}$ (not reported) both with the joint model and the true λ are essentially equal to the prior, while the plug-in posterior is concentrated on larger values. Under such conditions, even with the true linkage structure, we do not have any useful information for estimating the mea-

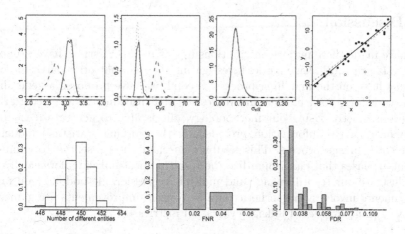

Fig. 4. Experiment II. Left upper panels: prior (dotdash lines) and posterior of $\beta, \sigma_{y|\bar{x}}, \sigma_{x|\bar{x}}$ with the joint record linkage and regression model (solid lines), the true linkage structure (dotted lines) and the plug-in approach (dashed lines). Right upper panel: estimated regression line and (x, y) pairs with the joint model (solid line and full circles) and the plug-in approach (dashed line and empty circles) Lower panels: posterior for $k(z)$, FNR and FDR.

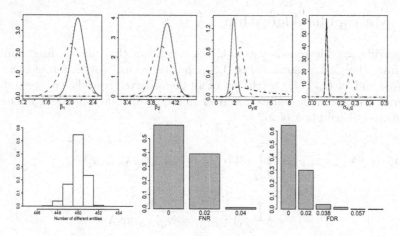

Fig. 5. Experiment III. Same caption as Fig. 3.

surement error variances due to the lack of duplicated x values. Thus, while the joint model correctly does not contrast the information provided by the prior, the presence of false matches creates (y, x) pairs that could be also explained by a larger measurement error of the covariates. We observe that the joint modeling of the record linkage and regression data improves the matching process as noted by the higher concentration of $k(z)$ (second row, left lower panel) around the true value of 450 and the lower FNRs and FDRs (second row, right lower panels) with respect to results obtained with the hit-and-miss model only.

5   Discussion

We have made three major contributions in this paper. First, we have proposed a Bayesian record linkage model investigating the role that prior partition models may have on the matching process. Second, we have proposed a generalized framework for record linkage and regression that accounts for the record linkage error exactly. Using our methodology, one is able to generate a feedback mechanism of the information provided by the working statistical model on the record linkage process. This feedback mechanism is essential to eliminate potential biases that can jeopardize the resulting post-linkage inference. Third, we illustrate our record linkage and multiple regression methodology on many experiments involving a synthetic data set, where improvements are gained in terms of standard record linkage evaluation metrics.

Acknowledgments. Steorts was supported by NSF-1652431 and NSF-1534412. Tancredi and Liseo were supported by Ministero dell' Istruzione dell' Universita e della Ricerca, Italia PRIN 2015.

Appendix

A   Metropolis Algorithm

We provide our Metropolis-within-Gibbs algorithm that allows direct simulation from the joint posterior. Let $\lambda_{(-ij)}$ be the vector λ with the element λ_{ij} removed and let $C_{j'} \setminus (ij)$ be the set of all the observed records with record (ij) removed, which conditionally on λ, refer to the population individual j' The full conditional distribution of λ_{ij} is

$$p(\lambda_{ij} = q | \lambda_{(-ij)}, \alpha, v) \propto \prod_{j'=1}^{N_{pop}} p(V_{C_{j'}} = v_{C_{j'}} | \alpha, \lambda) \, p(\lambda_{ij} = q | \lambda_{(-ij)})$$

$$\propto \prod_{j'=1}^{N_{pop}} \frac{p(V_{C_{j'}} = v_{C_{j'}} | \alpha, \lambda)}{p(V_{C_{j'} \setminus (ij)} = v_{C_{j'} \setminus (ij)} | \alpha, \lambda)} p(\lambda_{ij} = q | \lambda_{(-ij)}) \quad (8)$$

$$\propto \frac{p(V_{C_q} = v_{C_q} | \alpha, \lambda)}{p(V_{C_q \setminus (ij)} = v_{C_q \setminus (ij)} | \alpha, \lambda)} p(\lambda_{ij} = q | \lambda_{(-ij)}), \quad (9)$$

where $q = 1, \ldots N_{pop}$. In Eq. 8, set $\lambda_{ij} = q$, which implies that $C_{j'} = C_{j' \setminus ij}$ $\forall j' \neq q$. It follows that $\dfrac{p(v_{C_{j'}} | \alpha, \lambda)}{p(v_{C_{j'} \setminus ij} | \alpha, \lambda)} = 1$ $\forall j' \neq q$. When the population entity q represents an already existing cluster given $\lambda_{-(ij)}$, the above ratio can also be written as

$$\frac{p(V_{C_q} = v_{C_q}|\alpha, \lambda)}{p(V_{C_q\backslash(ij)} = v_{C_q\backslash(ij)}|\alpha, \lambda)}$$

$$= \prod_{l=1}^{p}\left[\alpha_l \theta_{l\,v_{ijl}} + (1-\alpha_l)\frac{\prod_{(i_h,j_h)\in C_q\backslash(ij)}\left((1-\alpha_l)\delta_{v_{i_hj_hl},v_{ijl}} + \alpha_l\theta_{lv_{i_hj_hl}}\right)}{p(V_{C_q\backslash(ij)\,l} = v_{C_q\backslash(ij)\,l}|\alpha, \lambda)}\right].$$

When the label q identifies a new cluster, the following simplification is possible:

$$\frac{p(V_{C_q} = v_{C_q}|\alpha, \lambda)}{p(V_{C_q\backslash(ij)} = v_{C_q\backslash(ij)}|\alpha, \lambda)} = \prod_{l=1}^{p}\theta_{l,v_{ijl}}.$$

Note that the posterior $p(\lambda, \alpha|v)$ is invariant with respect to the cluster labels and that we are only interested in the cluster composition. Thus, we can avoid simulating the entire population label distribution, and instead set $q \in \{1, \ldots, N\}$ (since there can be at most N clusters) and update λ_{ij} with the following:

$$q(\lambda_{ij} = q) = \begin{cases} \frac{p(V_{C_q} = v_{C_q}|\alpha, \lambda)}{p(V_{C_q\backslash(ij)} = v_{C_q\backslash(ij)}|\alpha, \lambda)}p(\lambda_{ij} = q|\lambda_{(-ij)}) & \text{if } q \text{ labels an observed cluster} \\ \prod_{j=1}^{p}\theta_{l,v_{ijl}}p(\lambda_{ij} = new|\lambda_{(-ij)})/(N - k_{(-ij)}) & \text{if } q \text{ labels a new cluster} \end{cases} \quad (10)$$

for $i = 1, \ldots, L, j = 1, \ldots, N_i$, where $k_{(-ij)}$ is the number of clusters without the label λ_{ij}. This way of updating the cluster assignment is standard when the CRP is used for a prior on the cluster assignments. In addition, the marginal likelihood of the cluster observations is known or can be easily calculated using a recursive formula, see for example [11,12].

To adapt the algorithm to the two different prior distribution of λ, note that, when q labels an observed cluster, the use of a uniform prior on λ implies that

$$p(\lambda_{ij} = q|\lambda_{-(ij)}) = 1/N_{pop} \quad \text{and} \quad p(\lambda_{ij} = new|\lambda_{-(ij)}) = (N_{pop} - k_{-(ij)})/N_{pop}.$$

With the PYP prior, the above mentioned probabilities are, respectively,

$$p(\lambda_{ij} = q|\lambda_{-(ij)}) = (n_q - \sigma)/(N - 1 + \vartheta) \quad p(\lambda_{ij} = new|\lambda_{-(ij)}) = (k_{-(ij)}\sigma + \vartheta)/(N - 1 + \vartheta)$$

where n_q here denotes the size of the cluster C_q without the entity λ_{ij}. Finally, when a uniform prior on the partition space is considered, one has

$$p(\lambda_{ij} = q|\lambda_{-(ij)}) \propto 1/(N_{pop})_{k_{(-ij)}} \quad \text{and} \quad p(\lambda_{ij} = new|\lambda_{-(ij)}) \propto (N_{pop} - k_{-(ij)})/(N_{pop})_{k_{(-ij)+1}}.$$

Finally, full conditional distributions of the components of α have a computationally manageable form using a recursive formula. In fact, assuming a standard Beta prior on each α_l, one obtains

$$p(\alpha_l|\lambda, v, \alpha_{-l}) \propto \prod_{j'=1}^{N} p(V_{C_{j'}l} = v_{C_{j'}l}|\alpha_l)\alpha_l^{p-1}(1-\alpha_l)^{q-1},$$

and a straightforward Metropolis step can be easily implemented.

B   Construction of PYP Priors

We now briefly describe adapting the PYP prior to our L database framework. Assume the first j records of the i-th database and all the records of the first $i-1$ databases are classified into $k_{i,j}$ clusters identified by the population labels $j'_1, \ldots, j'_{k_{i,j}}$ with sizes $n_1, n_2, \ldots, n_{k_{i,j}}$ respectively. Also, let $N_{i,j} = \sum_{l=1}^{i-1} N_l + j$ denote the total number of these records. Next, the label of the record $\lambda_{i,j+1}$ identifies a new cluster with probability

$$P\left(\lambda_{i,j+1} = \text{"new"}|\lambda_{1,1}, \ldots, \lambda_{i,j}\right) = \frac{k_{i,j}\sigma + \vartheta}{N_{i,j} + \vartheta},$$

where (ϑ, σ) are two parameters whose admissible values are $\sigma \in [0,1)$ with $\vartheta > -\sigma$ or $\sigma < 0$ with $\theta = m|\sigma|$ for some positive integer m. Moreover, $\lambda_{i,j+1}$ will assume an already observed label j'_g identifying a cluster with size n_g with probability

$$P\left(\lambda_{i,j+1} = j'_g|\lambda_{1,1}, \ldots, \lambda_{i_1,j_1}\right) = \frac{n_g - \sigma}{N_{i,j} + \vartheta} \quad g = 1, \ldots, k_{i,j}.$$

The above updating rule induces a prior on the set of the possible partitions of all the N records which can be written as [13]

$$P\left(z(\lambda) = z\right) = \frac{(\vartheta + \sigma)_{k-1,\sigma}}{(\vartheta + 1)_{N-1,1}} \prod_{g=1}^{k} (1 - \sigma)_{n_g-1,1},$$

where $\{n_1, \ldots, n_k\}$ are the cluster sizes of the partition z and $x_{r,s} = x(x + s) \cdots (x+(r-1)s)$. It can also be proved [13] that, under this prior, the expected value of $k(z)$ is

$$E(k(z)) = \sum_{i=1}^{N} \frac{(\vartheta + \sigma)_{(i-1)\uparrow}}{(\vartheta + 1)_{(i-1)\uparrow}} = \frac{\vartheta}{\sigma}\left[\frac{(\vartheta + \sigma)_{N\uparrow}}{\vartheta_{N\uparrow}} - 1\right]$$

and the variance is

$$Var(k(z)) = \frac{\vartheta(\vartheta + \sigma)}{\sigma^2}\frac{(\vartheta + 2\sigma)_{N\uparrow}}{\vartheta_{N\uparrow}} - \frac{\vartheta^2}{\sigma^2}\left(\frac{(\vartheta + \sigma)_{N\uparrow}}{\vartheta_{N\uparrow}}\right)^2 - \frac{\vartheta}{\sigma}\frac{(\vartheta + \sigma)_{N\uparrow}}{\vartheta_{N\uparrow}}$$

with $x_{s\uparrow} = \Gamma(x + s)/\Gamma(x)$. For more details, we refer to [19].

The above equations can be used for prior elicitation by fixing ϑ and σ in order to have $E(k(z))$ equal to a rough prior guess for the number of clusters and a specific amount of prior variability for $k(z)$. Moreover, in evaluating the asymptotic properties, [13] observes that as $N \to \infty$, $E(k(z))$ becomes infinite for non negative values of σ; on the other hand, if σ is negative, $k(z)$ is equal almost surely to m which thus takes the role of the size N_{pop} in a finite population framework.

C   Record Linkage Without Duplicate-Detection

We now consider record linkage of two databases without duplicate-detection. To consider this case, we simply modify the prior distribution on the λ's such that $\lambda_{ij_1} \neq \lambda_{ij_2} \; \forall j_1 \neq j_2$ and for $i = 1, 2$. In this case, clusters consist of at most two elements so that the distribution of the observed records v, conditional on λ and α, can be calculated analytically without exploiting the recursive formula.

If a uniform prior on the label space is assumed, the above conditioning is equivalent to assuming that the two databases are two simple random samples with replacement from a population of N_{pop} units. This is the same situation described in [18], where N_{pop} is assumed unknown. Assume that T denotes the number of common units between the two databases; then $k(z)$ is equal to $N_1 + N_2 - T$, where T follows a hypergeometric distribution

$$P(T = t) = \frac{\binom{N_1}{t}\binom{N_{pop}-N_1}{N_2-t}}{\binom{N_{pop}}{N_2}}, \quad \max\{0, N_1 + N_2 - N_{pop}\} \leq t \leq \min\{N_1, N_2\}.$$

From a computational perspective, the conditioning of the uniform prior does not imply substantial changes. In fact if a PYP prior is assumed, the standard record linkage framework can be tackled by imposing that $\lambda_{1j} = j$ for $j = 1, \ldots, N_1$ and that the units of the second database may only join a cluster composed by a single unit of the first database or create a new cluster, that is

$$p(\lambda_{2j+1} = q|\lambda_{11}\ldots,\lambda_{2j}) = \begin{cases} \frac{1-\sigma}{k_{2j}-j(1-\sigma+\vartheta)} & \text{if } q \leq N_1 \text{ and } n_q = 1 \\ 0 & \text{if } q \leq N_1 \text{ and } n_q = 2 \quad j = 0, 1 \ldots, N_2 - 1 \\ 0 & \text{if } q > N_1 \text{ and } n_q = 1, \end{cases}$$

and

$$p(\lambda_{2j+1} = new|\lambda_{11}\ldots,\lambda_{2j}) = \frac{k_{2j}\sigma + \vartheta}{k_{2j} - j(1-\sigma) + \vartheta} \quad j = 0, 1 \ldots, N_2 - 1$$

where $k_{20} = N_1$ and k_{2j} is the number of distinct elements considering the first database and the first j elements of the second database. Finally, notice that

$$p(\lambda_{21}, \ldots, \lambda_{2N_2}|\lambda_{11}, \ldots, \lambda_{1N_1}) = \frac{(1-\sigma)^{N-k_{2N_2}} \prod_{l=N_1+1}^{k_{2N_2}}(\sigma(l-1) + \vartheta)}{\prod_{l=1}^{N_2}(k_{2l-1} - (l-1)(1-\sigma) + \vartheta)} \frac{(N - k_{2N_2})!}{N!}.$$

This implies that the λ's are no longer exchangeable. This problem, although interesting from a theoretical perspective, does not cause computational issues.

The conditional prior probabilities for the Gibbs step updating of λ_{2j} to be used from Eq. (10) are

$$p(\lambda_{2j} = q|\lambda_{-(2j)}) \propto \begin{cases} (1-\sigma) & \text{if } q \leq N_1 \text{ and } n_q = 1 \\ 0 & \text{if } q \leq N_1 \text{ and } n_q = 2 \\ 0 & \text{if } q > N_1 \text{ and } n_q = 1, \end{cases}$$

and

$$p(\lambda_{2j} = new|\lambda_{-(2j)}) \propto (k_{-(2j)}\sigma + \vartheta) \prod_{l=j}^{N_2-1} \left[\frac{k_{2l} - l(1-\sigma) + \vartheta}{k_{2l} + 1 - l(1-\sigma) + \vartheta} \right].$$

D   Record Linkage Experiment

We provide the parameter settings for the record linkage experiments. For the case with duplicate detection, we considered the effect of the PYP prior for λ with $(\theta, \sigma) = (0.4, 0.98)$, $(2, 0.975)$, $(10, 0.965)$. These prior distributions have a common prior mean of $k(z)$ almost equal to 450; however, their respective variance are quite different. For the case of no duplicate detection, we consider the effect of the constrained PYP prior for λ with $(\theta, \sigma) = (1, 0.6)$, $(1, 0.725)$, and $(1, 0.86)$. These values of the hyper-parameters (θ, σ) produce prior means for the number of matches equal to 75, 50 and 25.

E   Regression Experiments

We elaborate on our three regression experiments. In the first experiment, we modify the data set by adding two columns with the pairs y and x generated from the model in Sect. 3.1, conditional on the true λ structure. For clusters with two records we simulate a single true value $\tilde{x}$ of the covariate from a normal distribution with zero mean and variance equal to $\sigma_{\tilde{x}}^2 = 9$. Then, conditionally on $\tilde{x}$, we generate two independent draws x from a normal distribution with mean $\tilde{x}$ and variance $\sigma_{x|\tilde{x}}^2 = 0.01$ and two corresponding independent draws y from a normal distribution with mean $\beta\tilde{x}$ with $\beta = 3$ and variance $\sigma_{y|\tilde{x}}^2 = 4$. Instead, the records without duplication are augmented with a single pair (y, x) that is generated conditionally on a single value $\tilde{x}$ following the same model of the duplicated records.

In the second experiment, we use the modified RL500 data set that consists of two databases. We then remove y from the second databases and x from the first database. Given the 50 entities with duplication, 28 belong to both the databases reporting the y variable on the first database and the x variable on the second database. Moreover, 9 entities only belong to first database with 2 duplicate records of y, and 13 entities only belong to the second database with 2 copies of x. In addition, we assume the same priors as in the first experiment.

In the third experiment, we modify the RL500 data set by generating data from a regression model with two covariates, where we assume $\beta_1 = 2$ and $\beta_2 = 4$, $\sigma_{y|\tilde{x}}^2 = 4$ and a diagonal covariance matrix $\Sigma_{x|\tilde{x}}$ with elements $\sigma_{x_1|\tilde{x}}^2 = \sigma_{x_2|\tilde{x}}^2 = 0.01$. We then split this data set into two databases of size 250, and then remove y from the second database and remove the two covariates from the first database. To mimic the case of record linkage without duplicate detection we arrange the two databases so that they share 50 entities without duplications within each databases.

References

1. Christen, P.: Data Matching: Concepts and Techniques for Record Linkage, Entity Resolution, and Duplicate Detection. Springer, Heidelberg (2012). https://doi.org/10.1007/978-3-642-31164-2
2. Copas, J., Hilton, F.: Record linkage: statistical models for matching computer records. J. R. Stat. Soc. A **153**, 287–320 (1990)
3. De Blasi, P., Favaro, S., Lijoi, A., Mena, R., Prunster, I., Ruggiero, M.: Are gibbs-type priors the most natural generalization of the dirichlet process? IEEE Trans. Pattern Anal. Mach. Intell. **37**(2), 803–821 (2015)
4. Goldstein, H., Harron, K., Wade, A.: The analysis of record-linked data using multiple imputation with data value priors. Stat. Med. **31**, 3481–3493 (2012)
5. Gutman, R., Afendulis, C.C., Zaslavsky, A.M.: A Bayesian procedure for file linking to analyze end-of-life medical costs. J. Am. Stat. Assoc. **108**, 34–47 (2013)
6. Gutman, R., Sammartino, C., Green, T., Montague, B.: Error adjustments for file linking methods using encrypted unique client identifier (eUCI) with application to recently released prisoners who are HIV+. Stat. Med. **35**, 115–129 (2016)
7. Hof, M., Zwinderman, A.: Methods for analyzing data from probabilistic linkage strategies based on partially identifying variables. Stat. Med. **31**, 4231–4242 (2012)
8. Kim, G., Chambers, R.: Regression analysis under incomplete linkage. Comput. Stat. Data Anal. **56**, 2756–2770 (2012)
9. Lahiri, P., Larsen, M.D.: Regression analysis with linked data. J. Am. Stat. Assoc. **100**, 222–230 (2005)
10. Liseo, B., Tancredi, A.: Bayesian estimation of population size via linkage of multivariate normal data sets. J. Off. Stat. **27**, 491–505 (2011)
11. MacEachern, S.N.: Estimating normal means with a conjugate style Dirichlet process prior. Commun. Stat.-Simul. Comput. **23**, 727–741 (1994)
12. Neal, R.M.: Markov chain sampling methods for Dirichlet process mixture models. J. Comput. Graph. Stat. **9**, 249–265 (2000)
13. Pitman, J.: Combinatiorial Stochastic Processes. Ecole d'Eté de Probabilités de Saint-Flour XXXII. LNM, vol. 1875. Springer, Berlin (2006). https://doi.org/10.1007/b11601500
14. Sadinle, M.: Detecting duplicates in a homicide registry using a Bayesian partitioning approach. Ann. Appl. Stat. **8**, 2404–2434 (2014)
15. Steorts, R.C.: Entity resolution with empirically motivated priors. Bayesian Anal. **10**, 849–875 (2015)
16. Steorts, R.C., Hall, R., Fienberg, S.E.: SMERED: a Bayesian approach to graphical record linkage and de-duplication. J. Mach. Learn. Res. **33**, 922–930 (2014)
17. Steorts, R.C., Hall, R., Fienberg, S.E.: A Bayesian approach to graphical record linkage and de-duplication. J. Am. Stat. Soc. (2016)
18. Tancredi, A., Liseo, B.: A hierarchical Bayesian approach to record linkage and population size problems. Ann. Appl. Stat. **5**, 1553–1585 (2011)
19. Yamato, H., Shibuya, M.: Moments of some statistics of pitman sampling formula. Bull. Inf. Cybern. **32**, 1–10 (2000)

Probabilistic Blocking
with an Application to the Syrian Conflict

Rebecca C. Steorts[1]([⊠]) and Anshumali Shrivastava[2]

[1] Department of Statistical Science, Affiliated Faculty, Computer Science,
Biostatistics and Bioinformatics, the Information Initiative at Duke (iiD),
and the Social Science Research Institute (SSRI), Duke University, Durham, USA
beka@stat.duke.edu
[2] Department of Computer Science, Rice University, Houston, USA
anshumali@rice.edu

Abstract. Entity resolution seeks to merge databases as to remove duplicate entries where unique identifiers are typically unknown. We review modern blocking approaches for entity resolution, focusing on those based upon locality sensitive hashing (LSH). First, we introduce k-means locality sensitive hashing (KLSH), which is based upon the information retrieval literature and clusters similar records into blocks using a vector-space representation and projections. Second, we introduce a subquadratic variant of LSH to the literature, known as Densified One Permutation Hashing (DOPH). Third, we propose a weighted variant of DOPH. We illustrate each method on an application to a subset of the ongoing Syrian conflict, giving a discussion of each method.

1 Introduction

A commonly encountered problem in statistics, computer science, and machine learning is merging noisy data sets that contain duplicated entities, which is known as entity resolution (record linkage or de-duplication). Entity resolution tasks are intrinsically difficult because they are quadratic in computational complexity. In addition, for such tasks to be accurate, one often seeks models that are robust to model-misspecification and also have low error rates (precision and recall). These criteria are both difficult to satisfy, and have been at the core of entity resolution research [3,8,9,26].

One way of approaching the computational complexity barrier is by partitioning records into "blocks" and treating records in different blocks as non-co-referent *a priori* [3,8]. There are several techniques for constructing a blocking partition. The most basic method picks certain fields (e.g., geography, or gender and year of birth) and places records in the same block if and only if they agree on all such fields. This amounts to an *a priori* judgment that these fields are error-free. This is known as traditional blocking, and is a deterministic scheme. Unlike traditional blocking, probabilistic schemes such as locality sensitive hashing (LSH) use all the fields of a record, and can be adjusted to ensure that blocks

J. Domingo-Ferrer and F. Montes (Eds.): PSD 2018, LNCS 11126, pp. 314–327, 2018.
https://doi.org/10.1007/978-3-319-99771-1_21

are manageably small. For example, [25] introduced data structures for sorting and fast approximate nearest-neighbor look-up within blocks produced by LSH.

This approach is fast and has high recall (true positive rate), but suffers from low precision (too many false positives). In addition, this approach is called *private* if, after the blocking is performed, all candidate records pairs are compared and classified into matches/non-matches using computationally intensive "private" comparison and classification techniques, e.g., see [4].

LSH has been recently proposed as a way of blocking for entity resolution, where one place similar records into bins or blocks. LSH methods are defined by a type of similarity and a type of dimension reduction [1]. Recently, [24] proposed clustering-based blocking schemes — k-means locality sensitive hashing (KLSH), which is based upon the information retrieval literature and clusters similar records into blocks using a vector-space representation and projections. (While KLSH had been used before within the information retrieval literature, this is the first known instance of its application to entity resolution [13]). [24] showed that KLSH gave improvements over popular methods in the literature such as traditional blocking, canopies [12], and k-nearest neighbors clustering. In addition, [21] showed that minwise hashing based approaches are superior to random projection based approaches when the data is very sparse and feature poor. Furthermore, computational improvements can be gained by using the recently proposed densification scheme known as densified one permutation hashing (DOPH) [21,22]. Specifically, the authors proposed an efficient substitute for minwise hashing, which only requires one permutation (or one hash function) for generating many different hash values needed for indexing. In short, the algorithm is linear (or constant) in the tuning parameters, making this algorithm very computationally efficient.

In this paper, we review traditional blocking methods that are deterministic, and describe why such methods are not practical. We then review scalable LSH methods for blocking. Specifically, we give two recent approaches an methods from above that are scalable to large entity resolution data sets – KLSH and DOPH. Since both methods are known to work well on toy examples, we illustrate both algorithms on a real data set taken from a subset of the Syrian conflict, which is likely to be more realistic of industrial sized data sets. We illustrate evaluation metrics of all methods and the computational run time.

2   Blocking Methods

Blocking is a computational tool used in entity resolution that allows one to place similar records into blocks or partitions using either a deterministic or probabilistic mechanism. We first review traditional blocking methods, and then review probabilistic blocking methods. We propose two probabilistic blocking methods for large scale blocking methods based upon LSH.

2.1   Traditional Blocking

Traditional blocking requires domain knowledge to pick out highly reliable, if not error-free, fields for blocking. While traditional blocking is intuitive and easy to implement, it has at least four drawbacks. The first one is that the resulting blocks may still be so large that linkage within them is computationally impractical. The second is that because blocks *only* consider selected fields, much time may be wasted comparing records which happen to agree on those fields but are otherwise radically different. The third is due to the fact that traditional blocking methods are by nature deterministic, and thus, must be changed for each application at hand. The fourth is that a deterministic method cannot be guaranteed to be private. Given that traditional blocking is impractical for many reasons, we refer readers to [24], and we focus instead on probabilistic types of blocking, namely variants LSH.

2.2   Variants of Locality Sensitive Hashing

In this section, we first provide terminology, known as shingling, that is essential for using LSH for blocking. Next, we describe how can one produce blocks using k-means LSH (KLSH). Then we introduce the notation of LSH, and the linear variant—Densified One Permutation Hashing (DOPH). Finally, we propose a weighted variant of DOPH (weighted DOPH).

2.3   Shingling

In entity resolution tasks, each record can be represented as a string of textual information. It is often convenient to represent each record instead by a "bag", "shingle" (or "multi-set") of length-k contiguous sub-strings that it contains. In this paper, we use a k-shingle (k-gram or k-token) based approach to transform the records, and our representation of each record is a set, which consists of all the k-contiguous characters occurring in record string.

As an illustration, for the record BAKER, TED, we separate it into a 2-gram representation. The resulting set is the following:

$$BA, AK, KE, ER, ER, TE, ED.$$

For example, consider Sammy, Smith, whose 2-gram set representation is

$$SA, AM, MM, MY, MS, SM, MI, IT, TH.$$

We now have two records that have been transformed into a 2-gram representation. Thus, for every record (string) we obtain a set $\subset \mathcal{U}$, where the universe $\mathcal{U}$ is the set of all possible k-contiguous characters.

2.4   KLSH

We explore a simple random projection method, KLSH, where the similarity between records is measured using the inner product of a bag-of-shingled vectors that are weighted by their inverse document frequency. We first construct a k-shingle of a record by replacing the record by a bag (or multi-set) of length-k contiguous sub-strings that it contains. After the shingles are created, the dimensionality of the bag-of-shingled vectors is then reduced using random projections and by clustering the low-dimensional projected vectors via the k-means algorithm. That is, the mean number of records per cluster is controlled by n/c, where n is the total number of records and c is the number of block-clusters.

2.5   Locality Sensitive Hashing

We now turn to LSH, which is used in computer science and database engineering as a way of rapidly finding approximate nearest neighbors [5,10]. Specifically, the variant of LSH that we utilize is scalable to large databases, and allows for similarity based sampling of entities in a subquadratic amount of time.

In LSH, a hash function is defined as $y = h(x)$, where y is the *hash code* and $h(\cdot)$ the *hash function*. A *hash table* is a data structure that is composed of *buckets* (not to be confused with blocks), each of which is indexed by a *hash code*. Each reference item (record) x is placed into a bucket $h(x)$.

More precisely, LSH is a family of function that map vectors to a discrete set, namely, $h : \mathbb{R}^D \rightarrow \{1, 2, \cdots, M\}$, where M is in finite range. Given this family of functions, similar points (records) are likely to have the same hash value compared to dissimilar points (records). The notion of similarity is specified by comparing two vectors of points (records), x and y. We will denote a general notion of similarity by $\mathrm{SIM}(x, y)$. In this paper, we only require a relaxed version LSH, and we define this below. For a complete review of LSH, we refer to [19]. Formally, a LSH is defined by the following definition below:

Definition 1 (*Locality Sensitive Hashing (LSH)*). Let x_1, x_2, y_1, $y_2 \in \mathbb{R}^D$ and suppose h is chosen uniformly from a family $\mathcal{H}$. Given a similarity metric, $\mathrm{SIM}(x, y)$, $\mathcal{H}$ is locality sensitive if $\mathrm{SIM}(x_1, x_2) \geq Sim(y_2, y_3)$ then $Pr_{\mathcal{H}}(h(x_1) = h(x_2)) \geq Pr_{\mathcal{H}}(h(y_1) = h(y_2))$, where $Pr_{\mathcal{H}}$ is the probability over the uniform sampling of h.

2.5.1   Minhashing

One of the most popular forms of LSH is minhashing [1], which has two key properties—a type of similarity and a type of dimension reduction. The type of similarity used is the Jaccard similarity and the type of dimension reduction is known as the minwise hash, which we now define.

Let $\{0, 1\}^D$ denote the set of all binary D dimensional vectors, while $\mathbb{R}^D$ refers to the set of all D dimensional vectors (of records). Note that records can be represented as a binary vector (or set) representation via a shingling representation More specifically, given two record sets (or equivalently binary vectors) $x, y \in \{0,1\}^D$, the Jaccard similarity between $x, y \in \{0,1\}^D$ is $\mathcal{J} = \dfrac{|x \cap y|}{|x \cup y|}$, where $|\cdot|$ is the cardinality of the set.

More specifically, the minwise hashing family applies a random permutation π, on the given set S, and stores only the minimum value after the permutation mapping, known as the *minhash*. Formally, the minhash is defined as $h_\pi^{min}(S) = \min(\pi(S))$, where $h(\cdot)$ is a hash function.

Given two sets S_1 and S_2, it can be easily shown that

$$Pr_\pi(h_\pi^{min}(S_1) = h_\pi^{min}(S_2)) = \frac{|S_1 \cap S_2|}{|S_1 \cup S_2|}, \tag{1}$$

where the probability is over uniform sampling of π. It follows from Eq. 1 that minhashing is a LSH family for the Jaccard similarity.[1]

2.6 DOPH

In this section, we introduce the linear variant of LSH, known as DOPH. Let K be the number of hash functions and let L be the number of hash tables. A (K, L) parameterized blocking scheme requires $K \times L$ hash computations per record. For a single record, this requires storing and processing hundreds (or even thousands) of very large permutations. This in turn requires hundreds or thousands of passes over each record. Thus, traditional minwise hashing is prohibitively expensive for large or moderately sized data sets. In order to cross-validate the optimal (K, L) tuning parameters, we need multiple independent runs of the (K, L) parameterized blocking scheme. This expensive computation is a major computational concern. To avoid this computational issue, we can use *one permutation* of the hash function, where $k = K \times L$ minhashes are made in one single pass over the data [21,22].

Due to sparsity of data vectors (from shingling), empty blocks (in the hash tables) are possible and destroy LSH's essential property [19]. To restore this, we rotate the values of non-empty buckets and assign a number to each of the empty buckets. Our KL hashed values are simply the final assigned values in each of the KL buckets. The final values were shown to satisfy Eq. 1, for any S_1, S_2, as shown in [21,22].

2.7 Weighted DOPH

Minhashing, however, only uses the binary information and ignores the weights (or values) of the components, which is important for entity resolution problems due to the unbalanced nature of the data (small amount of duplicate records). This is the reason why we observe slightly better performance for synthetic data

---

[1] In this paper, we utilize a shingling based approach, and thus, our representation of each record is likely to be very sparse. Moreover, [23] showed that minhashing based approaches are superior compared to random projection based approaches for very sparse data sets.

of LSH methods used in [24], one of which is based upon random projections. To explore this more broadly, we examine the power of minwise hashing for entity resolution, a situation where the data is quite unbalanced, while simultaneously utilizing the weighting of various components.

Suppose now x, y are non-negative vectors. For our problem, we are only interested in non-negative vectors because shingle based representations are always non-negative. We utilize the generalization of Jaccard similarity for real valued vectors in $\mathbb{R}^D$, Unlike the minhash, this variant is sensitive to the weights of the components, and is defined as

$$J_w = \frac{\sum_i \min\{x_i, y_i\}}{\sum_i \max\{x_i, y_i\}} = 1 - \frac{\|x - y\|_1}{\sum_i \max\{x_i, y_i\}}, \tag{2}$$

where $\| \cdot \|_1$ represents the ℓ_1 norm. Consistent weighted sampling [2,6,7,11] is used for hashing the weighted Jaccard similarity J_w. In our application to the subset of the Syrian dataset, we find minhash and weighted minhash give similar error rates, which can be seen in Sect. 4.

With DOPH, the traditional minwise hashing scheme is linear or constant in the tuning parameters. For the weighted version of minhashing, we propose a different way of generating hash values for weighted Jaccard similarity, similar to that of [2,6]. As a result, we obtain the fast and practical one pass hashing scheme for generating many different hash values with weights, analogous to DOPH for the unweighted case. Overall, we require only one scan of the record and only one permutation.

Given any two vectors $x, y \in \mathbb{R}^D$ as the shingling representation, we seek hash functions $h(\cdot)$, such that the collision probability between two hash functions is small. More specifically, this means that

$$Pr(h(x) = h(y)) = \frac{\sum_i \min\{x_i, y_i\}}{\sum_i \max\{x_i, y_i\}}. \tag{3}$$

Let δ be a quantity such that all components of any vector $x_i = I_i^x \delta$ for some integer I_i^x.[2] Let the maximum possible component x_i for any record be x and let M be an integer such that $x_i = M\delta$. Thus, δ and M always exist for finitely bounded data sets over floating points.

Consider the transformation $T : \mathbb{R}^D \to \{0,1\}^{M \times D}$, where for $T(x)$ we expand each component $x_i = I\delta$ to M dimensions and with the first I dimensions have value 1 and the rest value 0.

Observe that for vectors x and y, $T(x)$ and $T(y)$ are binary vectors and

$$\frac{|T(x) \cap T(y)|}{|T(x) \cup T(y)|} = \frac{\sum_i \min\{I_i^x, I_i^y\}}{\sum_i \max\{I_i^x, I_i^y\}}$$
$$= \frac{\sum_i \min\{I_i^x, I_i^y\}\delta}{\sum_i \max\{I_i^x, I_i^y\}\delta} = \frac{\sum_i \min\{x_i, y_i\}}{\sum_i \max\{x_i, y_i\}} \tag{4}$$

---

[2] The assumption holds when dealing with floating point numbers for small enough δ.

In other words, the usual resemblance (or Jaccard similarity) between the transformed $T(x)$ and $T(y)$ is precisely the weighted Jaccard similarity between x and y that we are interested in. Thus, we can simply use the DOPH method of [21,22] on $T(x)$ to get an efficient LSH scheme for weighted Jaccard similarity defined by Eq. 4. The complexity here is $O(KL + \sum_i I_i)$ for generating k hash values, a factor improvement over $O(k \sum_i I_i)$ without the densified scheme.

Often I_i is quite large (when shingling) and $\sum_i I_i$ is large as well. When $\sum_i I_i$ is large, [6] give simple and accurate approximate hashes for weighted Jaccard similarity. They divide all components x_i by a reasonably big constant so that $x_i \leq 1$ for all records x. After this normalization, since $x_i \geq 0$, for every x, we generate another bag of word x_S by sampling each x_i with probability $x_i \leq 1$. Then x_S is a set (or binary vector) and for any two records x and y, the resemblance between x_S and y_S sampled in this manner is a very accurate approximation of the weighted Jaccard similarity between x and y. After applying the DOPH scheme to the shingled records, we generate k different hash values of each record in time $O(KL + d)$, where d is the number of shingles contained in each record. This is a vast improvement over $O(KL + \sum_i I_i)$. Algorithm 1 summarizes our method for generating k different minhashes needed for blocking.

Algorithm 1. Fast KL hashes

Data: record x,
Result: KL hash values for blocking
$x_S = \phi$;
forall the $x_i > 0$ **do**
$\quad |\quad x_S \cup i$ with probability proportional to x_i;
end
return KL densified one permutation hashes (DOPH) of x_S

---

3 Evaluation Metrics

We evaluate each of our hashing methods below using recall and reduction ratio (RR). The recall measures how many of the actual true matching record pairs have been correctly classified as matches. There are four possible classifications. First, record pairs can be linked under both the truth and under the estimate, which we refer to as *correct links* (CL). Second, record pairs can be linked under the truth but *not* linked under the estimate, which are called *false negatives* (FN). Third, record pairs can be *not* linked under the truth but linked under the estimate, which are called *false positives* (FP). Fourth and finally, record pairs can be *not* linked under the truth and also *not* linked under the estimate, which we refer to as *correct non-links* (CNL). The vast majority of record pairs are classified as correct non-links in most practical settings. Then the true number links is CL + FN, while the estimated number of links is CL + FP. The usual definitions of false negative rate and false positive rate are

$$\text{FNR} = \frac{\text{FN}}{\text{CL+FN}}, \qquad \text{FDR} = \frac{\text{FP}}{\text{CL+FP}},$$

where by convention we take FDR = 0 if its numerator and denominator are both zero, i.e., if there are no estimated links. The recall is defined to be

$$\text{recall} = 1 - \text{FNR}.$$

The precision is defined to be[3]

$$\text{precision} = 1 - \text{FDR}.$$

The reduction ratio (RR) is defined as

$$\text{RR} = 1 - \frac{s_\text{M} + s_\text{N}}{n_\text{M} + n_\text{N}},$$

where n_M and n_N are the total of matched and non-matched records and the number of true matched and true non-matched candidate record pairs generated by an indexing technique is denoted with $s_\text{M} + s_\text{N} \leq n_\text{M} + n_\text{N}$. The RR provides information about how many candidate record pairs were generated by an indexing technique compared to all possible record pairs, without assessing the quality of these candidate record pairs. We also evaluate the methods using the precision, where precision calculates the proportion of how many of the classified matches (true positives + false positives) have been correctly classified as true matches (true positives). It thus measures how precise a classifier is in classifying true matches. This measure is useful if we wish to use hashing based approaches for entity resolution, however, as we show, we are not able to achieve both a high precision and recall (see Sect. 4). It's most important for a blocking method to have a high RR and recall because the entity resolution task can correct for potential problems that are represented with a low precision. On the other hand, the error summarized by the recall cannot be improved by an entity resolution task.

4 Application

We test the two blocking approaches on a subset of the ongoing Syrian conflict, where via the Human Rights Data Analysis Group (HRDAG), we have access to four databases from the Syrian conflict which cover roughly the same period, namely March 2011 – April 2014. In this section, we apply LSH based methods to the subset of the Syrian dataset. (We do not consider any methods in the literature that performed worse than KLSH in terms of RR and recall. See [20] for further details and experiments on traditional and other probabilistic blocking schemes.)

---

[3] Note that the precision for a blocking procedure is not expected to be high since we are only placing similar pair in the same block (not fully running an entity resolution procedure or de-duplication procedure, which would try and maximize both the recall and the precision).

4.1   The Syrian Data

The four data sources consist of the Violation Documentation Centre (VDC), Syrian Center for Statistics and Research (CSR-SY), Syrian Network for Human Rights (SNHR), and Syria Shuhada website (SS). Each database lists a different number of recorded victims killed in the Syrian conflict, along with available identifying information including full Arabic name, date of death, death location, and gender. Since the above information is collected indirectly, such as through friends and religious leaders, or traditional media resources, it comes with many challenges. For example, the data set contains natural biases, spelling errors, missing values in addition to duplication of those killed in the conflict. The ambiguities in Arabic names make the situation more challenging as there can be a large textual difference between the full and short names in Arabic. Such ambiguities and lack of additional information make blocking on this data set considerably challenging [18]. Owing to the significance of the problem, HRDAG has provided labels for a large subset of the data set. More specifically, five different human experts from the HRDAG manually reviewed pairs of records in the four data sets, classifying them as matches if referred to the same entity and non-matches otherwise. (More information regarding the Syrian data set can be found in Appendix A).

4.2   KLSH Applied to Syrian Data

We first apply KLSH to the subset of the Syrian data set, which greatly contrasts the empirical studies shown in [24]. The parameters to be set for KLSH are the number of random projections (p) and the number of clusters to output (k). Using this k-means approach to blocking, the mean number of records within a cluster can be fixed.

Figure 1 (left panel) displays the results of KLSH clustering applied on the subset of the Syrian database, where we plot the recall versus the total number of blocks. We set the number of random projections to be $p = 20$ and allow the shingles to vary from $k = 1, 2, 3, 4$. This figure shows that a 1-shingle always achieves the highest recall. We notice that using a 1-shingle, a block size of 100, the recall is 0.60, meaning that 40% of the time the same two records are split across blocks.

4.3   DOPH Applied to Syrian Data

Due to the poor results achieved by KLSH for the Syrian data set, we apply minhashing using both the unweighted and weighted DOPH algorithm to the full Syrian database using shingles 2—5. We illustrate that regardless of the number of shingles used, the recall and RR are close to 1 as illustrated in Fig. 2. Furthermore, using unweighted DOPH, we see that a shingle of three overall is most stable in having a recall and RR close to 0.99 as illustrated in Fig. 3. Using weighted DOPH, we see that a shingle of two or three overall is most stable in having a recall and RR close to 0.99. In terms of computational run time, we

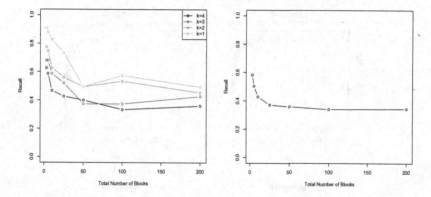

Fig. 1. Left: KLSH on subset of Syria database (20,000 records) using p = 20. Right: KLSH on entire Syrian database using p = 20 and k = 1. One can see that the recall is very poor compared with previous approaches applied using KLSH, and thus, the method is not suitable for blocking on this particular data set.

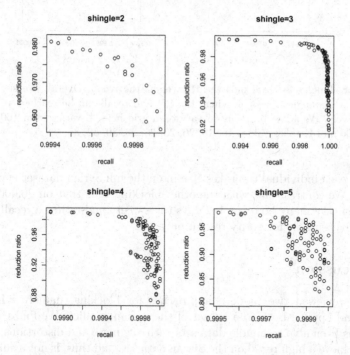

Fig. 2. For shingles 2–5, we plot the RR versus the recall. Overall, we see the best behavior for a shingle of 3, where the RR and recall can be reached at 0.98 and 1, respectively. We allow L and K to vary on a grid here. L varies from 100–1000 by steps of 100 and K takes values 15, 18, 20, 23, 25, 28, 30, 32, 35.

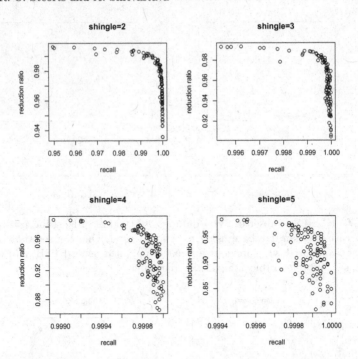

Fig. 3. For shingles 2–5, we plot the RR versus the recall. Overall, we see the best behavior for a shingle of 2 or 3, where the RR and recall can be reached at 0.98 and 1, respectively. We allow L and K to vary on a grid here. L varies from 100–1000 by steps of 100 and K takes values 15, 18, 20, 23, 25, 28, 30, 32, 35.

note that each individual run takes 10 min on the full Syrian dataset and 100 GB of RAM. We contrast this with the other blocking runs that on 20,000 records from Syria take many hours or 1–2 days (or a week) and return a recall and RR that is unacceptable for entity resolution tasks.

5   Discussion

We have reviewed two modern approaches for blocking, namely KLSH and DOPH and applied both to a subset of the Syrian conflict. We find that while KLSH has been able to handle data sets with low noise and distortions, it is not able to achieve a high recall on the Syrian data set, and thus, is not a suitable for entity resolution for data sets that have similar levels of noise as in the Syrian data set. On the other hand, DOPH performs well given the sparsity and noisy levels on the observed data at hand, and appears to be an excellent, stable, and scalable choice for the blocking step in an entity resolution task. This merits further investigations with scalable variants of LSH for entity resolution tasks.

Acknowledgments. We would like to thank HRDAG for providing the data and for helpful conversations. We would also like to thank Stephen E. Fienberg and Lars Vilhuber for making this collaboration possible. Steorts's work is supported by NSF-1652431 and NSF-1534412. Shrivastava's work is supported by NSF-1652131 and NSF-1718478. This work is representative of the author's alone and not of the funding organizations.

A   Syrian Data Set

In this section, we provide a more detailed description about the Syrian data set. As already mentioned, via collaboration with the Human Rights Data Analysis Group (HRDAG), we have access to four databases. They come from the Violation Documentation Centre (VDC), Syrian Center for Statistics and Research (CSR-SY), Syrian Network for Human Rights (SNHR), and Syria Shuhada website (SS). Each database lists each victim killed in the Syrian conflict, along with identifying information about each person (see [17] for further details).

Data collection by these organizations is carried out in a variety of ways. Three of the groups (VDC, CSR-SY, and SNHR) have trusted networks on the ground in Syria. These networks collect as much information as possible about the victims. For example, information is collected through direct community contacts. Sometimes information comes from a victim's friends or family members. Other times, information comes from religious leaders, hospitals, or morgue records. These networks also verify information collected via social and traditional media sources. The fourth source, SS, aggregates records from multiple other sources, including NGOs and social and traditional media sources (see http://syrianshuhada.com/ for information about specific sources).

These lists, despite being products of extremely careful, systematic data collection, are not probabilistic samples [14–16,18]. Thus, these lists cannot be assumed to represent the underlying population of all victims of conflict violence. Records collected by each source are subject to biases, stemming from a number of potential causes, including a group's relationship within a community, resource availability, and the current security situation.

A.1   Syrian Handmatched Data Set

We describe how HRDAG's training data on the Syrian data set was created, which we use in our paper.

First, all documented deaths recorded by any of the documentation groups were concatenated together into a single list. From this list, records were broadly grouped according to governorate and year. In other words, all killings recorded in Homs in 2011 were examined as a group, looking for records with similar names and dates.

Next, several experts review these "blocks", sometimes organized as pairs for comparison and other times organized as entire spreadsheets for review. These experts determine whether pairs or groups of records refer to the same individual victim or not. Pairs or groups of records determined to refer to the same individual are assigned to the same "match group." All of the records contributing to a single "match group" are then combined into a single record. This new single record is then again examined as a pair or group with other records, in an iterative process.

For example, two records with the same name, date, and location may be identified as referring to the same individual, and combined into a single record. In a second review process, it may be found that record also matches the name and location, but not date, of a third record. The third record may list a date one week later than the two initial records, but still be determined to refer to the same individual. In this second pass, information from this third record will also be included in the single combined record.

When records are combined, the most precise information available from each of the individual records is kept. If some records contain contradictory information (for example, if records A and B record the victim as age 19 and record C records age 20) the most frequently reported information is used (in this case, age 19). If the same number of records report each piece of contradictory information, a value from the contradictory set is randomly selected.

Three of the experts are native Arabic speakers; they review records with the original Arabic content. Two of the experts review records translated into English. These five experts review overlapping sets of records, meaning that some records are evaluated by two, three, four, or all five of the experts. This makes it possible to check the consistency of the reviewers, to ensure that they are each reaching comparable decisions regarding whether two (or more) records refer to the same individual or not.

After an initial round of clustering, subsets of these combined records were then re-examined to identify previously missed groups of records that refer to the same individual, particularly across years (e.g., records with dates of death 2011/12/31 and 2012/01/01 might refer to the same individual) and governorates (e.g., records with neighboring locations of death might refer to the same individual).

References

1. Broder, A.Z.: On the resemblance and containment of documents. In: Proceedings of 1997 Compression and Complexity of Sequences. IEEE, pp. 21–29 (1997)
2. Charikar, M., Chen, K., Farach-Colton, M.: Finding frequent items in data streams. In: Widmayer, P., Eidenbenz, S., Triguero, F., Morales, R., Conejo, R., Hennessy, M. (eds.) ICALP 2002. LNCS, vol. 2380, pp. 693–703. Springer, Heidelberg (2002). https://doi.org/10.1007/3-540-45465-9_59
3. Christen, P.: A survey of indexing techniques for scalable record linkage and deduplication. IEEE Trans. Knowl. Data Eng. **24**, 1537–1555 (2012)
4. Christen, P., Gayler, R., Hawking, D.: Similarity-aware indexing for real-time entity resolution. In: Proceedings of the 18th ACM Conference on Information and Knowledge Management, pp. 1565–1568 (2009)
5. Gionis, A., Indyk, P., Motwani, R., et al.: Similarity search in high dimensions via hashing. In: Very Large Data Bases (VLDB), vol. 99, pp. 518–529 (1999)
6. Gollapudi, S., Panigrahy, R.: Exploiting asymmetry in hierarchical topic extraction. In: Proceedings of the 15th ACM International Conference on Information and Knowledge Management. ACM, pp. 475–482 (2006)

7. Haeupler, B., Manasse, M., Talwar, K.: Consistent weighted sampling made fast, small, and easy. Technical report (2014). arXiv:1410.4266

8. Herzog, T., Scheuren, F., Winkler, W.: Data Quality and Record Linkage Techniques. Springer, New York (2007). https://doi.org/10.1007/0-387-69505-2

9. Herzog, T., Scheuren, F., Winkler, W.: Record linkage. Wiley Interdisc. Rev.: Comput. Stat. **2** (2010). https://doi.org/10.1002/wics.108

10. Indyk, P., Motwani, R.: Approximate nearest neighbors: towards removing the curse of dimensionality. In: STOC, Dallas, TX, pp. 604–613 (1998)

11. Ioffe, S.: Improved consistent sampling, weighted minhash and L1 sketching. In: ICDM, Sydney, AU, pp. 246–255 (2010)

12. McCallum, A., Nigam, K., Ungar, L.H.: Efficient clustering of high-dimensional data sets with application to reference matching. In: Proceedings of the Sixth ACM SIGKDD International Conference on Knowledge Discovery and Data Mining, pp. 169–178. ACM (2000)

13. Paulevé, L., Jégou, H., Amsaleg, L.: Locality sensitive hashing: a comparison of hash function types and querying mechanisms. Pattern Recogn. Lett. **31**, 1348–1358 (2010)

14. Price, M., Ball, P.: The limits of observation for understanding mass violence. Can. J. Law Soc./Revue Canadienne Droit et Société **30**, 237–257 (2015a)

15. Price, M., Ball, P.: Selection bias and the statistical patterns of mortality in conflict. Stat. J. IAOS **31**, 263–272 (2015b)

16. Price, M., Gohdes, A., Ball, P.: Documents of war: understanding the Syrian conflict. Significance **12**, 14–19 (2015)

17. Price, M., Klingner, J., Qtiesh, A., Ball, P.: Updated statistical analysis of documentation of killings in the Syrian Arab Republic. United Nations Office of the UN High Commissioner for Human Rights (2013)

18. Price, M., Klingner, J., Qtiesh, A., Ball, P.: Updated statistical analysis of documentation of killings in the Syrian Arab Republic. United Nations Office of the UN High Commissioner for Human Rights (2014)

19. Rajaraman, A., Ullman, J.D.: Mining of Massive Datasets. Cambridge University Press, Cambridge (2012)

20. Sadosky, P., Shrivastava, A., Price, M., Steorts, R.C.: Blocking methods applied to casualty records from the syrian conflict (2015). arXiv preprint arXiv:1510.07714

21. Shrivastava, A., Li, P.: Densifying one permutation hashing via rotation for fast near neighbor search. In: Proceedings of The 31st International Conference on Machine Learning, pp. 557–565 (2014)

22. Shrivastava, A., Li, P.: Improved densification of one permutation hashing. In: Proceedings of the 30th Conference on Uncertainty in Artificial Intelligence (2014)

23. Shrivastava, A., Li, P.: In defense of minhash over simhash. In: Proceedings of the Seventeenth International Conference on Artificial Intelligence and Statistics, pp. 886–894 (2014)

24. Steorts, R.C., Ventura, S.L., Sadinle, M., Fienberg, S.E.: A comparison of blocking methods for record linkage. In: Domingo-Ferrer, J. (ed.) PSD 2014. LNCS, vol. 8744, pp. 253–268. Springer, Cham (2014). https://doi.org/10.1007/978-3-319-11257-2_20

25. Vatsalan, D., Christen, P., O'Keefe, C.M., Verykios, V.S.: An evaluation framework for privacy-preserving record linkage. J. Privacy Confidentiality **6**, 3 (2014)

26. Winkler, W.E.: Overview of record linkage and current research directions. Technical report, U.S. Bureau of the Census Statistical Research Division (2006)

Spatial and Mobility Data

SwapMob: Swapping Trajectories for Mobility Anonymization

Julián Salas[1]([✉]), David Megías[1], and Vicenç Torra[2]

[1] Internet Interdisciplinary Institute (IN3),
CYBERCAT-Center for Cybersecurity Research of Catalonia,
Universitat Oberta de Catalunya (UOC), Barcelona, Spain
{jsalaspi,dmegias}@uoc.edu
[2] School of Informatics, University of Skövde, Skövde, Sweden
vtorra@his.se

Abstract. Mobility data mining can improve decision making, from planning transports in metropolitan areas to localizing services in towns. However, unrestricted access to such data may reveal sensible locations and pose safety risks if the data is associated to a specific moving individual. This is one of the many reasons to consider trajectory anonymization.

Some anonymization methods rely on grouping individual registers on a database and publishing summaries in such a way that individual information is protected inside the group. Other approaches consist of adding noise, such as differential privacy, in a way that the presence of an individual cannot be inferred from the data.

In this paper, we present a perturbative anonymization method based on swapping segments for trajectory data (SwapMob). It preserves the aggregate information of the spatial database and at the same time, provides anonymity to the individuals.

We have performed tests on a set of GPS trajectories of 10,357 taxis during the period of Feb. 2 to Feb. 8, 2008, within Beijing. We show that home addresses and POIs of specific individuals cannot be inferred after anonymizing them with SwapMob, and remark that the aggregate mobility data is preserved without changes, such as the average length of trajectories or the number of cars and their directions on any given zone at a specific time.

1 Introduction

With the pervasive use of smartphones and the location techniques such as GPS, GSM and RFID, the opportunities to deliver content depending on current user location have increased. Location Based Services (LBS) provide considerable advantages such as allowing users to benefit from live location-based information for transportation, recommendations of places of interest, or even the opportunity to meet friends in nearby locations. Such location-based data can be useful also for intelligent transportation systems, in which vehicles may serve as sensors for collecting information about traffic jams, weather, and road conditions.

J. Domingo-Ferrer and F. Montes (Eds.): PSD 2018, LNCS 11126, pp. 331–346, 2018.
https://doi.org/10.1007/978-3-319-99771-1_22

However, revealing users' locations may have some privacy risks. If the data is linked to the real identities it may reveal personal preferences (e.g., sexual, political or religious orientation), or it may be used for inferring habits and know the time when a person is at home or away. To avoid such inconveniences, a variety of anonymization techniques have been developed to hide the identity of the user or her exact location, e.g., [26].

Moreover, as Giannotti et al. mention in [10], big data (in particular trajectory data) may be used to understand human behavior through the discovery of individual social profiles, by the analysis of collective behaviors, spreading epidemics, social contagion, and to study the evolution of sentiment and opinion; however, trusted networks and privacy-aware social mining must be pursued and methods for protection and anonymization for such data must be developed to enforce the data subjects' rights and promote their participation.

2   Related Work

Different solutions have been proposed for anonymizing trajectories in data publishing. Abul et al. [1], propose the (k, δ)-anonymity model, which consists on publishing a cylindrical volume of radius δ that contains the trajectory of at least k moving objects. Note that this idea is an extension of the concept of k-anonymity for databases [22].

Terrovitis and Mamoulis [27] consider a discrete spatial domain, e.g., spatial information is given in terms of addresses in a city map. Hence, the user trajectories are expressed as sequences of POIs. They present the use case of the RFID cards from the Octopus[1] company in Hong Kong, which collects the transaction history of its customers. The company may want to publish sequences of transactions by the same person as trajectories, for extracting movement and behavioral patterns. However, if a given user, Alice, uses her card to pay at different convenience stores that belong to the same chain (e.g., convenience stores), that company may reidentify Alice if her sequence of purchases is unique in the published trajectory database.

A similar approach in [18] is obtained by transforming sequences by adding, deleting, or substituting some points of the trajectory, while preserving also frequent sequential patterns [2] obtained by mining the anonymized data.

In [13,14], Hoh et al. discuss the use of mobility data for transportation planning and traffic monitoring applications to provide drivers with feedback on road and traffic conditions. For modelling the threats to privacy in such datasets, they assume that an adversary does not have information about which subset of samples belongs to a single user, however by using multi-target tracking algorithms [19] subsequent location samples may be linked to an individual that is periodically reporting his anonymized location information.

In [13] they consider the attack of deducing home locations of users by leveraging clustering heuristics used together with the decrease of speed reported by

---

[1] http://www.octopuscards.com/.

GPS sensors. Then, propose data suppression techniques by changing the sampling rate (e.g., from 1 min to 2, 4 and 10) for protecting from such inferences.

In [14], in order to prevent adversaries from tracking complete individual paths, they propose an algorithm that perturbs slightly the trajectories of different individuals (to make them closer) in such a way that the adversary may not be able to follow which segment of the path corresponds to which user by using multi-target tracking algorithms. This is done with a constraint on the Quality of Service, which is expressed as the mean location error between the actual and the observed locations. They argue that adequate levels of privacy can only be obtained if the density of users is sufficiently high.

This is closely related to [3] in which Mix Zones are introduced, these are spatial areas on which users' location is not accessible, hence when users are simultaneously present on a mix zone, their pseudonyms are changed. This procedure is performed to difficult the linkage of the incoming and outgoing path segments to the same specific user.

They design a model for location privacy protection that aims to preserve the advantages of location aware services while hiding their identities from the applications that receive the users' locations. The existence of a trusted middleware system (or sensing infrastructure) is assumed and the applications register their interest in a geographic space with the middleware, such space is called application zone. Examples of such application zones are hospitals, universities or supermarket complexes, in general it could be any open or closed space.

The regions in which applications cannot trace user movements are called mix zones, and the borders between a mix zone and an application zone are called boundary lines. Applications do not receive traceable user identities, they receive pseudonyms that allow communication between them. Such communication passes through the trusted intermediary and the pseudonyms of users change when they enter a mixed zone.

In order to measure location privacy, Beresford and Stajano [4] define the anonymity set as the group of people visiting the mix zone during the same time period. However, as the boundary and time when a user exits a mix zone is strongly correlated to the boundary and time when the user enters it, such information may be exploited by an attacker, therefore they use the information theoretic metric that Serjantov and Danezis [24] proposed for anonymous communications which considers the varying probabilities of users sending and receiving messages through a network of mix nodes.

This is modeled in [4] as a movement matrix in which they record the frequency of ingress and egress points to the mix zone at several times. Then, a bipartite weighted graph is defined in which vertices model ingress and egress pseudonyms and edge weights model the probability that two pseudonyms represent the same underlying person. Therefore, a maximal cost perfect matching of these graphs represents the most probable mapping among incoming and outgoing pseudonyms.

However, since the solution to many restricted matching problems (such as this one) is NP-hard [25], Beresford and Stajano [4] describe a method for achieving partial solutions.

An approach that does not consider middleware to obtain location privacy is proposed in Chap. 9 from [11]. It consists of a system with an untrusted server and clients communicating in a P2P network for privacy preserving trajectory collection. The aim of their data collection solution is to preserve anonymity in any set of data being stored, transmitted or collected in the system. This is achieved by means of k-anonymization and swapping. Briefly, the protocol consists of the clients recording their private trajectories, cloaking them among k similar trajectories and exchanging parts of those trajectories with other clients in the P2P network. However, the final step (the data reporting stage) clients send anonymous partial trajectories to the server, that have been generated in such a way that the server can filter all the synthetic trajectory data that has been generated for cloaking during the process, and recover the original trajectory.

One of the advantages of performing trajectory anonymization on the user side, as in [20], is that the anonymization process is no longer centralized. Thus data subjects gain control, transparency and more security for their data.

For a brief overview of privacy protection techniques and a discussion of k-anoymity and differential privacy models in different frameworks, cf. [21].

In [7], a differential privacy model for transit data publication is considered, using data from the Société de Transport de Montréal (STM). The data are modeled as sequential data in a prefix tree that represents all the sequences by grouping the sequences with the same prefix into the same branch. Their algorithm takes a raw sequential dataset D, a privacy budget ϵ, a user specified height of the prefix tree h and a location taxonomy tree T, and returns a sanitized dataset $\tilde{D}$ satisfying ϵ-differential privacy. For measuring utility, in the STM case, sanitized data are mainly used to perform two data mining tasks, count query and frequent sequential pattern mining [2].

Other ϵ-differentially private mechanism for publishing trajectories called SDD (Sampling Distance and Direction) can be found in [15]. They focus on ship trajectories with known starting point and terminal point. And consider that two trajectories T and T' with the same number of positions are adjacent if they differ at exactly one position excluding the starting point and the terminal point.

In [28], a differentially private algorithm for location privacy is proposed, following a discussion on the (in)applicability of differential privacy in a variety of settings, such as [6,16]. Their algorithm considers temporal correlations modeled as a Markov chain and proposes the "δ-location set" to include all probable locations (where the user might appear). The authors argue that, to protect the true location, it is enough to hide it in the δ-location set in which any pairs of locations are not distinguishable. However, they leave the problem of protecting the entire trace of released locations as future work.

In this paper, we present an anonymization method considering that the data are dynamic, the rate at which the information is collected is not constant, and the databases are being generated as the data is received.

3   Proposed Method: SwapMob

We propose a method for anonymization of mobility data by swapping trajectories, which works in a similar way as the mix zones but in a non-restricted space.

Our algorithm (SwapMob) simulates an online P2P system for exchanging segments of trajectories. That is, when two users are near they interchange their partial trajectories, see Sect. 3.1. In this way, all users' trajectories are mixed incrementally, and the moving users keep generating segments of trajectories that are being swapped. In the end, each trajectory retrieved is made of small segments of trajectories of different individuals, who have met during the day, as depicted in Fig. 1. Hence, the relation between data subjects and their data is obfuscated while keeping a precise aggregated data, such as the number of users in each place at each time and the locations that have been visited by different anonymous users.

We formalize our method after a brief explanation of previous definitions and assumptions.

3.1   Definitions

We assume that we have a database in which the i-th observation is a tuple (ID_i, lat_i, long_i, t_i) that consists of the individual's identifier (ID_i), the latitude (lat_i), longitude (long_i) and timestamp (t_i).

Then, the trajectory T_x of an individual x will consist of all the observations with identifier x ordered by their timestamps t_i. These can be represented as $T_x = (x_1, x_2, \ldots, x_m)$ if there are m observations for individual x.

We say that *two individuals meet* or their trajectories cross (on points x_i and y_j) if they have been co-located. We denote this by $x_i \approx y_j$. Note that being co-located depends on thresholds for proximity (χ) and time (τ), since the sampling rate of positions is not regular nor constant. Moreover two persons cannot be in the exact same place at the same time.

We define a *matching* as a maximal subset of pairs of elements of a set.

We denote by $Sw(T)$ the resulting trajectory after all swaps have been applied to T. Next, we define the following two primitives for our algorithm: *generate random matching* and *swap*.

1. *Swap:* Given two trajectories $T_x = (x_1, \ldots, x_i, x_{i+1}, \ldots)$ and $T_y = (y_1, \ldots, y_j, y_{j+1}, \ldots)$ that meet in points x_i and y_j, a swap of T_x with T_y at points x_i and y_j results in $Sw(T_x) = (y_1, \ldots, y_j, x_{i+1}, \ldots)$ and $Sw(T_y) = (x_1, \ldots, x_i, y_{j+1}, \ldots)$.

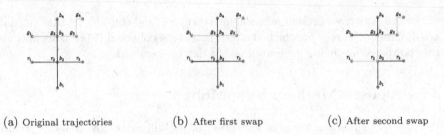

(a) Original trajectories   (b) After first swap   (c) After second swap

Fig. 1. Three trajectories before and after swapping (Color figure online)

2. *Generate random matching:*
   Given a set of elements $S = s_1, s_2, \ldots, s_m$, we generate a random matching by making pairs of the first $m/2$ with the following $m/2$ numbers, followed by a random permutation of all numbers m.

   Note that, in case that the number of elements m is odd, to generate a matching we must leave out one element and that all possible random matchings can be generated following our procedure.

Crossing Paths and Swapping. We propose a model such that two peers get in contact (meet) if they have been co-located on a similar timestamp depending on parameters of proximity χ and time τ.

Next, we simulate SwapMob protocol by swapping the users IDs when the users have passed close enough. We calculate the set of users that get in contact in a given time interval, and choose a random matching among them when they are even and a matching of all but one, when they are odd. Here, the swapping is carried out in a pairwise manner, but it could be done as a permutation such as in [4].

Note that changing pseudonyms (IDs) is equivalent to swapping the partial trajectories.

In Fig. 1, we present an example of three simple trajectories crossing T_r, T_g, T_b. We assume that they are moving from left to right and upwards, $T_r = (r_1, r_2, r_3)$, $T_g = (g_1, g_2, g_3, g_4)$ and $T_b = (b_1, b_2, b_3, b_4)$. Note that we are also assuming that the blue trajectory meets the red trajectory first ($b_2 \approx r_2$) and then the green trajectory ($b_3 \approx g_2$). In this tiny example, we can see how the iterative swaps preserve parts of the trajectory intact, but at the end each trajectory has parts of many others, such as the green one which ends having a segment of the blue trajectory, a segment of the red and a segment of its original trajectory $Sw(T_g) = (r_1, r_2, b_3, g_3, g_4)$.

3.2 SwapMob Anonymizer

We follow a similar architecture to the one in [13] in which a Trusted Third Party (TTP) knows the vehicles identities but can not access sensor information (such

---

Algorithm 1. Offline algorithm for swapping trajectories

Input: Trajectory Database. Thresholds for time τ and proximity χ.
Output: Swapped trajectories identifiers $Sw(T_i)$.
Partition the timestamps $t = \bigcup \tau_j$ in intervals of length τ
for *each pair of registers i, j in interval τ_j* **do**
    if $dist(l_i, l_j) < \chi$ **then**
        add i, j to close records list (possible swaps) S_{τ_j} at the given time
        interval.
    end
end
generate random matching with possible swaps in S_{τ_j}
order all swaps in $\bigcup S_{\tau_j}$ by timestamp
for *each pair $i \approx j$ in $\bigcup S_{\tau_j}$* **do**
    swap T_i with T_j
end
return *Swapped trajectories $Sw(T_i)$*

---

as position and speed); and a Service Provider (SP) knows the sensor measures but not the identities. Further, the SP calculates which records are close to each other without knowing to which individual they belong and communicates them to the TTP (in this case SwapMob anonymizer) such that it can swap their identities without knowing at which location they were.

This is achieved in the following way (See Fig. 2):

1. Users communicate with SwapMob, sending their sensor data (M) encrypted with the public key (K_{SP}) of SP. SwapMob keeps the number of register (i), which user has sent it (u_i), its current pseudonym (ID_i), the timestamp (t_i) and the encrypted sensor data $E(M_i, K_{SP})$, which includes their encrypted location (l_i).
2. SwapMob sends the vector ($i, t_i, E(M_i, K_{SP})$) to the SP, who decrypts $E(M_i, K_{SP})$ and keeps a buffer of data on interval τ_j that contains all timestamps between timestamp t_j and t_{j+1} and has length τ, that is $\tau_j = \{t : t_j < t < t_{j+1}\}$.
3. SP sends the set S_{τ_j} of registers that were at distance less than the predefined threshold χ during the interval of time τ_j back to SwapMob, more formally $S_{\tau_j} = \{i, i' : d(l_i, l_{i'}) < \chi \text{ and } t_i, t_{i'} \in \tau_j\}$. SwapMob calculates the swaps and stores the users and swapped IDs list, that is, for every record i SwapMob keeps the corresponding swapped id $Sw(ID_i)$ and the user (u_i) to which such pseudonym corresponds.
4. Finally, every given period of time which could be daily, weekly or monthly, SwapMob reports the list of ($i, Sw(ID_i)$) to SP.

The authentication data integrity of the communications can be guaranteed with a hash-based message authentication code.

In this way, SP obtains the measures of all sensors M in real-time (Step 2), and at the end of the day also gets the anonymized trajectories of the users

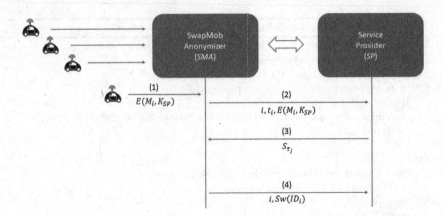

Fig. 2. Architecture of our system

that generated them (Step 4). Even, though SP knows which records belong to S_{τ_j} (Step 3), SP does not know to which other record they have been swapped during period τ_j, and by the iterative swappings it gets even harder to associate them to a specific user.

At the same time, SwapMob only knows the users, the timestamps at which they have crossed, and the reported trajectories are already anonymized by SwapMob (Step 4).

Our system, can be applied for the use case proposed in [3], by defining a set of swap zones (similar to the mix zones) and adding the restriction that the swapping cannot be performed outside such places. Then, the spatio-temporal trajectories of users between such swap zones could be monitored in an anonymous and precise way.

However, there will still be some differences. Namely, the swap zone that we consider is the entire application zone, whereas in [3] a user entering a mix zone can be distinguished from another user emerging from the same zone if the size of the mix zone is too large.

This same argument justifies that the distance and time parameters, χ and τ must not be too large either in our algorithm, otherwise swapping could not be credible.

3.3   Protecting Against Reidentification

It is well known that de-identification does not necessarily means anonymization. The same attributes that are used for extracting knowledge, may be used for pointing to a specific individual, and uniquely relating his/her data to her real identity.

Other notions of privacy are defined depending on the context, which may be of statistical databases [9], networks [33], or geo-located data.

By identifying the POIs of an individual, it is possible to infer his habits (e.g., does sport, travels a lot), the locations that he visits frequently (may be related to political or religious beliefs) or even related to health (clinics, hospitals). This may also be used to infer his schedule, predict his future locations, and learn his past locations and possibly his personal relations by observing frequent or periodic co-location. Moreover, such habits and locations can be easily used to reidentify the individuals behind the data. As it has been proven on previous anonymity studies on anonymity of home/work location.

Regarding this topic, Golle and Partridge studied in [12] workers who revealed their home and work location with noise or rounding on the order of a city block, a kilometer or tens of kilometers (census block, census tract, or county) and showed that the sizes of the anonymity set were respectively 1, 21 and 34,980. That is, when the data granularity was on the order of a census block, the individuals were uniquely identifiable, and for granularities on the order of census track or county, they were protected within sets of size 21 or 34,980. In [31], Zang and Bolot inferred the top N locations of a user from call records and correlated such information with publicly-available side information such as census data. Then, they showed that the top 2 locations likely correspond to home and work location and that the anonymity sets are drastically reduced if an attacker infers them.

Therefore, for protecting the individuals against reidentification, is crucial to protect their home addresses and POIs, to provide them with minimum guarantees of keeping them anonymous. Swapped data may not allow for following a specific individual and his whereabouts, and thus, this will not permit personalization or individual classification, which are ways of protecting their privacy.

A different approach regarding the possibility of reidentification and the (im)possibility of protection, is in [17], where they measure the uniqueness of human mobility traces depending on their resolution and the available outside information, assuming that an adversary knows p random spatio-temporal points. Then, they coarsen such data spatially and temporally to find a formula for uniqueness depending on such parameters.

We argue that SwapMob preserves anonymity by dissociating the segments of trajectories from the subject that generated them.

An attacker may know several spatio-temporal points of an individual that uniquely identify him. However, to link a register in the anonymized database to such an individual, the points known by the attacker must belong to the same trajectory after swapping. In most cases, the attacker will not learn the entire trajectory information since the published trajectory is made of segments from many different individuals. Of course, when publishing the trajectories, it should be noted that they have been generated by SwapMob, and the anonymization may be reversed if the SwapMob Anonymizer and the Service Provider (see Fig. 2) share their information for guaranteeing accountability.

3.4   Utility of Swapped Data

In this paper we are assuming that the interest of using data anonymized by SwapMob is for making mobility maps and predictions that may be useful for intelligent transportation systems and for planning in a city. As Hoh and

Gruteser proposed in [14], pre-specified vehicles could periodically send their locations, speeds, road temperatures, windshield wiper status and other information to the traffic monitoring facility. These statistics can provide information on the traffic jams, average travel time or the quality of specific roads, and can be used for traffic light scheduling and road design.

Furthermore, the sensors do not necessarily have to be attached to vehicles, they could be carried on mobile phones, and the utility of using the individuals for sensing is preserved, since all their sensor data, including all their movements and timestamps (in aggregate) are kept intact by SwapMob.

In [5], a real-time urban monitoring platform and its application to the City of Rome was presented, they used a wireless sensor network to acquire real-time traffic noise from different spots, GPS traces of locations from 43 taxis and 7268 buses, and voice and data traffic served by each of the base transceiver stations from a telecom company in the urban area of Rome. These are few examples of sensor that could be carried by individuals, anonymized and transmitted to a service provider via SwapMob.

Another example is the offline mining in [29] representing the knowledge from taxi-drivers as a landmark graph could be done with SwapMob anonymized data. A landmark is defined as a road segment that has been frequently traversed by taxis, and a directed edge connecting two landmarks represents the frequent transition of taxis between the two landmarks. This graph is then used for traffic prediction and for providing a personalized routing service.

In general, lossless maps of flows in the city can be obtained by using Swap-Mob at several aggregation levels and for different timestamps.

4   Empirical Evaluation

We tested our algorithm on the T-drive dataset [29,30] which contains the GPS trajectories of 10,357 taxis during the period of Feb. 2 to Feb. 8, 2008 within Beijing. The total number of points in this dataset is about 15 million and the total distance of the trajectories reaches to 9 million km. It is important to note that not all taxis appear every day and not all report their positions at the same interval. The average sampling interval is about 177 s and 623 m. Each line contains the following data: taxi id, date time, longitude, latitude.

4.1   Reidentification by POIs and Home Location

Recall that our main privacy motivations are to protect locations related to people's habits and also the association of the trajectories to specific individuals.

We show in the next subsections how to infer points of interest of an individual such as his home location. Then, we show that such locations cannot be inferred after applying SwapMob to the data.

Inferring Points of Interest and Home Location. In [32], if an individual remains at less than 200 m from a given point during at least 20 min, then it is considered a stay point (or point of interest) thus the two thresholds used for detecting them are time $\tau = 20$ and distance $\chi = 200$.

In general, for obtaining points of interest, we discretize the space in square cells of 111.32 m (or 0.001 decimal degrees) and count which are the most populated cells for each individual, considering that the most populated one should contain the home location.

This is similar to [8] that discretize the world into 25 by 25 km cells and define the home location as the average position of check-ins in the cell with the most check-ins, see also [23].

In [3], Beresford and Stajano tested this kind of attack on real location data from the Active Bat, consisting in following the trajectory of a pseudonym to a "home" location (in this case the user's desk in the office) and successfully de-anonymize all the users by correlating where does any given pseudonym spend most of its time and who spends more time than anyone else at any given desk.

We validated this assumption empirically by looking at the distribution of timestamps for such set of locations, which is greater than 5% and increases consistently from 20 h until 6 h where it reaches its peak and decreases below 5% around 10 h, see Fig. 3. Also, if we assume that those with more than 5% of relative frequency, are the correct home locations, we obtain that the correct home locations amount for 83.7% of the total, a very similar percentage to the one obtained in [8] by manual inspection which was 85% accuracy.

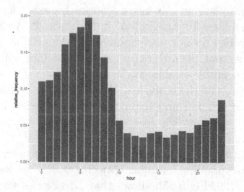

Fig. 3. Frequency histogram of hours at deduced home locations, as a percentage of total number of measurements for each hour.

Home Locations After Swapping. We validated the results of our anonymization technique by trying to infer users home locations after swapping. Considering that the home location is the most relevant POI, we can argue that these experiments prove that not only home locations will be protected by our method, but all POIs.

For carrying out the swapping process, we assumed that two taxis were co-located if they passed at distance at most 111 m approximately ($\chi = 0.001$) in a 1-min interval ($\tau = 60$). Note that this is about 6 and 3 times less the average sampling interval for distance and time in the dataset.

We found out that the inferred location after swapping was always different from the real one, for all except for the 51 trajectories that did not swapped. However, we inspected some of them and observed that they didn't swapped probably due to the fact that they are outside Beijing, as we can see in Fig. 4.

Fig. 4. Some trajectories that did not swapped

4.2   Reidentification by Linkage

In this section we simulate an adversary who knows exact locations and times-tamps, and tries to reidentify a trajectory in the dataset. We also show that the anonymized trajectories do not always intersect the original ones. This means that, even in the case that the adversary can link the data points that he knows, he may not learn the entire original trajectory.

In Fig. 5a we represent the empirical cumulative distribution function of the intersection between original and anonymized trajectories, and in Fig. 5b the percentage of each trajectory disclosed depending on how many exact spatio-temporal points an adversary knows.

Figure 5a shows that 84% of all the anonymized trajectories intersect in less than 1/4 their corresponding original trajectory, 68% in less than 1/10 and 28% in less than 1/100. Figure 5b shows that adversaries knowing as many as 10 precise spatio-temporal points, are still not able to reidentify 58% of the population, and even when they are able to find the corresponding anonymized trajectory to the one that they are attacking, they will not learn more than 50% of the original trajectory in 95% of the cases. This is in contrast to the results on [17], in which 4 spatio-temporal points are enough to uniquely characterize 95% of the traces, and the traces considered the most difficult to identify can be characterized with 11 points.

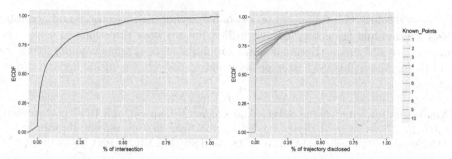

(a) intersection between original and anonymized (b) attackers capability of inferring trajectories
trajectories   from known locations

Fig. 5. Empirical cumulative distributions

5   Conclusions

We have defined and tested a novel algorithm for real-time mobility data
anonymization that consists on swapping trajectory segments. In contrast to
the k-anonymity or differential privacy models for trajectory anonymization, the
proposed method does not modify the data, but its association to specific individuals, and it is performed on real time, without the need of having the entire
dataset. The proposed protocol tackles both identity and location privacy, and
our data model can be adapted to protect either single trajectory positions, as
they lose the relation to the individual who has generated the data, or the whole
trajectories, since they are mixed among many different peers.

We show that is not possible to infer correctly the home locations after the
anonymization and, also, that an adversary who knows exact points of the trajectory is not able to use them for reidentification, because in most cases they
no longer correspond to the anonymized trajectory. And, even in the improb-
able case that the adversary correctly relates the anonymized trajectory with
the original, we have shown that he cannot infer the entire trajectory but just a
small part of it.

We have simulated our protocol with an offline algorithm, although, the
protocol could be run in real time in which data is transmitted by user devices to
our anonymizer that communicates and collaborates with a server. By changing
the anonymizer for a group protocol, the protocol could provide security against
collusion between the service provider and the anonymizer.

It must be pointed out that swapping cannot be carried out when an individual does not cross anyone in her path. Hence, the proposed technique will
not anonymize the individuals who do not cross anyone in their daily activity.
However, it is not very common for an individual to spend too much time without meeting someone or going out from home. Moreover, such individuals can
be kept outside the database without compromising its utility. The use case considered is for obtaining aggregate mobility data and exact count queries, which
neither k-anonymity or differential privacy can provide.

Nevertheless, this comes at the cost of modifying the trajectories and possibly losing individual trajectory mining utility. Future work directions to solve this issue are to add the restriction of non-swapping streets or non-swapping zones for improving the utility to better preserve entire trajectories inside a given street or zone.

Acknowledgments. Julián Salas acknowledges the support of a UOC postdoctoral fellowship. This work is partly funded by the Spanish Government through grant TIN2014-57364-C2-2-R "SMARTGLACIS", and Swedish VR (project VR 2016-03346).

References

1. Abul, O., Bonchi, F., Nanni, M.: Never walk alone: uncertainty for anonymity in moving objects databases. In: Proceedings of the 2008 IEEE 24th International Conference on Data Engineering, ICDE 2008, pp. 376–385. IEEE Computer Society, Washington, DC (2008). https://doi.org/10.1109/ICDE.2008.4497446
2. Agrawal, R., Srikant, R.: Mining sequential patterns. In: Proceedings of the Eleventh International Conference on Data Engineering, ICDE 1995, pp. 3–14. IEEE Computer Society, Washington, DC (1995). http://dl.acm.org/citation.cfm?id=645480.655281
3. Beresford, A.R., Stajano, F.: Location privacy in pervasive computing. IEEE Pervasive Comput. **2**(1), 46–55 (2003). https://doi.org/10.1109/MPRV.2003.1186725
4. Beresford, A.R., Stajano, F.: Mix zones: user privacy in location-aware services. In: Proceedings of the 2nd IEEE Annual Conference on Pervasive Computing and Communications Workshops (PERCOMW 2004), pp. 127–131 (2004)
5. Calabrese, F., Colonna, M., Lovisolo, P., Parata, D., Ratti, C.: Real-time urban monitoring using cell phones: a case study in Rome. IEEE Trans. Intell. Transp. Syst. **12**(1), 141–151 (2011)
6. Chatzikokolakis, K., Andrés, M.E., Bordenabe, N.E., Palamidessi, C.: Broadening the scope of differential privacy using metrics. In: De Cristofaro, E., Wright, M. (eds.) PETS 2013. LNCS, vol. 7981, pp. 82–102. Springer, Heidelberg (2013). https://doi.org/10.1007/978-3-642-39077-7_5
7. Chen, R., Fung, B.C., Desai, B.C., Sossou, N.M.: Differentially private transit data publication: a case study on the montreal transportation system. In: Proceedings of the 18th ACM SIGKDD International Conference on Knowledge Discovery and Data Mining, KDD 2012, pp. 213–221. ACM, New York (2012). http://doi.acm.org/10.1145/2339530.2339564
8. Cho, E., Myers, S.A., Leskovec, J.: Friendship and mobility: user movement in location-based social networks. In: Proceedings of the 17th ACM SIGKDD International Conference on Knowledge Discovery and Data Mining, KDD 2011, pp. 1082–1090. ACM, New York (2011). http://doi.acm.org/10.1145/2020408.2020579
9. Danezis, G., et al.: Privacy and data protection by design - from policy to engineering. Technical report, ENISA (2015)
10. Giannotti, F., et al.: A planetary nervous system for social mining and collective awareness. Eur. Phys. J. Spec. Top. **214**(1), 49–75 (2012). https://doi.org/10.1140/epjst/e2012-01688-9
11. Gidófalvi, G.: Spatio-temporal data mining for location-based services. Ph.D. thesis, Faculties of Engineering, Science and Medicine Aalborg University, Denmark (2007)

12. Golle, P., Partridge, K.: On the anonymity of home/work location pairs. In: Tokuda, H., Beigl, M., Friday, A., Brush, A.J.B., Tobe, Y. (eds.) Pervasive 2009. LNCS, vol. 5538, pp. 390–397. Springer, Heidelberg (2009). https://doi.org/10. 1007/978-3-642-01516-8_26

13. Hoh, B., Gruteser, M., Xiong, H., Alrabady, A.: Enhancing security and privacy in traffic-monitoring systems. IEEE Pervasive Comput. 5(4), 38–46 (2006)

14. Hoh, B., Gruteser, M.: Protecting location privacy through path confusion. In: Proceedings of the First International Conference on Security and Privacy for Emerging Areas in Communications Networks, SECURECOMM 2005, pp. 194–205. IEEE Computer Society, Washington, DC (2005). https://doi.org/10.1109/ SECURECOMM.2005.33

15. Jiang, K., Shao, D., Bressan, S., Kister, T., Tan, K.L.: Publishing trajectories with differential privacy guarantees. In: Proceedings of the 25th International Conference on Scientific and Statistical Database Management, SSDBM, pp. 12:1–12:12. ACM, New York (2013). http://doi.acm.org/10.1145/2484838.2484846

16. Kifer, D., Machanavajjhala, A.: No free lunch in data privacy. In: Proceedings of the 2011 ACM SIGMOD International Conference on Management of Data, SIGMOD 2011, pp. 193–204. ACM, New York (2011). http://doi.acm.org/10.1145/1989323. 1989345

17. de Montjoye, Y.A., Hidalgo, C.A., Verleysen, M., Blondel, V.D.: Unique in the crowd: the privacy bounds of human mobility. Sci. Rep. 3 (2013)

18. Pensa, R.G., Monreale, A., Pinelli, F., Pedreschi, D.: Pattern-preserving k-anonymization of sequences and its application to mobility data mining. In: PiLBA (2008). https://air.unimi.it/retrieve/handle/2434/52786/106397/ ProceedingsPiLBA08.pdf$#$page=44

19. Reid, D.B.: An algorithm for tracking multiple targets. IEEE Trans. Autom. Control 24, 843–854 (1979)

20. Romero-Tris, C., Megías, D.: User-centric privacy-preserving collection and analysis of trajectory data. In: Garcia-Alfaro, J., Navarro-Arribas, G., Aldini, A., Martinelli, F., Suri, N. (eds.) DPM/QASA -2015. LNCS, vol. 9481, pp. 245–253. Springer, Cham (2016). https://doi.org/10.1007/978-3-319-29883-2_17

21. Salas, J., Domingo-Ferrer, J.: Some basics on privacy techniques, anonymization and their big data challenges. Math. Comput. Sci. (2018)

22. Samarati, P., Sweeney, L.: Generalizing data to provide anonymity when disclosing information (abstract). In: Proceedings of the Seventeenth ACM SIGACT-SIGMOD-SIGART Symposium on Principles of Database Systems, PODS 1998, p. 188. ACM, New York (1998). http://doi.acm.org/10.1145/275487.275508

23. Scellato, S., Noulas, A., Lambiotte, R., Mascolo, C.: Socio-spatial properties of online location-based social networks. In: ICWSM 2011, pp. 329–336 (2011)

24. Serjantov, A., Danezis, G.: Towards an information theoretic metric for anonymity. In: Dingledine, R., Syverson, P. (eds.) PET 2002. LNCS, vol. 2482, pp. 41–53. Springer, Heidelberg (2003). https://doi.org/10.1007/3-540-36467-6_4. http://dl.acm.org/citation.cfm?id=1765299.1765303

25. Tanimoto, S.L., Itai, A., Rodeh, M.: Some matching problems for bipartite graphs. J. ACM 25(4), 517–525 (1978). http://doi.acm.org/10.1145/322092.322093

26. Terrovitis, M.: Privacy preservation in the dissemination of location data. SIGKDD Explor. Newsl. 13(1), 6–18 (2011). http://doi.acm.org/10.1145/2031331.2031334

27. Terrovitis, M., Mamoulis, N.: Privacy preservation in the publication of trajectories. In: Proceedings of The Ninth International Conference on Mobile Data Management, MDM 2008, pp. 65–72. IEEE Computer Society, Washington, DC (2008). https://doi.org/10.1109/MDM.2008.29

28. Xiao, Y., Xiong, L.: Protecting locations with differential privacy under temporal correlations. In: Proceedings of the 22Nd ACM SIGSAC Conference on Computer and Communications Security, CCS 2015, pp. 1298–1309. ACM, New York (2015). http://doi.acm.org/10.1145/2810103.2813640

29. Yuan, J., Zheng, Y., Xie, X., Sun, G.: Driving with knowledge from the physical world. In: Proceedings of the 17th ACM SIGKDD International Conference on Knowledge Discovery and Data Mining, KDD 2011, pp. 316–324. ACM, New York (2011). http://doi.acm.org/10.1145/2020408.2020462

30. Yuan, J., et al.: T-drive: driving directions based on taxi trajectories. In: Proceedings of the 18th SIGSPATIAL International Conference on Advances in Geographic Information Systems, GIS 2010, pp. 99–108. ACM, New York (2010). http://doi.acm.org/10.1145/1869790.1869807

31. Zang, H., Bolot, J.: Anonymization of location data does not work: a large-scale measurement study. In: Proceedings of the 17th Annual International Conference on Mobile Computing and Networking, MobiCom 2011, pp. 145–156. ACM, New York (2011). http://doi.acm.org/10.1145/2030613.2030630

32. Zheng, Y., Zhang, L., Xie, X., Ma, W.Y.: Mining interesting locations and travel sequences from gps trajectories. In: Proceedings of the 18th International Conference on World Wide Web, WWW 2009, pp. 791–800. ACM, New York (2009). http://doi.acm.org/10.1145/1526709.1526816

33. Zhou, B., Pei, J., Luk, W.: A brief survey on anonymization techniques for privacy preserving publishing of social network data. SIGKDD Explor. Newsl. 10(2), 12–22 (2008). http://doi.acm.org/10.1145/1540276.1540279

Safely Plotting Continuous Variables on a Map

Peter-Paul de Wolf$^{(\boxtimes)}$ and Edwin de Jonge

Statistics Netherlands, The Hague, The Netherlands
pp.dewolf@cbs.nl

Abstract. A plotted spatial distribution of a variable is an interesting type of statistical output favored by many users. Examples include the spatial distribution of people that make use of child care, of the amount of electricity used by businesses or of the exhaust of certain gasses by industry. However, a spatial distribution plot may be exploited to link information to a single unit of interest. Traditional disclosure control methods and disclosure risk measures can not readily be applied to this type of maps. In previous papers [5,6] we discussed plotting the distribution of a dichotomous variable on a cartographic map. In the present paper we focus on plotting a continuous variable and derive a suitable risk measure, that not only detects unsafe areas, but also contains a recipe to repair them. We apply the risk measure to the spatial distribution of the energy consumption of enterprises to test and describe its properties.

Keywords: Cartographic map · Disclosure risk · Spatial distribution
Continuous variable

1 Introduction

The use of spatial mapping of (statistical) information is becoming more popular with the increasing availability of easy tools to produce cartographic plots. Since humans often are visually oriented, spatial mapping helps in understanding data. Indeed, often policy makers are using and asking for maps to explore which places in their cities need special attention with regard to their domain of interest.

In our previous papers [5,6] we concentrated on plotting spatial distributions of dichotomous variables. We effectively plotted the probability of occurrence of a phenomenon such as making use of youth care. We defined disclosure risk measures as well as utility measures for that specific situation, translating the traditional measures for frequency count tables to measures as function of areas

---

The views expressed in this paper are those of the authors and do not necessarily reflect the policy of Statistics Netherlands.

The authors like to thank Rob van de Laar and Anne Miek Kremer for reviewing an earlier version of this paper.

J. Domingo-Ferrer and F. Montes (Eds.): PSD 2018, LNCS 11126, pp. 347–359, 2018.
https://doi.org/10.1007/978-3-319-99771-1_23

or locations. The utility was defined in terms of hot-spots, reflecting the intended use by policy makers. A nice example of the use of hot-spots is e.g., described in [1].

Spatial mapping can obviously be used to display the spatial distribution of a numeric variable such as the mean energy consumption per enterprise. Such maps can be very useful tools for policymakers. See e.g., http://www.esru.strath.ac.uk/EandE/Web_sites/12-13/SmartCities/index.html where a project is presented that aims to make use of Geographical Information Systems (GIS) mapping techniques to analyse the spatial and temporal distribution of energy consumption throughout a manageable area of Glasgow in support of a range of possible decision-makers, to identify opportunities for future energy networks.

National statistical offices also provide a lot of regional and spatial information that can be visualized on a map. Regarding energy consumption statistics, Statistics Netherlands e.g. produces figures on energy consumption that are used on http://www.nationaleenergieatlas.nl/kaarten, a site about national issues concerning energy (consumption of electricity and gas, location of solar systems, windmills, etc.).

In this paper we use the term 'map' in the sense of a cartographic map or at least of a map that can relate physical locations to units in the (target) population. We will describe a disclosure risk measure as a function of area for the situation of displaying a numerical variable on a map. That measure will be based on the traditional measures for magnitude tables (see e.g., Chap. 4 of [4]).

Applying disclosure control techniques to publications of statistics, is a trade off between reducing the disclosure risk and maximizing utility, see e.g., [3,4]. In the current paper we mainly focus on the disclosure risk aspects. We will only briefly mention some aspects of utility.

We apply the proposed spatial disclosure risk measure to real-life data on energy consumption. To estimate the spatial distribution of energy consumption, we make use of a 'simple' estimator (mean energy consumption on predefined grid cells) and of a kernel density type estimator. For kernel density type estimators, information and applications can be found in e.g., [2,7,8]. We show how the disclosure risk measure evolves with changing the zoom level of the spatial mapping.

2   How to View Map-Data

Plotting a distribution on a (cartographic) map can be done in different ways. A traditional way of plotting spatial data is by making use of predefined administrative regions or a predefined gridding. One typically calculates the average value of a variable of interest for each region and colors the region area with a corresponding color in the map. This results in a so-called choropleth, with colored regions, as an example see left hand side of Fig. 1.

The mean values for the regions can be regarded as cell-values in a table and thus the 'standard' risk measures like a $p\%$-rule can be used to identify 'risky' cells, i.e., 'risky' regions. To deal with such 'risky' regions, one would typically

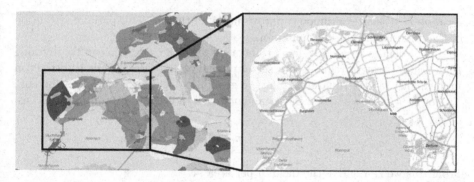

Fig. 1. Losing information when zooming in, due to disclosure risk Source: http://www.nationaleenergieatlas.nl/kaarten

suppress such a region (i.e., not color that region) or merge regions into larger areas that do no longer violate the p%-rule.

A disadvantage of this method is that each region gets a uniform color. I.e., whenever there are sub-regions that contain quite different values, they are not visible, unless zooming is allowed, sensible and available. Allowed in the sense that disclosure issues are not too restrictive (hence e.g., showing nothing at large zoom-factors, see e.g., Fig. 1), sensible in the sense that the administrative regions can be subdivided into nested, smaller administrative regions and available in the sense that for each zoom-level the average should have been computed beforehand. To overcome some of the just mentioned disadvantages, alternatively one can construct a kernel-type estimator of the spatial distribution of the variable of interest.

In the current paper we focus on the spatial distribution of a continuous variable, that is continuous in its value, not in its spatial distribution. In practice the locations of the target population units are discrete. Our method uses a spatial density estimator which *suggests* that the spatial distribution is continuous: each location on the map has a 'value' for the variable. This should increase usability, as displaying a spatial density can be seen as a middle ground between a coarse choropleth and a sparse and noisy location plot.

2.1 Spatial Distribution

We will restrict ourselves to spatial distributions defined on $\mathbb{R}^2$. It is convenient to write the (target) population as $\mathcal{U} = \{r_1, \ldots, r_N\}$ with $r_i = (x_i, y_i)$ the representation of element i of the population by its coordinates (x_i, y_i). I.e., $\mathcal{U}$ is a set of points in $\mathbb{R}^2$. This reflects the notion that the location is an *identifying* variable. In official statistics, the r_i often coincides with the locations of the target population of houses or of enterprises. Assume furthermore that we have measurements on each location r_i of phenomenon g, e.g. energy consumption, and denote these measurements by $g_1, \ldots, g_N$. We construct a continuous function $g(r) : \mathbb{R}^2 \to \mathbb{R}$ being a spatial distribution, representing the value of g at any location r. It is an continuous approximation of the discrete spatial distribution of the underlying, finite, population.

Let $\mathcal{A}$ denote an area, defined to be a subset of $\mathbb{R}^2$. An area can be anything from administrative regions to a set of points: $\mathcal{A}$ can be a building, a street, a municipality, a grid square, a county, a general polygon or a collection thereof.

For an area $\mathcal{A}$ we define the total amount of g (e.g. energy consumption) as

$$G(\mathcal{A}) = \int_{\mathcal{A}} g(r) \, dr$$

From this we can derive the mean g per area $\mathcal{A}$

$$\bar{G}_{\mathrm{a}}(\mathcal{A}) = G(\mathcal{A}) \big/ \|\mathcal{A}\| \tag{1}$$

where $\|\mathcal{A}\|$ denotes the size of area $\mathcal{A}$. Alternatively we can derive the mean g per unit in area $\mathcal{A}$ as

$$\bar{G}_{\mathrm{u}}(\mathcal{A}) = G(\mathcal{A}) \big/ \sum_{i=1}^{N} \mathbb{1}(r_i \in \mathcal{A}) \tag{2}$$

where $\mathbb{1}(B)$ equals 1 if B is true and 0 if B is false. Note that in these definitions we tacitly assume that the denominators in (1) and (2) are positive. In case a denominator is zero, the mean would be 'undefined'.

Using administrative areas in Eq. (2) would coincide with the ideas behind the more traditional way of plotting distributions as described in the beginning of this section. However, we now can more generally derive the mean g for *any* region on a map.

Zooming. An attractive feature of plotting spatial distributions is the ability to zoom in on specific areas. One usually starts with an overview impression of the spatial distribution by looking at a larger region (e.g., country level) and then zoom in on specific regions with special characteristics (e.g., neighborhoods). For this feature, there are (at least) two ways of dealing with the displayed spatial distribution. In the first place one could calculate/estimate the spatial distribution at a certain zoom-level and keep the thus obtained values fixed when zooming into lower level regions. A second possibility is to calculate the spatial distribution for several zoom-levels: when zooming into lower level regions, more detailed estimates could be calculated.

A disadvantage of the first option is that one has to decide at which zoom level the spatial distribution will be calculated. The second option feels more natural, especially when the mean energy consumption is plotted: zooming in to more detailed regions assumes one is interested in more detailed mean values as well. However, this could obviously conflict with the idea of protecting individual units. Thus fixing a maximum zoom level would be a first requirement for the second option.

2.2   Disclosure Risk

The recent discussion about the Strava fitness app[1] shows that the location of a population unit could sometimes be considered *sensitive* information: plotting the spatial distribution of 'running' people revealed the 'secret' location or shape of military compounds. In the current paper however we will restrict ourselves to the situation where location is 'only' an *identifying* variable.

Disclosure risk is related to individual units, whereas a spatial distribution is a function of location. This means we have to make a link between the spatial distribution and the risk measure. To simplify the discussion, we assume that the full population is observed.

Disclosure Scenarios. If we want to assess disclosure risk, we first need to specify possible disclosure risk scenarios. In case we plot a distribution of a variable on a map, several aspects play a role:

- The location of a population unit is very identifying.
- The possibility to zoom in on a map makes it easy to pinpoint the exact location of a population unit.
- The value of the variable at a certain location may be sensitive information.
- The spatial characteristics of the target population (e.g., where to find densely or sparsely populated regions) implies how identifiable a population unit is.

Based on those aspects, we have the following disclosure scenario in mind:

Definition 1. *Basic disclosure scenario for plots of spatial distributions*
An attacker first locates 'hot-spots': regions of high value of the spatial distribution. He then zooms in at that region until he can recognize/locate individual units from the population. Finally, he links the value of the spatial distribution to those individual units.

Two subscenarios can be distinguished: in case the attacker is not a population unit with a location in the hot-spot of interest, he is called an *external* attacker. In case the attacker is a population unit inside the hot-spot of interest, he is called an *internal* attacker. In the latter case the internal attacker can use information on his own contribution to derive more accurate information on another unit in the hot-spot compared to an external attacker. ◇

Risk Measure. In the traditional approach, each administrative region is regarded as a table cell and thus a $p\%$-rule can be applied to each administrative region. In case of a continuous spatial distribution, it makes no sense to consider each individual location as a table cell. By representing the units in the population by their locations r_i, we have no area related to each individual unit.

Indeed, there appear to be no 'natural' areas to be checked for a concentration rule like the $p\%$-rule. Note that 'natural' areas of two distinct units could be

---

[1] See e.g., http://www.abc.net.au/news/science/2018-01-29/strava-heat-map-shows-military-bases-and-supply-routes/9369490.

overlapping or actually coincide. In practice, the 'natural' region could be taken within a predefined set of regions or within unions of smallest allowable grid cells.

For a region $\mathcal{A}$ we calculate the total value $G(\mathcal{A})$ of variable g. As concentration rule for that area, we check for each unit in area $\mathcal{A}$ whether its value can be estimated within $p\%$ of its true value using $G(\mathcal{A})$, either by another unit in $\mathcal{A}$ (if there is an internal attacker) or by a unit not in $\mathcal{A}$ (an external attacker). In case $\mathcal{A}$ contains only a single unit, this leads to the requirement that the integrated distribution over $\mathcal{A}$ should be more than $p\%$ away from the true value of that single contribution. In case there are two or more observation in area $\mathcal{A}$, we treat $\mathcal{A}$ to be a table cell and apply the $p\%$-rule to that area. Summarizing, we define the following concentration based risk measure

$$
\mathsf{R_C}(\mathcal{A}; p) = \begin{cases} 0, & \text{if } \mathcal{A} \cap \mathcal{U} = \emptyset & \text{(3a)} \\ (1 + p/100)g_{i_1} - G(\mathcal{A}), & \text{if } \mathcal{A} \cap \mathcal{U} = \{r_{i_1}\} & \text{(3b)} \\ (1 + p/100)g_{i_1} + g_{i_2} - G(\mathcal{A}), & \text{otherwise} & \text{(3c)} \end{cases}
$$

with $g_{i_1} = \max(g_i : r_i \in \mathcal{A})$ (i.e., the largest value in area $\mathcal{A}$) and $g_{i_2} = \max(g_i : r_i \in \mathcal{A} \text{ and } r_i \neq r_{i_1})$ (i.e., the second largest value in area $\mathcal{A}$) and say that an area is not safe to be published if $\mathsf{R_C}(\mathcal{A}; p) > 0$. Cases (3b) and (3c) refer to the situations where the attacker is external or internal to the area $\mathcal{A}$ respectively.

Traditionally when publishing mean values on administrative areas, (some form of) group disclosure is discussed: whenever the variation of the individual values in the area is small, the mean value is a good estimate of each individual contribution. For an attacker to be able to use that notion, he should have some idea about the variation of the values over the area. In case of e.g., the value of apartments in flats, this may be easy to determine. In case of energy consumption this may be less easy. In the current paper we will not further address the group disclosure issue.

Properties of the Risk Measure. The definition of risk measure is obviously meaningful in case $G(\mathcal{A}) = \sum_{i:r_i \in \mathcal{A}} g_i$. Indeed, in that case it coincides with the $p\%$-rule applied to a table cell defined by area $\mathcal{A}$. However, in case the spatial distribution is estimated using a 2-dimensional kernel density estimator (kde, see e.g., [5]), in practice we could get one of the strange situations that

$$
\hat{G}_{\mathsf{kde}}(\mathcal{A}) < g_{i_1} \quad \text{or} \quad \hat{G}_{\mathsf{kde}}(\mathcal{A}) > 0 \text{ while } \mathcal{A} \cap \mathcal{U} = \emptyset
$$

and the risk measure (3) does not make sense. It could be an option to use a kernel density estimator for the values of individual contributions as well, e.g., use an estimated $\hat{g}_{i_1}$, but that may pose other anomalies as we will show in Sect. 3.

Note that, in theory, for each individual unit j we could find an area $\mathcal{A}_j^*$ such that the total energy consumption over that area is within $p\%$ of its true contribution g_j. However, for an attacker it would be impossible to find that area without knowing the true value beforehand.

Risk Measure in Practice. The risk measure in (3) is defined for any arbitrary area $\mathcal{A}$. Ideally the risk should be calculated using the 'natural' area that relates to an individual unit of the (target) population. In practice, it turns out to be difficult to define or derive such a 'natural' area. Therefore it is appropriate to use a predefined set of areas to be checked for disclosure: a set of grid cells, unions of grid cells or other 'logical' areas (e.g., the area of the building where the business in question resides).

In case an area turns out to be unsafe according to the risk measure, one has to adjust the way the information is plotted on the map. It is now no longer necessary to 'suppress' the spatial distribution over that area: we adjust the value of the spatial distribution such that the integration over $\mathcal{A}$ is more than $p\%$ away from the value of that single contribution. This can be viewed as an analogue to adding noise to the table cell representing area $\mathcal{A}$. Essential in this approach is that the area $\mathcal{A}$ is indeed a 'logical' or 'natural' area.

Hence, a practical approach would be that the risk measure $R_C(\mathcal{A}; p)$ is to be calculated for each area in a predefined set of areas and (the parameters of) the estimator should be adjusted such that the risk measure would be non-positive for all areas in that set.

3   Use Case: Energy Consumption of 'Westland'

3.1   Data Set

To test the defined disclosure risk $R_C(\mathcal{A}; p)$, we apply it to a data set describing the energy consumption of enterprises. Note that the energy consumption of dwellings also is an interesting variable, which seemingly could be analyzed and presented in the same spatial distribution, but it is wise to treat dwellings and enterprises as different populations. First of all, the energy consumption levels are very different, so the consumption of dwellings is barely visible when plotted on the same scale as enterprises. Secondly, the spatial density of the location of dwellings is much higher then that of enterprises, resulting in different requirements for spatial resolution.

In this example we will restrict our map to a small part of the Netherlands called 'Westland' in which many commercial greenhouses are situated as well as some enterprises in the Rotterdam industrial area. We will look at the spatial distribution of mean electricity consumption per enterprise. The data set contains the undisclosed micro-data of the electricity consumption of enterprises. Furthermore, a published aggregated data set is available in which disclosure control methods have been applied using a $p\%$-rule on the zip-code's (*postcode's*) most detailed level.

Figure 2 shows the map visualization for the published data. The map is attractive but the spatial distribution is difficult to spot, because the data is aggregated per zip-code and the colorization is done per building. This visualization stresses the building density and not so much the distribution of the mean energy consumption. For example in the lower left part of Fig. 2 the high mean energy consumption is barely visible, because of the low spatial density of enterprises.

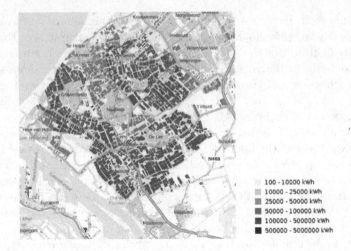

100 - 10000 kWh
10000 - 25000 kWh
25000 - 50000 kWh
50000 - 100000 kWh
100000 - 500000 kWh
500000 - 5000000 kWh

Fig. 2. Published mean energy consumption of enterprise (per zip-code) Source: http://
www.nationaleenergieatlas.nl/kaarten

3.2 Visualizing Mean Electricity Consumption

Spatial distributions are typically visualized using so called heat maps. Heat
maps are well known to the general public, because they are often used to display
precipitation or temperature zones in weather forecasts.

A heatmap has several important visual features:

- It has a *resolution* or *cell size*, restricting its spatial granularity at which
  details are shown.
- It uses a sequential *color scale* that is used to map the value at each location
  to a color. Such a color scale is designed to be perceived monotonic, i.e. a
  higher value has a 'higher' color. For this data set, we use a logarithmic color
  scale since energy consumption appears to be a log-normal phenomenon.
- It may provide a smoothed version of reality in which spatial distribution is
  presented continuous: in stead of restricting the display of values to the exact
  location of spatial objects, it may color the surrounding area, comparable to
  how regional statistics are cartographically displayed.

A naive approach to visualizing spatial distribution and producing a heat map is
to maximize resolution and to skip spatial smoothing. Figure 3 plots the original
data, plus three increasingly smoothed versions in the same (5-step-color) scale
to make them comparable. It furthermore shows that spatial patterns often are
more apparent when smoothing is used. This is similar to how uni-variate density
estimators work: they tend to reduce statistical noise and reveal the underlying
distribution. Moreover, maximizing resolution is at odds with disclosure risk as
we will see now.

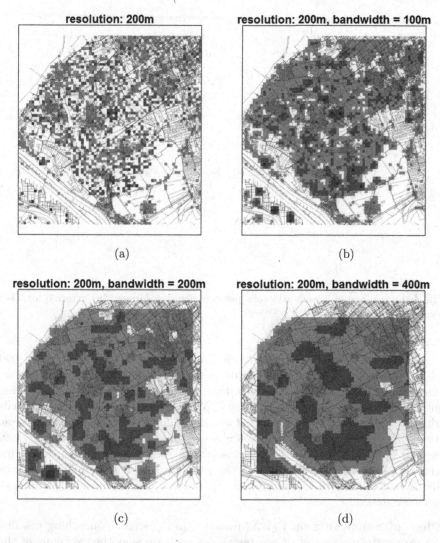

resolution: 200m

(a)

resolution: 200m, bandwidth = 100m

(b)

resolution: 200m, bandwidth = 200m

(c)

resolution: 200m, bandwidth = 400m

(d)

Fig. 3. Spatial distribution with increasing smoothing on data defined at 200 m × 200 m blocks (The legend has been removed in these examples to prevent disclosure of sensitive values): (a) without smoothing; (b)–(d) kde estimator with increasing bandwidth from 100 m to 400 m.

Effect of Resolution on R_C. Resolution is both important for visualization as well as for the disclosure risk. The more detail, the more utility, but also the higher the risk of being disclosed. Since rectangular grids are used to create heatmaps we will calculate $R_C(\mathcal{A}_i; p)$ for each grid cell $\mathcal{A}_i$. We assess R_C for the example data set on rectangular grids of various cell sizes and calculate the percentage of unsafe cells relative to the number of non-empty cells, to see its dependency on cell size (resolution).

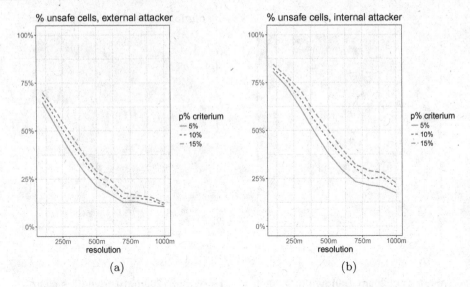

Fig. 4. Percentage of unsafe grid cells per cell size (resolution): (a) external attacker scenario; (b) internal attacker scenario.

Figure 4 shows that a detailed resolution (100 m to 200 m) makes many grid cells unsafe, which is sensible since this level of detail approaches the area of an enterprise: many grid cells contain in that case one or two enterprises, making both disclosure (sub)scenarios successful. When the cell size increases, most cells are safe. Figure 4a shows the percentage of unsafe grid cells when the *external* attacker scenario is applied for three different values of p: 5, 10 and 15. In Fig. 4b the corresponding lines for the *internal* attacker scenario are higher, since this scenario places extra restrictions on the risk measure. For example it enforces that a grid cell has to contain at least three enterprises to make it possible to be considered safe.

Effect of Smoothing on R_C. As reasoned in [5], spatial smoothing has disclosure control properties: it creates values that are smoothed versions of the observed values and makes pinpointing the location of individual values more difficult. To assess the effect of smoothing on disclosure risk $R_C(\mathcal{A}_i; p)$, the same area $\mathcal{A}_i$ is used as in the rectangular grid, but a different estimated risk $\hat{R}_{C,h}(\mathcal{A}_i; p)$ is used:

$$
\hat{R}_{C,h}(\mathcal{A}; p) = \begin{cases} 0, & \text{if } \mathcal{A} \cap \mathcal{U} = \emptyset & (4a) \\ (1 + p/100)\hat{g}_{i_1,h} - \hat{G}_h(\mathcal{A}), & \text{if } \mathcal{A} \cap \mathcal{U} = \{r_{i_1}\} & (4b) \\ (1 + p/100)\hat{g}_{i_1,h} + \hat{g}_{i_2,h} - \hat{G}_h(\mathcal{A}), & \text{otherwise} & (4c) \end{cases}
$$

with $\hat{G}_h(\mathcal{A})$, $\hat{g}_{i_1,h}$ and $\hat{g}_{i_2,h}$ kernel density estimators with bandwidth h for $G(\mathcal{A})$, g_{i_1} and g_{i_2} respectively. Fixing p at a constant value (e.g. 5), the dependency of $\hat{R}_{C,h}(\mathcal{A}_i; p)$ on h can be assessed by calculating the percentage of unsafe grid cells as function of h. Figure 5a shows the effect for various cell sizes of increasing bandwidth h (100 m–1500 m) for the external attacker scenario. The horizontal dashed line shows the percentage of zip-code areas (PC6 = postal code at most detailed level) that is considered unsafe according to the $p\%$ rule. For cell size 200 m and more, smoothing indeed decreases disclosure risk and allows publishing a (much) higher percentage of grid cells than the percentage of published zip codes in the current practice of Statistics Netherlands. The internal attacker scenario in Fig. 5b is interesting: for cell size 200 m and more, smoothing decreases disclosure risk, although a lot slower than in the external attacker scenario, but for cell size of 100 m, the risk increases with h. This may partially be the effect of using an *estimated* disclosure risk, including estimates of the largest and second largest individual contributions instead of the true values.

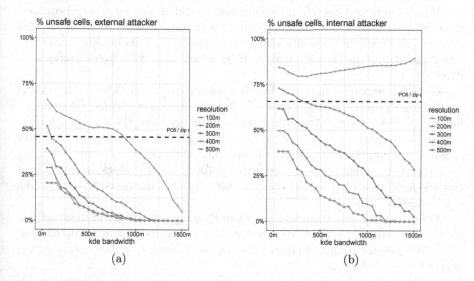

(a)   (b)

Fig. 5. Percentage of unsafe grid cells per bandwidth (smoothness): (a) external attacker scenario; (b) internal attacker scenario.

4   Discussion and Future Work

In this paper we derive and describe a disclosure risk measure $R_C(\mathcal{A}; p)$ for displaying continuous data on cartographic maps. The risk measure is a function of an area $\mathcal{A}$. Ideally, the risk measure is calculated for a set of 'natural' areas. With 'natural' area we indicate an area that is logically connected with an individual unit in the population. This could be the area of the building where a business resides or the land that is associated with a farm. To simplify the procedure we did not choose a 'natural' area, but assessed the risk measure for rectangular grids. We plotted the mean energy consumption per enterprise on the grid without smoothing as well as with different levels of smoothing using a 2-dimensional kernel density estimator.

The first observation we make is that smoothing often helps in finding and understanding spatial patterns. Moreover, smoothing in itself provides some disclosure protection by smearing out the energy consumption over a (larger) neighborhood of each business.

As discussed in Sect. 2.2 the disclosure risk measure might show strange behavior when using a kernel density estimator: the estimated total energy consumption of area $\mathcal{A}$ may be smaller than the largest contribution to that area or there may be a positive total energy consumption even in case there is no business present in that area. Trying to avoid this problem, we estimated the individual energy contributions using a kernel density estimator as well.

Starting with small grid cells of $100\,\text{m} \times 100\,\text{m}$, we found that a large number of those grid cells contain no business at all and almost no grid cell contains three or more businesses. Using estimated individual contributions in the risk measure, the number of unsafe grid cells increased with increasing bandwidth when considering the scenario with an internal attacker. This feels counter intuitive.

A possible explanation is that by using separate estimates for the largest and second largest contributor in a grid cell, the ordering might change: the estimate for the second largest contribution might be larger than the estimate for the largest contribution. Unfortunately we did not have time to develop estimates of individual contributions taking the ordering into account.

Increasing percentage of unsafe cells for $100\,\text{m} \times 100\,\text{m}$ blocks suggests that this resolution does not reflect the 'natural' areas of businesses: it seems to be too detailed.

As future work we will go into more detail on remarks we made about adjusting the kernel density estimator such that the plot is directly safe for all 'natural' areas. We will discuss how to define 'natural' areas and show how to adjust the kernel type estimator.

References

1. Chainey, S., Reid, S., Stuart, N.: When is a hotspot a hotspot? A procedure for creating statistically robust hotspot maps of crime. In: Kidner, D., Higgs, G., White, S. (eds.) Innovations in GIS 9: Socio-Economic Applications of Geographic Information Science, pp. 21–36. Taylor and Francis, Abingdon (2002)
2. Danese, M., Lazzari, M., Murgante, B.: Kernel density estimation methods for a geostatistical approach in seismic risk analysis: the case study of Potenza Hilltop Town (Southern Italy). In: Gervasi, O., Murgante, B., Laganà, A., Taniar, D., Mun, Y., Gavrilova, M.L. (eds.) ICCSA 2008. LNCS, vol. 5072, pp. 415–429. Springer, Heidelberg (2008). https://doi.org/10.1007/978-3-540-69839-5_31
3. Duncan, G.T., Stokes, S.L.: Disclosure risk vs. data utility: the RU confidentiality map as applied to topcoding. Chance **17**(3), 16–20 (2004)
4. Hundepool, A., et al.: Statistical Disclosure Control. Wiley Series in Survey Methodology. Wiley, Hoboken (2012). ISBN 978-1-119-97815-2

5. de Jonge, E., de Wolf, P.-P.: Spatial smoothing and statistical disclosure control. In: Domingo-Ferrer, J., Pejić-Bach, M. (eds.) PSD 2016. LNCS, vol. 9867, pp. 107–117. Springer, Cham (2016). https://doi.org/10.1007/978-3-319-45381-1_9
6. de Wolf, P.P., de Jonge, E.: Location related risk and utility. Presented at UNECE/Eurostat Worksession Statistical Data Confidentiality, 20–22 September, Skopje (2017). https://www.unece.org/fileadmin/DAM/stats/documents/ece/ces/ge.46/2017/3_LocationRiskUtility.pdf
7. Worton, B.J.: Kernel methods for estimating the utilization distribution in home-range studies. Ecology **70**(1), 164–168 (1989)
8. Zhou, Y., Dominici, F., Louis, T.A.: A smoothing approach for masking spatial data. Ann. Appl. Stat. **4**(3), 1451–1475 (2010). https://doi.org/10.1214/09-AOAS325

Author Index

Printed in the United States
By Bookmasters